FUNDAMENTALS OF LEAN MANUFACTURING

Don T. Phillips
Berna E. Tokgoz

"Fundamentals of Lean Manufacturing" by Don T. Phillips and Berna E. Tokgoz. ISBN 978-1-951985-71-4.

Published 2020 by Virtualbookworm.com Publishing Inc., P.O. Box 9949, College Station, TX 77842, US.

Dedication

This book is dedicated to Henry Ford. He was a Lean Engineer before the terms Lean Manufacturing and Lean thinking were ever birthed.

Born: July 30, 1863, Springwells Township, Michigan
Died: April 7, 1947, Fair Lane

Acknowledgements

There are many individuals who have contributed to this book on Lean Engineering. Dr. Mark Spearman and Dr. Wallace Hopp first awakened interest in the field of Factory Physics and without their pioneering work of establishing a scientific body of knowledge to unify Manufacturing Systems Analysis, the content of this book would not be placed in its proper perspective. Their work in awakening the engineering community to the corrupting influence of variance has not gone unnoticed. Of Course, there are numerous pioneers of the Toyota Production System which must be acknowledged including Taiichi Ohno and Shigeo Shingo. Early lean pioneers such as Bob Hall, Kenichi Sekine, Dr. Dick Schonberger and Yasuhiro Monden must also be recognized, along with the work and publications of Dr. Bernard Schroer and Michael Baudin. Each were Lean Engineers in their own way. Concepts and ideas have been modified, presented and summarized which stem directly from Dr. Rajan Suri, Dr. John M. Nicholas, Dr. Dave Sly of ProPlanner and of course Mark Spearman and Wallace Hopp.

Admiration and respect are offered to James P. Womack and Daniel T. Jones for their landmark publications in Lean Thinking, and to John Krafcik who coined the term *Lean.*We join these pioneers in believing that Lean thinking and Lean Manufacturing are here to stay.

Since Dr J. T Black of Auburn University and Dr Don T. Phillips of Texas A& M University collaborated to coin the term ***Lean Engineer*** in 2014 this Engineering Discipline has been defined within the framework of Industrial Engineering, Mechanical Engineering, Quantitative Business and Industrial Technology programs. A program dedicated to Lean Principles has been started at Ohio State University.

Regrettably, Dr J T Black of Auburn University passed away in 2018. His wisdom, wit and pioneering work in Lean Manufacturing Cells will not be forgotten by those who knew and loved him.

We only hope that in some small way we have captured the DNA of a new engineering discipline called ***Lean Engineering*** in this book and clarified key concepts in Lean Manufacturing.

Preface

Perhaps we should stop looking for the sharpest needle in the haystack
And find one that we can sew with... March &Siman, Organizations

The motivation for constructing this textbook evolved from over 30 years of teaching and industrial experience in Manufacturing Systems Design and Analysis. As the years passed, it became abundantly clear that traditional mass production lines were not suited to the flexibility and versatility required to competitively compete in an evolving international marketplace. In order to frame this observation and describe how it came about, we need to briefly describe the evolution of modern manufacturing systems.

Manufacturing is not a new endeavor; it has been around since ancient times. Biblical and historical records record extended battles between ancient empires. These encounters required thousands of bows, arrows, knives, swords and shields. We know that the Romans manufactured these items in Rome to equip their legions. In those days, manufacturing was a craft activity. Skilled individuals made these and other things using crude tools and no automation. This continued for over 3000 years. In the late 18th century and the early 19th century all of this began to change. Exploiting the wheel and the concept of leveraging power by using gears, a new manufacturing paradigm slowly emerged. Using the power of water from streams, manufacturing plants began to automate by generating power from falling water. Pulleys and gears were connected by crude homemade belts and the cottage industries of individual skilled craftsman gave way to groups of powered machines and the job shop emerged. These initial efforts at automation were pioneered and perfected in the Eastern United States where water power was plentiful and creative men were out to make their fortune. The world came to see......and the world adopted these methodologies. Water power reached its zenith and its destiny when James Watt (1726-1819) invented the first modern steam engine. This paved the way for the locomotive, ocean liners and powerful machinery. The old had passed away, and the new had emerged. This was the *First Industrial Revolution,* called the American Armory System by the historians.

The first true Manufacturing engineer was undoubtedly Henry Ford. Henry Ford pursued a vision of creating an automobile that almost any American could afford, and at the same time operate without failure for long periods of time. Initially, Henry Ford was influenced by two key principles: (1) standardization and interchangeability of parts and (2) specialization of labor (Turner, Mize and Case). As far as we can tell, Adam Smith (1776) and Charles Babbage (1832) were among the first to recognize the efficiency and productivity of division of labor. The advantages and superiority in mass production of part standardization and interchangeability were pioneered by Eli Whitney (1785-1793) when he standardized parts which were used to build a cotton gin.

Fueled by these two key principles, Henry Ford became the progenitor of the *Second Industrial Revolution.* Almost single handedly, he designed and built the first Mass Production Assembly line and built millions of Model T (1908-1927) and Model A (1927-1931) Ford vehicles. As a testimony to his genius and engineering skills, many of these vehicles are still running today.

"I will build a car for the great multitude. It will be large enough for the family, but small enough for the individual to run and care for. It will be constructed of the best materials, by the best men to be hired, after the simplest designs that modern engineering can devise. But it will be so low in price that no man making a good salary will be unable to own one – and enjoy with his family the blessing of hours of pleasure in God's great open spaces" Henry Ford, 1908

Although Henry Ford was very successful, he had a fatal flaw in his thinking. Obsessed with standardization, he purposely eliminated any variety or customization of mass produced vehicles. One of his famous quotations was: *"You can buy any color of Model A that you choose, as long as it is black".* Reluctant or unable to respond to the individual customer, his pioneering efforts were soon to come to an end. Almost simultaneously, entrepreneur name William C. Durant arrived on the scene.

In 1908, Durant recognized that the automobile was here to stay, but that a customer was going to demand variety and customization. He formed General Motors Corporation, and the *Second Industrial Revolution* was underway. Durant set up several automotive divisions (Buick, Pontiac, Cadillac and Chevrolet) and within each division he produced one line of automobiles, but with variety in several different models. General Motors pioneered the Mass production of mixed-model automobiles and the concepts of high speed assembly lines. A new engineer emerged which we now call a manufacturing engineer, but they were mostly Mecanical engineers. This new breed of manufacturing engineers and manufacturing practices were fueled by the work of Frederick F. Taylor (Scientific management) and Frank & Lillian Gilbreath (Motion and time Study). Both were concerned about the best way to perform a task (time and cost). Henry Gantt developed charts to schedule these factors. These people can rightly be considered to be the first Industrial Engineers. The world's first Industrial engineering Degree was established at Penn State University in 1909 in the Department of Mechanical engineering.

While mass production lines were being pioneered by General Motors, there was a dark cloud which hung over world peace. In 1914 WWI started and lasted until 1918. Mass production of war machines such as guns, tanks and planes were built upon existing knowledge and much success was achieved. However, modern mass manufacturing and paced assembly lines were not begun until 1941 when the United States entered WWII (1939-1945). Between 1941 and 1945 there were enormous strides taken in the mass production of war machines. After WW II, these methodologies and production innovations began to find their way into general American Manufacturing. Following WW II, America emerged as the world leader in manufactured goods and services. Led by the aerospace and automotive industries, mass production continued to evolve and produce at higher rates. During this era, Industrial Engineering played a large role in finding better ways to do things, designing more efficient systems and optimizing key parameters. The work of Charles Babbage, Henry Metcalf, Henry Gantt, Frederick Taylor and the Gilbreaths (Frank & Lillian) established Industrial Engineering as the *productivity people.* Between 1923-1932 Elton Mayo conducted his now famous *Hawthorne Experiments* at the Western Electric Company. Mayo directed a series of studies which scientifically established that there were close and undeniable connections between the physical conditions, mental attitudes and welfare of the workforce to productivity, quality, and a host of other factors. These factors were directly related to corporate profits. These experiments expanded the scope of Industrial engineering to include Human factors and Ergonomics, and had a dramatic influence on Taylor and the Gilbreaths. George Danzig developed Linear Programming at Stanford University in 1939, and proved its usefulness in solving a wide range of practical problems. During WW II, the field of Operations Research emerged in England under a group called *Blackett's Circus*. The Industrial Engineering community adopted Operations Research into their program of study, and modern Industrial Engineering emerged.

The Second Industrial Revolution was not instantaneous, but spanned a period of over 50 years, including two world wars. The evolution of assembly line manufacturing started with Henry Ford's production system and is currently best typified by the high speed assembly lines of General Motors Corporation, Ford Motor Company and the Chrysler Corporation. In this era, the Mass Production System reached unprecedented capabilities to produce a wide variety of automobiles. However, as previously discussed, the Achilles heel of American Automobile Manufacturing is the lack of flexibility and versatility to rapidly change from one set of finished products to another. This may seem paradoxical in contrast of the previous statement that a wide variety of automobiles are produced, but manufactured variety is actually built into the mass production systems. A typical American automobile manufacturer will spend up to 3 years designing future automobiles, and to change mass production from one year to the next requires a complicated and expensive changeover sometimes spanning 6 months. Modern Mass production Lines often involve massive *supermachines* and highly automated processes, blended with some manual tasks. This is a highly automated form of Henry Ford's flow line for building and producing modern automobiles in extremely large volumes (400,000 to 500,000 per year). The enabling technologies are repeat-cycle, dedicated machines with automated material handling devices and sophisticated control systems. Although a wide variety of automobile configurations are produced, within any one family of cars there is a large degree of standardized basic configurations and interchangeability of parts.

As the modern Mass Production System evolved, it became necessary to develop and deploy highly sophisticated Production Planning and Control Systems. In 1964 Joseph Orlicky introduced Materials Requirement Planning (MRP) as a tool to combat emerging international competition. In 1983 Oliver Wight transformed MRP into Manufacturing Resource Planning or MRP II. As the modern computer matured into an incredible computational platform and the PC emerged, the race to implement computerized production planning and control was underway. By 1975, MRP was implemented in 150 companies. By 1981, this had risen to over 8000 companies. MRP II brought master scheduling, rough-cut capacity planning, and material requirements planning under one computational umbrella. By 1989, about one third of the software industry software sold to American industrywas MRP II ($1.2 billion worth of software). However, major problems began to surface. First, modern MRP II systems can require a large support staff to manage its capabilities. Second, there has been a failure to fully integrate factory floor status into the planning framework because of cost and time. Third, such a wide degree of computer control tends to dehumanize the factory worker. Skilled labor with years of experience began to be supplanted by computerized instructions. When things went bad (frequently)and work flow priorities disrupted the planned schedule, expeditors were sent to resolve the problem. Worse, because expensive high speed mass production lines were driven to

100 % utilization, any disruption to product flow caused a corresponding failure in planned production. The stage was set for a new Industrial Revolution to emerge.

The *Third Industrial Revolution* came from an unpredictable and surprising source. After WWII ended, the Japanese Toyota automobile manufacturing company was ready to take on the world supremacy of American Automobile Manufacturing;but there was a lack of sophisticated computer software and hardware to build upon. Further, labor was plentiful and comparatively cheap. Finally, the Japanese economy was not ready to support the sales required to support another General Motors Corporation. In the early 1950's Taiichi Ohno conceived and designed an entirely new manufacturing system concept that was destined to change the entire manufacturing world. His system became known as the *Toyota Production System* (TPS), and was rooted in five(5) basic concepts: (1) Just-in-time production… only manufacture what is needed, in the quantity needed, at the correct point in time (2) Implement pull vs. push production strategies (3) Control production with simple, manual control strategies which became known as Kanban (4) convert all Mass Production lines into a Linked-Cell Manufacturing System (L-CMS) with Mixed Model Final Assembly (MMFA). Finally, and most important (5) Aggressively and continuously vigorously attack any form of waste (Muda) in the entire manufacturing system from the supply chain to final product delivery. The methodologies, strategies and tools to accomplish this transformation are contained in this textbook.

Ohno found perhaps the one man in Japan that could make the TPS concept a reality. His name was Shigeo Shingo. When history records the Third Industrial Revolution, the name of Shigeo Shingo will rank with those of Henry Ford, Frederick Taylor and Lillian Gilbreth. Shingo was a Japanese Industrial Engineer. If Ohno was the conceptual force behind TPS, Shingo was the driving force behind its implementation. Every Industrial Engineer should read his books on the Toyota Production System. This new design took years to develop and implement, but by the mid-1960's the world began to take notice. In 1982 Shoenberger visited Japan and wrote a landmark textbook called *Japanese Manufacturing Techniques*. In 1990, J. T. Black published his first book describing manufacturing cells called the *Design of the Factory with a Future*. In 1991 Womack, Jones and Roos published a book called The *Machine that Changed the World.* The term lean production introduced in this book was coined by John Krafcik. After extensive trips to Japan, Womack and Jones rocked the entire industrial world by publishing a textbook called *Lean Thinking* in 1996. The term *Lean* stuck, and TPS became known as the *Lean Production System*. Academicians such as Hall, Monden and Black soon joined the chase; and understanding the transformation to Lean Production began to emerge. The concepts of U-shaped, Lean Production Cells were published by Black and his colleagues (Black and Hunter, 2003). Lean concepts are now being applied to service systems, hospitals, insurance companies, small job shops and large manufacturing companies. One of the first companies to jump on the Lean bandwagon was Harley-Davidson, who reported spectacular results in the 1980's.

Around 2008, JT. Black and Don T. Phillips met and discussed the need for an exciting new breed of manufacturing engineer… one who would be thoroughly trained and equipped to lead Lean implementation across a wide variety of applications. It was agreed that perhaps the best engineering discipline from which this new engineer might emerge was Industrial Engineering. This decision was by no means self-serving: The modern Industrial Engineer is trained to continuously improve the intersection of people, scarce resources and equipment. The TPS and current Lean transformations require extensive knowledge of this intersection, with specific training in systems design and analysis, statistics, human factors and ergonomics, production planning and control, facilities layout and design, supply chain management and warehousing/distribution/material handling. However, Black and Phillips soon realized that while modern Industrial engineering provided a strong foundation, there were significant educational changes which needed to take place. Traditional IE skills need to be augmented with the methodologies of aggressive variance reduction, the principles of Factory Physics, Lean cell design, setup time reduction, decoupler design, Poka-yoke insertion, Kanban control, full preventative maintenance, total quality management and zero defects, 5 S methodologies and WIP reduction strategies: Worker paced production and not machine paced production, push versus pull strategies and many other new skills including the intersection of 6-sigma methods. This new breed of Industrial Engineer deserved a new name: We call this new engineer a ***Lean Engineer***. We will be quick to state that the new Lean Engineer does not supplant or replace the traditional Industrial Engineer, but rather creates a new career path and game-changing engineering discipline with specialized training in Lean Concepts. This concept prompted us to construct and self-publish a book in 2014 called *Lean Engineering: The Future has Arrived* to use in teaching Industrial Engineering students about these new Lean concepts. This book was used as a launching pad for further educational material at Texas A& M University and Auburn University. Although not publicized, it was noticed by the Shingo Institute in Salt Lake City and was awarded the International Book-of-the-Year in 2015. This book unified Lean Thinking into one book and defined a new breed of Manufacturing Engineer: ....***The Lean Engineer***. That book represented only an initial effort to define and equip this new engineer with a core set of tools. In 2016, Dr. Don t. Phillips joined Lamar University in Beaumont, Texas. After classroom testing these educational materials at Texas A&M University, Auburn University and Lamar University the stage was set to write this textbook, In a real sense, we seek to convince the discipline of Industrial Engineering to embrace, support and educate this new Lean Engineer. Succinctly stated, this new baby needs a home. We propose….. now we plead…. that Lean Engineering and Lean Thinking be unified under Industrial Engineering and taught as a new, emerging discipline.

Many academic institutions, including Texas A&M University and Auburn University, have introduced stand-alone Lean course(s). While this has created a general awareness of Lean concepts, a unified body of knowledge and the unique design aspects of Lean Systems design have not yet emerged. The truth is also that Lean Thinking and Lean Concepts are no longer confined to automobile manufacturing. These concepts have been found to be universally effective in reducing wastes, improving system performance, increasing quality and simply resulting in higher profits. Lean is now surfacing as a transformative theory in all forms of production, service and mass production systems. In a recent informal survey by the authors, 61% of all manufacturers reported that they were either planning to implement Lean or were in the process of doing so. This is not a fad nor is it a temporary fix…. It is the *Third Industrial Revolution*. It demands to be unified under the title of Lean Engineering. The authors are convinced that when some Engineering discipline decides to do this… the world will beat a path to their door. But much needs to be done; this text is only a meager beginning.

As efforts to integrate Lean concepts, Industrial engineering and 6-sigma emerge into a unified engineering discipline; it will become clear that the main benefit will be to create a deeper understanding of the unified power in these now mostly disjoint but complementary disciplines. The first step is not simply to integrate these concepts into a cohesive program of study, but to transform and change the very culture of manufacturing engineering. If 60%-70% of all manufacturing companies have begun this cultural change, how can we not also do so?

We are simply hopeful that this book will help articulate and define the very unique nature of Lean Transformation and Lean Systems Design, creating a pathway to *Factories with a Future* and *Companies with a new Concept*.

Don T. Phillips and Berna E. Tokgoz
Lamar University Beaumont, Texas
January, 2021

Thoughts and Things…….

Table of Contents

Chapter 1
Introduction to Lean Engineering

Changing any system will always take longer than you might think. — *JT. Black*

KISS...Keep It Simple Stupid, or at least as simple as possible. — *Don T. Phillips*

1.1 What is Lean Engineering?

In the 21st Century, we are witnessing in America the restructuring of manufacturing systems. In the early years of this century, manufacturing was on the decline, and assembly was on the rise. Extensive outsourcing to manufacturers outside the United States became preferred and commonplace. Outdated tax laws and cheaper labor rates caused a major shift in American manufacturing strategies. The entire manufacturing and assembly enterprise in Europe...led by the Japanese... became the focus of intensive cost saving activities and waste reduction. The restructuring of the manufacturing enterprise and the supply chain which supports that enterprise was underway to produce goods and services at less cost, with higher quality in less time with higher customer satisfaction. Underlying these goals were the principles of lean engineering, six-sigma engineering and industrial engineering. Leading these innovative practices was a Japanese company called Toyota, who had risen from meager beginnings to the world leader in automobile manufacturing. Toyota pioneered a new manufacturing system design and operation called the *Toyota Production System* or TPS. It was soon adopted by other foreign manufacturers and quickly caused a major shift in commodity pricing.

Since its origin in the early 1950s this new system has acquired many names; including Just-in Time (JIT), World-Class-Manufacturing, Stockless Production, Zero Inventory and Integrated Manufacturing Production System (IMPS). The most commonly used name in Europe for this new system of operation was called the *Lean Production System.* In spite of spectacular success in Toyota Manufacturing, the western would was slow to adopt this new philosophy and was largely ignored until a book called *The Machine Which Changed the World* was published by Womack and Jones in 1990.

The term *Lean Production* stands in opposition and contrast to traditional *Mass Production* systems. In any manufacturing system, products, goods and services are produced by transforming raw materials to finished goods and products by a set of manufacturing process involving people, materials and machines. The Lean Manufacturing System systematically transforms *raw materials* into *final product(s)* through a system of *Value Added* activities which minimizes cycle time and waste. It is important that we distinguish the *Manufacturing System* from the *Manufacturing Enterprise*.

It is interesting to note that the term *Lean Thinking* has come to be synonymous with *Lean Manufacturing*. Lean thinking is simply the mental process of aggressively and continuouslytransforming a traditional manufacturing system into one which minimizes work-in-progress, reduces cycle time, improves product quality and reduces waste (*muda*).

Table 1.1
Names and Origin of Lean Manufacturing

Lean Production	Womack and Jones (In book the Machine that Changed the World)
The Toyota Production System	Mondon, Ohno, and Shingo (Toyota Motor Company)
Integrated Pull Manufacturing System	AT&T
Minimum Inventory Production System	Westinghouse Corporation
MAAN-Material as Needed	Harley-Davidson
Just in Time System	Dick Schonberger (In book Japanese Manufacturing Techniques)
World Class Manufacturing	Dick Schonberger (In book World Class Manufacturing)
ZIPS- Zero Inventory Production System	Omark and Hall (In book Zero Inventory)
Quick Response Manufacturing	The Appeal Industry (The Toyota Sewing System)
Stockless Production	Hewlett-Packard Corporation
Kanban System	Many Japanese and American Companies
The New Production System	SuzukiMotor Company
Continuous Flow Manufacturing (CFM)	Many American Companies
The Linked Cell Manufacturing System (L-CMS)	Black (Auburn University)
One-Piece Flow	Sekine
The Integrated Manufacturing Production System - IMPS	Black
The Lean Engineer	Black and Phillips

While *Continuous Improvement* is certainly a necessary and ongoing goal of Lean Manufacturing, it fails to capture the whole range of design, implementation and sustainment activities required to create a true *Lean Enterprise System*. The management, control and efficient operation of such a system requires a set of unique managerialand engineering skills which are part of what we will call a *Lean Engineer*.

The Lean Engineer does not represent a completely new body of knowledge, but a unique combination of several traditional and some new areas of knowledge. This book is an attempt to capture the tools, techniques and operational activities necessary to transform a traditional system or enterprise into a Lean System or enterprise. The force required to do this is embodied in what we now call a *Lean Engineer*.

1.2 What is an Enterprise System?

An *Enterprise System* supports and manages the *Manufacturing System* where the actual value-added activities take place on the shop floor. Figure 1.1 provides a high-level view of the management and control functions which constitute a typical *Enterprise System*. The enterprise system is a self-contained or distributed system which includes all of the manufacturing and operational components to produce a final product. In today's global economy, this system may be distributed among various plant locations or international subassembly operations. It may be a complete product or a subassembly. These operational components are coordinated, managed and controlled by a *Supply Chain Management System.* The elements in a typical manufacturing system are automated or semi-automated machine tools, workstations, tooling and fixtures to support the manufacturing workstation; and human resources to interact with the material handling, warehousing and distribution systems to support product flows. The arrangement, functions, physical locations, flexibility and versatility of these elements are largely determined by the *Manufacturing System Design.*

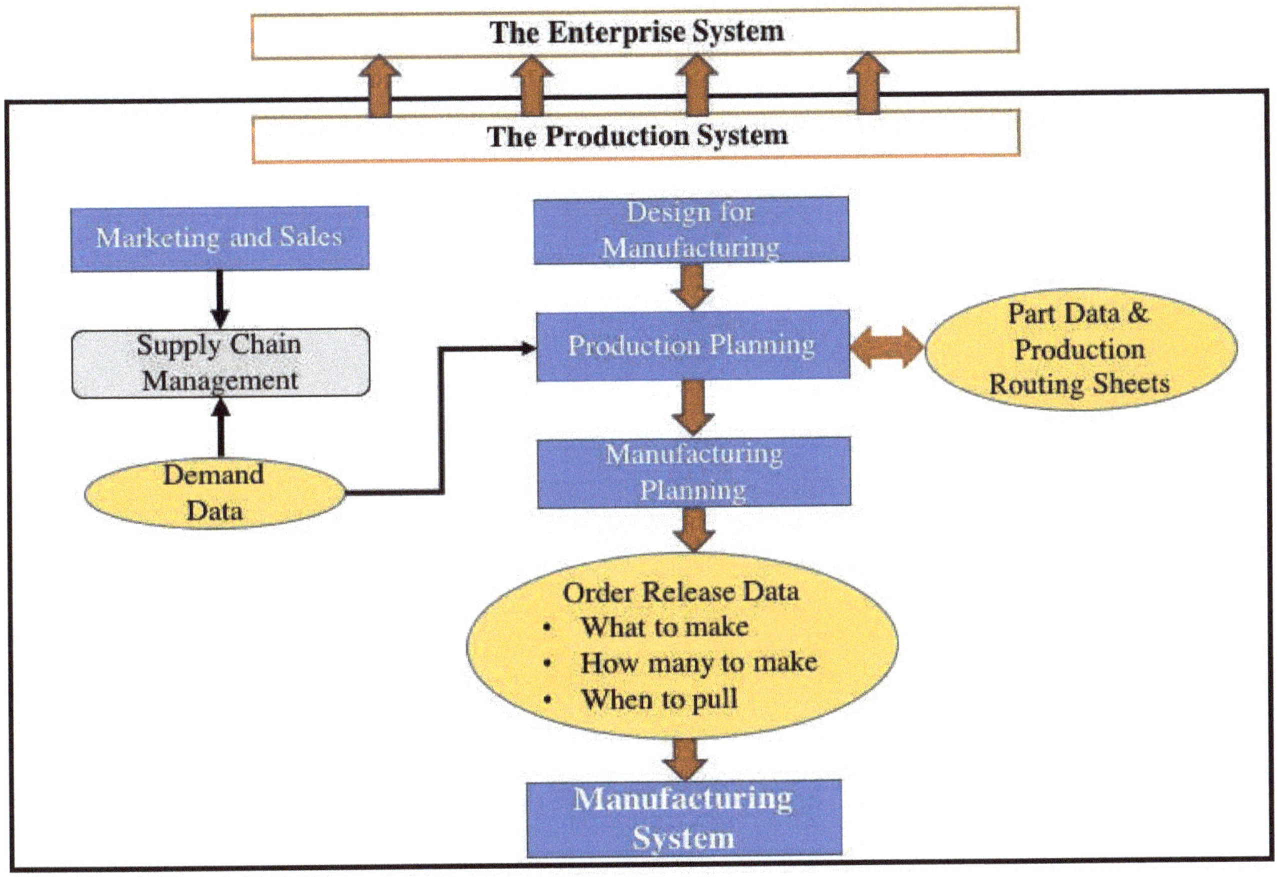

Figure 1.1
The Enterprise System

The collection of processes and people that actually produce a final product is called a *Manufacturing System.* A typical manufacturing system is shown in Figure 1.2and is characterized by a complex arrangement of *physical elements*described by measurable operating parameters (Black, 1991). Typical operating parameters are machine availability, throughput time, cycle time and output rates (or production rates). The relationship of necessary manufacturing components and their complexity determine how efficiently a system can be operated and controlled. System control not only involves individual process steps, but also involves the entire enterprise system. How well each process and element of the system harmoniously exists and supports one another largely affects the profit margin, efficient use of scarce resources and satisfaction of stated system goals and objectives. The entire manufacturing system must be continuously coordinated and controlled to efficiently move raw materials (work-in-process) throughout the system, schedule people, processes and customer orders, manage inventory levels, ensure high product quality, maintain target production rates and minimize system costs.

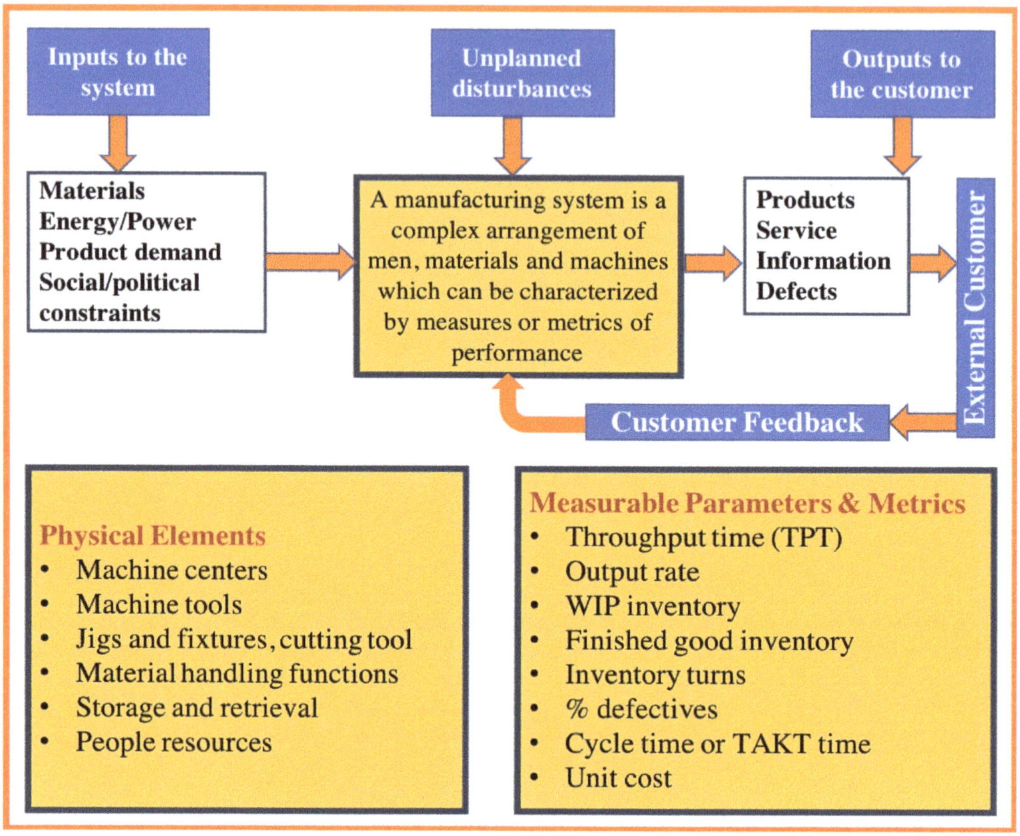

Figure 1.2
A Manufacturing System

The manufacturing system receives inputs that include materials, information and energy. The integrated system is a complex set of elements and people that include a wide variety of machines, tooling, material handling equipment and labor pools. The entire system exists to serve *customers*; both internal and external. Individual machine centers within the system represent *internal customers.* The *external customer* is the person or entity who consumes goods and services produced by the manufacturing system. Internal customers create value as raw materials are transformed into final product. External customers (usually) pay money for the delivery of these final goods and services. There is (usually) only one reason for a competitive, production system to exist… and that is to generate profit (revenue). In a traditional manufacturing system, *revenue* or profit is usually determined as a reasonable corporate goal.

$$Revenue = Sales\ Price - Cost\ of\ Production$$

After desired revenue is set by corporate managers,*the cost of manufacturing or the cost of production* of a company productis then determined.

The cost of manufacturing/production can be determined with some accuracy. The Sales Price of a final product is then set to achieve the desired revenue. This traditional way of thinking is a recipe for corporate disaster in a competitive sales environment. If the sales price is too high, the consumer will look elsewhere for a comparable product. In the Lean Engineering world, a subtle but powerful alternative is used.

$$Sales\ Price = Revenue + Cost\ of\ Production$$

In this world view, if desired revenue is set at say 30% profit, then the path to lower unit cost to the consumer….and hence higher sales if a quality product is produced…. is to *lower the cost of production*. This is the real driving force behind Lean Engineering. Lean Engineering and its associated methodologies are directed to *cycle time reduction, minimum in-process inventories* and *waste reduction*. In the Lean world, all forms of waste are aggressively attacked and systematically reduced or eliminated. The result is a shorter throughput time, increased production rates, tighter system control and a higher quality product. This book will provide a framework and definition of how to achieve these goals, which we call *Lean Engineering*.

1.3 Framing the Lean Engineer

All manufacturing systems are complex, dynamic, and subject to variation. This means that they must be designed and built to constantly accommodate change. Many inputs previously described cannot be fully controlled by management, and the typical response to variance and disturbances are counteracted by constantly manipulating, changing and modifying order release, sequencing, scheduling, and material availability. Changing any of these basic functions of a manufacturing system can only create *more* variability in a system. Lean Engineering seeks to *level production, minimize disruptions, reduce variability* and *create a stable manufacturing environment.*

In order understand the difficulty in designing or changing existing systems using modeling and analysis tools the following list will provide insight.

- A manufacturing system is usually very complex, difficult to rigorously define and exhibit different goals in different areas;
- Accurate data describing system operational behavior is usually either nonexistent or difficult to easily obtain. Even when system data exists, it may be inaccurate, out of date or too obscure to analyze and use without considerable effort;
- Interactive behavior and operational relationships may be awkward to express in analytical terms and exist in nonlinear form. Hence, many standard modeling tools cannot be applied without oversimplification;
- The physical size of large manufacturing systems may inhibit detailed modeling and analysis;
- Real-world systems rarely reach "steady state" behavior, and dynamically change through time. Many external forces such as environmental parameters can change system behavior;
- All forms of accurate systems analysis are subject to modeling errors (inaccuracy or lack of sufficient detail); errors of omission (missing data and operational logic) and errors of commission (failure to properly use all of the data).

Because of these and other difficulties, systems simulation analysis has emerged as the most important tool for detailed systems analysis and for manufacturing system design. Of course, higher levels of modeling and analysis such as queuing networks often provide extremely valuable insight. As new Lean Engineering modeling and analysis capabilities emerge, systems analysis will be more effective.

1.4 Critical Control Functions in a Manufacturing System

Ideally, manufacturing systems control should be vertically integrated, with each level of system control being composed of horizontally integrated subsystems dedicated to appropriate tasks and corporate goals. Each level of control is vertically integrated to support the goals and objectives of the entire system. The most critical control functions are *production control, inventory control, quality control and machine tool control.* While the system as an integrated entity will have a number of goals and objectives (cost control, profit, etc.) these system goals are often sought to be realized by optimizing selected bits and pieces of existing subsystems. This will never work in a complex, interactive system. Real system control functions require complex information analysis from the lowest levels of operation to the highest. All decisions and operational policies must reflect both local (machine level) and global (corporate level) impact. To control any system:

- The boundaries, constraints and interactive dependencies must be clearly identified.
- System response and behavior to any system change must be identified.
- System behavioral objectives (cost, profit, span times, throughput rates, etc,) must be linked to subsystem operational rules

The rules and laws which describe, and capture system behavior are difficult to discover and explain. Recent landmark work in this area is beginning to produce meaningful methodologies and procedures (*Factory Physics* by Hopp and Spearman for example). Accurate rules and laws of behavior must accurately link the behavior of a system to perturbations or changes in both input and behavioral characteristics. The entire field of lean analysis is directed to this goal.

1.5 Trends in Manufacturing Systems Design

In subsequent chapters of this book, we will devote considerable time describing the design of Lean manufacturing systems. The behavior and efficacy of both *greenfield*systems and *existing* systems can be greatly influenced by the principles of Lean Engineering. However, the field of systems engineering has been partially clouded if not destroyed by academic researchers; principally those in Operations Research. Academicians have historically looked at manufacturing systems as an application *playground* for optimization theory.They have developed specific, powerful analysis tools to optimize elements of Manufacturing systems under the disguise of *Systems Analysis*. This is a certain recipe for failure in an interactive, staged manufacturing system. A good example of this is improving workstation activities without regard to whether the workstation is a *Bottleneck Operation* or not. We will discuss this ill-fated philosophy later in this book. In order to sustain this approach, many important nonlinearities, complex systems interactions and stochastic behavior have been *assumed* away. The exception to this rule seems to be a subset of researchers applying simulation modeling techniques. However, even a system's (subsystem) simulation model only provides a snapshot in steady state. It is correct to say that a manufacturing system *never*exists in steady-state. This is not to say that simulation models are not useful or incorrect: In fact, they are sometimes the only modeling tool which can be applied to the analysis of a complex, stochastic system.

In almost all cases, the operational and production system has often been assumed to be one which has is constant to a large degree, and not something which requires and has been designed for change. As the reader proceeds through this book, it will become clear that Lean Thinking and Lean Engineering is a path that *requires* and *demands* system behavioral, operational and managerial functions constantly change in such a way that a system is designed and operated to respond to constant change. This new way of systems design and operation has been fueled by the following trends:

- Many manufacturing systems now operate on an international, global scale. There are two aspects of this trend that should be noted.
    - As manufacturers and producers of goods and services build plants around the world, the new manufacturing systems design must function in places with different cultures, currency ad languages.
    - Outsourcing United States manufacturing functions to other high-tech, low-cost countries like Indonesia, Brazil and China will continue.

These two facts do not negate the reason to embrace Lean Thinking and to apply the principles of Lean engineering .... Rather they demand that waste elimination, WIP control and more effective system control policies be adapted to survive and be cost competitive.

- The proliferation of variety in products has resulted in a significant decrease in manufacturing build quantities. Smaller production lot sizes are necessary to facilitate increases in product options and variety; level production; quickly respond to customer demand; and support manufacturing versatility and agility.
- The use of modern, exotic varieties of raw materials such as composites and plastics have created a need for old manufacturing processes and equipment to be seamlessly replaced by new, modern processes. New manufacturing systems must be designed to support rapid system change, and existing systems must be reengineered to support versatility and agility. In fact, many manufacturing systems such as those in the semiconductor and automobile industry feel that they must respond to new technologies to even exist.
- The United States over the past decade has experienced a continual decrease in both the skills and desire for workers to embrace a career in advanced manufacturing.This trend has resulted in increased automation and the emergence of "workerless warehouses". As a consequence, direct labor hours are no longer considered the basis for calculating cost per unit and direct labor cost has dropped to 5%-10% of total cost in almost every modern factory today.
- Product reliability has increased in response to competition. Product liability (lawsuits) from use, misuse and abuse of products has also increased.
- To remain competitive, the total time from product design to finished product has had to decrease. This requires a more flexible manufacturing system design.

- The continued growth of computer usage has led to more tightly integrated product design and manufacturing system design. The use of 3D simulations, CAD/CAM application and rapid virtual prototyping has significantly impacted the design/manufacturing cycle time.
- Ergonomics and worker safety continue to grow in importance as worker absenteeism and compensation costs have escalated.
- Green Manufacturing philosophies and social pressures to become environmentally friendly are rapidly becoming driving forces. Zero wastes and disposable products will continue to present themselves as significant product design issues.
- Typically, 50% of total manufacturing costs come from the materials used (direct cost) and almost 35%-40% from overhead and indirect cost.The total cost for a typical unit of product is shown in Figure 1.4. For example, if a small model car cost $20,000 in a modern automated factory, then $10,000 would be for direct cost and $7,000-$8,000 in overhead and indirect costs. Viewed another way, on an assembly line producing 400 cars per 8-hour shift, we might need 1,600 workers. Each worker works on each car for 1 minute. If the worker is paid $20 per hour, the cost per laborer per car is $.30. Direct labor cost per car would be $480 or about 5% of the manufacturing cost.

The response to customer demands for cheaper-better-faster is translated into lower profits and higher costs unless Lean Manufacturing principles are applied throughout the product development life cycle. Reducing unit cost while increasing value; delivering superior quality products; meeting on-time delivery and literally guaranteeing customer satisfaction has forced American companies to adopt Lean strategies to become more nimbler and more responsive. Either using original TPS strategies; adopting TPS strategies to system design and operation; or developing new Lean Methodologies are enabling companies to bring competitive products to market cheaper, faster and on-time by matching production rates to final assembly, employing just-in-time strategies, implementing pull systems and smoothing production. These same strategies have resulted in increased flexibility and the capability to quickly respond to varying product demands and product mix.

1.6 Five Basic Manufacturing System Designs

All factory configurations may look different, but there are only four basic manufacturing system designs (MSDS); four classic configurations and a new linked cell Lean manufacturing system design. The four classic system configurations are (1) the job shop (2) the flow shop (3) the project shop and (4) continuous processing. The new design is a (5) *Lean Linked Cell* system. The linked cell design is a new Lean manufacturing concept composed of manufacturing parts, assemblies and subassemblies by U-shaped cells; all feeding a mixed model final assembly. It is conceivable that some might view the high-speed assembly line as another type of manufacturing system which is a specialized and typically highly automated flow shop, but this too simplistic as we will soon learn. Table 1.2 lists a few examples of each system. Table 1.3 summarizes the key characteristics of each type of Manufacturing System.

Table 1.2
Types and Examples of Manufacturing Systems

Type of Manufacturing System	Service System	Product Sales
Job shop	Auto repair, hospital, restaurant, insurance	Machine shop, metal fabrication, custom jewelry, machine tool producer
Flow shop or flow line	Hospital, cafeteria, custom shirts	TV manufacturer, auto assembly line, pre-fab housing, hot water heaters
Project shop	Movie set, Broadway play	Aircraft carrier, lamination chamber
Continuous processing	Telephone company, cable television	Chemical plant, oil refinery
Linked-cell lean shop	Subway sandwiches, car wash, upholstery shop	Automobile manufacturing, clothing manufacturer, pump manufacturer

Table 1.3
Characteristics of Manufacturing Systems

Component	Job Shop	Flow Shop	Project Shop	Continuous Processing	Lean Mfg.
Types of machines	Flexible, general purpose	Single purpose, single function	General purpose, mobile, manual & automated	Specialized, high technology	Simple, customized & single cycle machines
Layout	No particular order or pattern	Product flow	Fixed position	Immovable locations	Linked U-shaped reconfigurable cells
Setup & changeovers	Long, variable & frequent	Long, expensive & complex	Product specific, virtually none after design	Rare & expensive	Short, frequent
Operators or skilled personnel	Highly skilled, multifunction, versatile	Specialized, highly trained, dedicated to process step	Specialized, highly skilled	Skill level varies; processing, packaging and engineers	Multifunctional, cross trained, mobile and skilled
Inventory (WIP)	Very low	Large with buffer storage	Variable but usually high	Very high but in tanks, trucks, rail cars or ships	As small as possible
Process routings	By order and by machine, no fixed routing	Fixed and prioritized	No such concept but stages of	Many pipeline and different storage	Fixed with cells by part type and

			work are critical	locations/Modes to route through	final assembly
Production Control	Flow according to due date and machine availability once work is released	Follows a process routing sheet	Work completed in planned stages	Release and process as soon as possible. Usually capacity limited	Controlled by a Kanban signal strategy
Order release	Release order as soon as possible upon demand. No forecast	MRP or an order scheduling system based upon forecasted demand	No such concept except to control subassemblies & modules	No such concept as compared to piece-part manufacturing	Based upon a combination of actual demand and forecasts
Unit flow	One at a time	In large lots	No analogy to traditional mfg	Continuous	One-at-a-time or small lots
Push vs. Pull	Push	Push	Push & Pull	Push	Pull

1.6.1 The Job Shop

The distinguishing feature of a job shop is its flexibility. In a typical job shop, small and even unit lot sizes are produced to specific customer orders. Because the job shop must perform a wide variety of manufacturing operations, several different manufacturing processes and basic machine types are usually found, sometimes in a random fashion as new capabilities are added. General purpose and highly flexible machines are required, and each manufacturing operation is usually done by highly skilled labor. Process steps are frequently characterized by a wide range of work content, and standardization is usually difficult if not impossible.

As the demand for certain parts increase, machines usually become grouped according to the general manufacturing capabilities and characteristics represented by each work group. For example, one area might contain lathes; other milling machines; other grinders, etc. A finished product (part) will have a unique sequence of operations to be performed in a serial fashion, which usually results in highly inefficient product flow; and scheduling rarely exists. As the demand for certain parts increases, jobs may be manufactured and moved in medium sized lots or batches. The job shop has traditionally been used to produce small quantities of unique products in small batches.

A typical job shop is shown in Figure 1.3. A particular part may enter or leave the same functional area many times in a re-visitation sequence. The process flow chart for products in a job shop often looks like what is sometimes called *spaghetti flow*. *Routing sheets or travelers* are used to specify particular routing sequence with operating instructions. In a typical job shop, there may be thousands of different product routing sheets. Material handlers, forklift trucks or *pusher dogs* are often used to manually transfer a part from one machine to another. Sophisticated sequencing and scheduling rarely exists. A typical Job shop is shown in Figure 1.3. A job shop becomes extremely difficult to manage as the company grows, resulting in long throughput times, large in-process inventory and unpredictable delivery dates. Charles Carter, an engineer who worked for Cincinnati Milacron, performed a study where he tracked several parts and part types as they moved through the factory. He found that parts moving from start to finish spent about 90% of their time waiting or moving from one location to another, and only about 5% of their time is actual value-added processing (raw material transformation). Further studies revealed that even when a part was at a machining operation; only 30%-50% of *value-added* time was actually material transformation.

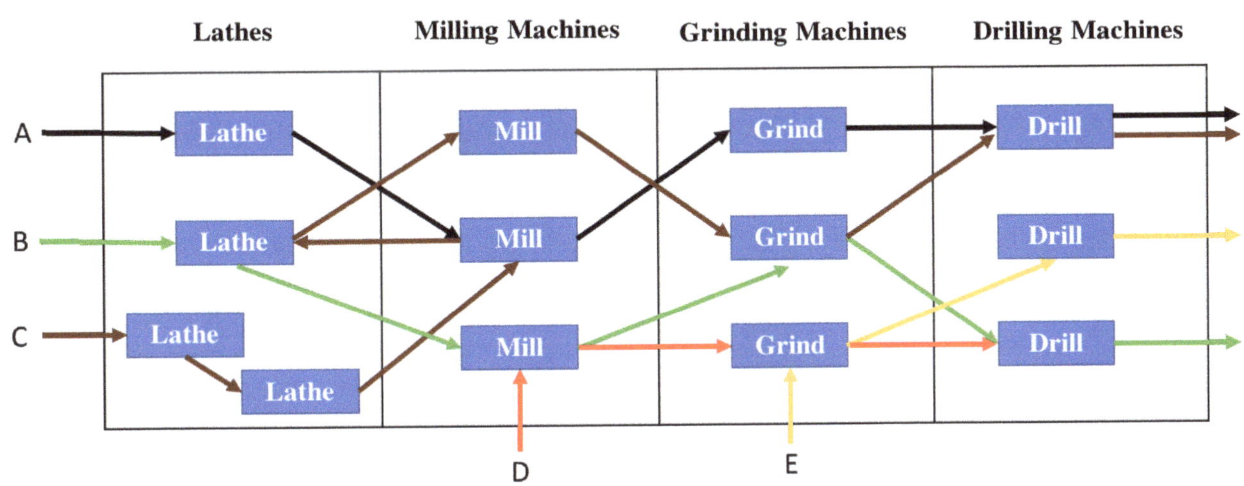

Figure 1.3
A Typical Job Shop

The remaining percentage of time was spent waiting on set-ups, tool changing, loading and unloading parts and inspection. Thus, the total time actually adding value was only 2%-3% of the product cycle time. These figures have since been verified many times, and even seem to hold in other type of manufacturing or assembly operations.

In recent years, programmable controllers and CNC/DNC capabilities have created standardization within certain classes of manufactured parts. The modern job shop still produces a lot of low volume parts, but by grouping orders and building make-to-stock, larger lot sizes or batch production can be achieved. Manufacturing lot sizes are often determined by the well-known Economic Lot Size (EOQ)equation (Wilsons EOQ). This is due to the simplicity of EOQ model calculations and not due to the efficacy of these models. The basic assumptions of EOQsimply cannot be met in a job shop. In later Chapters, we will show that a basic pull-strategy with load leveling yield far superior operating systems.

Basic batch and large lot production are usually used to satisfy continuous customer demand or relatively stable yearly demand patterns for an item. Because the production capability of any one machine often exceeds customer demand over a given period of time, most job shops build-to-stock for each item. The machine is continually being reconfigured to build value into a product at a particular step in its routing sheet. The corruptive effects of this policy are insidious and often financially large. It is necessary to build large quantities of work-in-process or make-to-stock in-process inventory from which products are pulled as needed. Perhaps worse, each part can experience delays at each processing step as the machine needed is being set up and being changed over.

In the case of DNC or CNC machines, these tools are often designed for set-up time reduction which should produce higher production rates. It has also been the goal to use higher speeds and feeds with multiple cutting tools. For example, automatic lathes capable of using many stored machine tools with computerized changeover will far outperform manually operated lathes. These machines are typically equipped with specialized jigs and fixtures, which increase precision, accuracy and output rates. However, such superiority over manually operated machines comes with a price. If the tooling required to manufacture a particular part is not part of the machine tool set on, the part cannot be produced. If tool sets are added to mitigate this problem, they may be highly underutilized and subject to obsolescence. One of the authors of this book was asked to perform an analysis of such a system with many automated machining centers because the cost of tooling was beginning to bankrupt the company and other alternatives were needed. The company was literally going broke trying to maintain a large number of automated cutting tools, jigs and fixtures. Many domestic products are made in small to medium sized job shops. Such manufacturing systems are often called machine shops, specialty shops, foundries or press working shops to name a few. It is unfortunate that a wide variety of names exist, because they all exhibit similar serious problems. The old adage is true… *a rose by any other name is still a rose.*

It is estimated that as much as 75% of all domestic manufacturing in the United States is done in job shops with between 5 and 100 machines with production lot sizes of 5-100 pieces per order. Hence, traditional and modern job shop operations constitute a large part of our gross national product. All forms of job shop operations can greatly be benefitted by the allocation of lean thinking and lean engineering.

1.6.2 The Flow Shop

When product demand is fairly stable and is of sufficient volume to support large lot or batch size production, a *flow shop* configuration becomes a superior manufacturing option to a job shop. A flow shop emerges when a single product or a group of products can be made on identical or similar dedicated machines at each step in a routing sequence. Traditionally, the flow shop is arranged such that the manufacturing line is serial and sequential. Flow shops arranged in such a fashion can be very long with many different types of machines arranged such that the output of one machine proceeds directly to the next machine in the manufacturing sequence. A flow shop is shown in Figure 1.4.

Flow shops are built for production runs which are larger than a Job shop and stable through time. Parts are transported between machines by material handlers or conveyors depending upon the volume of raw material or work-in-process being transferred and the manufacturing lot sizes. This form of manufacturing or assembly system can exhibit very high production rates and higher levels of automation than a job shop. A single flow line may have highly specialized equipment dedicated to the manufacture of a particular product. A flow shop might be designed to produce parts from cradle to grave, subassemblies or intermediate part types. The typical capital investment in machines or machining operations is usually very high. The extra cost of a flow line is justified by higher production rates and a large return on investment. In most cases, the sophisticated manual skills in a typical job shop are transferred to automated or semi-automated assembly/manufacturing operations. Ideally, items or parts flow through the system either one at a time or in small lots. The time that each item spends at each process station may vary considerably.

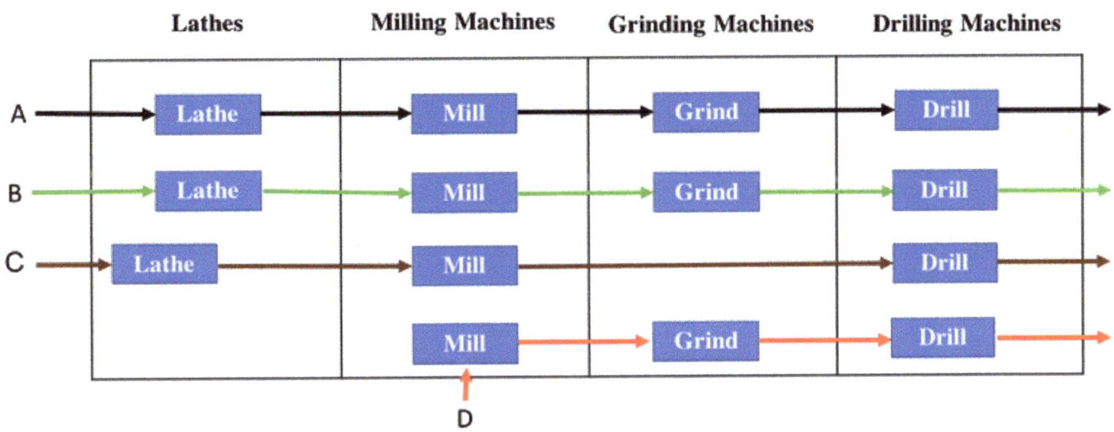

Figure 1.4
A Flow Shop

As product demand becomes fairly stable in large identical production runs, the flow shop will give way to the modern *mass* or *high-speed assembly line*. A prototype example of a high-speed assembly line is the modern automobile assembly line. A typical rate of production might be 60 cars per hour or one car per minute. This of course implies that any one car cannot spend more than one minute per car at any one machine or workstation. If material handling times, setups, etc. exceed 1 minute...there must be excess in process inventory to sustain the rest of production. Worse, this in-process inventory must be regularly replenished. This is often done on an idle shift or by subcontracting. To be economically feasible, high-speed production lines must be *balanced ,operate continuously over long periods of time*, and *never run out of raw materials......* all of which contribute to production variance and increased costs.The capital investment for a typical automobile assembly line is in excess of $1.5 billion. These manufacturing systems are designed to produce very large volumes per year, and design changes usually follow a 12-month cycle. Part of this phenomenon is the insatiable appetite of the consumer for more sophisticated and attractive vehicles on a yearly basis. It is unlikely that due to capital investment costs that the process could be radically changed less frequently.

Paradoxically,automobile consumers expect and demand many variations in color, add-ons and style. So, to meet the demands of a free economy and be competitive in an international market, automotive assembly lines need to be designed to exhibit *flexibility* and *versatility*, while retaining the capability to mass produce. In the flow line manufacturing system, the processing and assembly workstations are arranged in accordance to the product's sequence of operations. Workstations or machines are arranged in a serial production line with only one type of operation being performed at each processing step. Duplicate workstation (parallel machines) might be added to balance the line and perform the same function. The entire line is designed to produce a product or subassembly of the same type or from the same family of products. In most cases, there are little or no changeover and setup operations, and if there are any, they may be long and complicated. Hence, basic line integrity and functionality is difficult and expensive to change.

1.6.3 Mass Production

A mass production line was pioneered by Henry Ford. It is an extension of a flow line to produce more products in shorter periods of time due to extensive automation and control. Mass production facilities require a steady demand of product over relatively long period of time. As demand dictates a shift from batch mode to high volume production, mass production lines emerge. Mass production lines are usually fed by other high-volume flow lines which insert key subassemblies into the main line.

A variety of approaches have been used to develop machines and machine centers which are highly effective in mass assembly lines. Line efficiency and effectiveness usually depends in a large degree to how closely coupled engineering and product design are to the mass production line. Standardization of products, methods and methodologies are key to balanced line operation. If a particular part is fairly standardized and can be manufactured in large quantities, specializedmachines with a minimum of *touch labor* can be designed and implemented. It should be noted that the emergence of mass production lines with little or no skilled labor intervention has created havoc in many old corporate planning systems. Key performance metrics and product costing can no longer be based upon direct labor hours per part. This has given rise to *Activity Based Costing* (ABC) and other new accounting schemes. Direct labor costing methods are archaic and simply do not apply to high speed, highly automated production lines.

Mass Production: Ford & Automobile Assembly Line in 1913

- **Huge gains in productivity**
  - **Complete automobile chassis every 93 minutes**
  - **Previously:728 minutes**
- **Car prices dropped**
- **Millions of ordinary people start purchasing automobiles**

American Henry Ford (1863-1947) facilitates the factory conveyor system – each worker performs a specialized task

Figure 1.5
An Old Ford Mass Production Line

1.6.4 The Product Shop (Fixed Position Layout)

In some types of manufacturing systems, the product must remain in the same geographical area, and product transformation must be done in a fixed location. The modern jet passenger plane is a good example. All raw materials, people, processing, machining, assembly and inspection operations must be brought to the product site as shown in Figure 1.6

Figure 1.6
A Product Shop

In the late 1920's Henry Ford built his automobiles using a product shop. Skilled laborers with specialized tools and raw materials were brought to a single point of assembly. Some years later, Henry Ford used a skid drawn on by horses to move the car-in-progress along a straight line to the same (now stationary) set of workers and raw materials. This was the first modern high-speed production line.

The product shop might be considered an archaic, historical production strategy today but in fact it still exists in many forms. Large ships, aircraft, dams, machines cruise ships and skyscrapers are just a few product shop examples. A typical product shop may in fact contain elements of job shops, flow shops and mass production within its boundaries. Subassemblies and components may even use mass production lines. All product shops are fertile ground for Lean Engineering principles.

Clearly, the product shop is used to produce very large and often massive products. However, this is not always true. Some products must be manufactured in a single, fixed location because of environmental concerns. For example, a plating operation. Work is usually scheduled by project management techniques like the *Critical Path Method (*CPM) or the *Project Evaluation and Review Technique* (PERT). These methods use precedence diagrams and specialized analysis algorithms to sequence work and maintain project schedules (time and cost). Project shops are usually very complex, labor intensive and very expensive, often producing only a single product for millions of dollars, like the space shuttle.

1.6.5 Continuous Processing

Lean engineering and lean practices have not been widely used in the continuous processing industries, but significant cost savings and waste reduction have recently been reported (King and Floyd). In continuous processing, products usually flow in large quantities measured in barrels of product or thousands of gallons. Hence, these systems are often referred to as *Flow Production Systems.* Oil refineries, chemical processing plants and Biofarma production facilities are good examples of Continuous processing, although each of the previous examples of continuous processing actually have mass production components associated with many final products (cans of oil, plastic products, bottles of pills, etc.). In reality, most continuous processing facilities are not strictly processing products continuously, but rather in a mixture of high-volume flow lines. Movement of liquids, gasses and powders do flow continuously and require extremely complicated transformations of raw materials (distillation columns and separators)

Figure 1.7
An Oil Refinery

It is also true that facilities such as oil refineries are the least flexible of all mass production systems, but paradoxically very efficient with large output rates. The control systems are very complex, and work-in-process is hard to measure and discretize. Lean Engineering in these types of systems requires specialized knowledge in Chemical engineering, Materials engineering and Metallurgical engineering in many cases. There is an interesting intersection between the Lean engineer who is attempting to execute waste reduction and the Chemical engineer who is usually assigned the task of managing and controlling the system

1.6.6 The Lean Linked-Cell Manufacturing System (L-CMS)

The Toyota Production System (TPS) has developed a radical new approach to the design and operation of almost any piece-part manufacturing system and many other types of manufacturing systems as well. In a linked-cell manufacturing system, all parts, subassemblies and some finished products are each made in a U-Shaped Manufacturing Cell such as that shown in Figure 1.7. Mass production is designed to be versatile and flexible to respond to both new product and changing demand. Every Manufacturing area is reconfigured to reduce time in the system, The p*roduct final assembly line*is changed over to a *mixed model final assembly line* so that the daily production rates for erratic customer requirements can be *leveled* and *smoothed.* Production leveling, also called *production smoothing* is a set of Lean Engineering principles for reducing waste and guaranteeing that each process step produces at the same output rate. When demand is relatively constant production leveling can easily be accomplished, but when customer demand fluctuates, flexible and versatile manufacturing strategies with optimal WIP must be developed.

Each part or subassembly is combined to configure a final product by linking U-Shaped Production Cells to one another such that the output of any one cell might be the input to another downstream cell. Production rates in each cell are set by downstream demand, which in turn is driven by customer demand. A Linked-Cell System is shown in Figure 1.8.

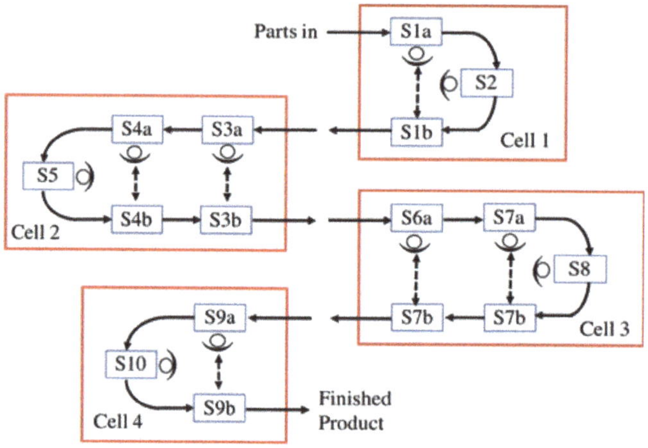

Figure 1.8
Multiple U-shaped Cells Functioning as One Production Line

The operations are either manual which require smaller capital investment or automated with simpler control functions than that usually found in CNC or DNC machining centers. All tasks and machines are performed by cross-trained, walking workers. Multiple workers are assigned to multiple tasks to achieve worker equality and line balance. Lean Cell workers are often self-balanced. Any job shop layout like that shown in Figure 1.7 can be redesigned into a U-shaped manufacturing cell such as those shown in Figure1.8.

U-shaped Lean manufacturing cells walk among designed machines with a minimum of transit or walking time. The operators usually walk in pre-assigned clockwise or counterclockwise loops with no crossing patterns. This U-shaped design lends itself well to standard operations sheets and worker flexibility. The implementation strategy to form these cells and the methodology used to rapidly adjust to changing production rate requirements will be discussed in Chapters 2 - 4. The design and operation on such systems are largely the goal of this book.

After U-shaped cells are implemented, the cells can then be linked to other sub-assembly points or final assembly with a production control system called *Kanban* which is based upon a *pull* strategy versus a *push* strategy. This will be explained in Chapter 6. Traditional production and implementation design rules are used to configure final assembly (Chapter 8). Key proprietary processes are easily protected and hidden within cell structures.In a typical linked-cell system, a predetermined level of inventory is maintained within each cell and between cells to minimize work-in progress. These inventory levels are tightly controlled by *Kanban* cards to match the desired output rate and quantity of goods produced. The system is also configured to compensate for any disturbances or variation which cannot be controlled or eliminated. Kanban card systems are discussed in Chapter 6.

Let us try to quickly summarize the key characteristics of a Lean Engineering designed production system with L-CMS production cells and compare these characteristics to other alternative configurations. It may be useful to once again examine Table 1.3.*Lean Cell design* produces parts in a highly flexible way, meaning that cell output and product mix can be altered quickly to respond to customer demand and market fluctuations. One of the myths of lean manufacturing is that it can only be applied successfully to production lines which manufacture a single product with fairly steady demand. Nothing could be further from the truth. U-shaped flexible cells are *not* dependent upon either machine location in a U-shaped line nor on machine cycle times. Indeed, families of similar parts can easily be produced in cells using one-piece flow, cross trained work, simple single-cycle machines, flexible work-holding devices, and rapid changeover tooling systems. These capabilities lead to small or unit lot size production and product-mix load leveling. Inter-cell inventory is tightly controlled by Kanban strategies and small *inventory supermarkets* (See Chapter 6). Quality is controlled within the cell, and less worker error is experienced due to training and worker experience in executing assigned tasks. Worker pride and morale is increased. It should be noted that no intra-cell sequencing or scheduling required, and no attempt is made to maximize the utilization of the machines… only the workers and a balanced workload. Cell throughput time and output rate is determined by the number of workers (See Chapter 2) once adequate production capacity is created.
A major advantage of Lean linked cells is that the basic design facilitates *one-piece flow* (OPF) and production flexibility.

OPF is the capability to move parts through the system one-at-a-time. Each machine in a cell executes a single processing step in the production sequence. Each part is checked carefully before moving to the next processing step (machine) in sequence. Volume flexibility is achieved by worker-machine separation and the instant reallocation of workers to machines. Contrary to popular belief, mixed model production in L-CMS is simple to implement as long as setup and changeover times are minimized (SMED). Simple or complex parts can be made in a properly configured cell.

Putting process steps together in a U-shaped cell allow both design engineers and skilled workers to see how parts are made and to immediately uncover production limitations and operational problems. Cell arrangement and multifunctional workers also create an opportunity to uncover any quality problems. In fact, *zero defects* are a mandate rather than a choice. The design engineer can more easily configure and implement future design changes. This is truly designing for manufacturing.

1.7 Evolution of Factory Designs

There was a time when there were no factories as we define them today. From earliest times until about 1850, most manufacturing was done by single, highly skilled workers in home workshops. This is often referred to as *cottage industries*. These early artisans were watch makers, blacksmiths, tool makers and wagon makers, etc. Any machines used for manufacturing were manually powered. In the mid-1800's the first integrated factories began to emerge. Automated machinery was being discovered, and large quantities of raw materials could be moved by locomotives or through rivers and lakes. These were typical of a wide range of *enabling technologies* which emerged to drive larger scale production. The first factory designs also evolved because power was being made available by new methods. According to Amber and Amber (1962) the major enabling technological innovation that led to modern factories was the ability to power machines on a reliable and consistent basis. The evolution of modern power systems was fueled by the emergence of steel mills. Producing and machining hard metals such as steel required significantly more power than wood or softer metals. Initially, factories were placed in locations dictated by where water could be used to turn large water wheels that drove large gears.

1.7.1 The First Factory Evolution

Closely following the availability of unlimited power was the concept of interchangeable parts by Eli Whitney and others. The first factories were built in New England next to rivers, where the water flowing downstream would turn large waterwheels. Machines were run by connecting waterwheels to large gears that drove long shafts which powered machines by connecting leather belts to gearing mechanisms. Later, the waterwheels were replaced by a steam engine, which enabled factories to be built away from waterways. Eventually, steam engines were replaced by large generators and electrical lines connected to the factory. Finally, electric motors were put on each machine in different areas or locations. The initial manufacturing configuration was job shop built around manufacturing functionality. The earliest factorieswere steel mills and gun manufacturers. Companies like Colt and Winchester invented high speed production and the race was on. Initial American gun manufacturing factories became famous worldwide ad people from all over the world came to see what was called the *American Armory System*. These were the beginnings of the *First Factory Evolution.*

1.7.2 The Evolution of the Job Shop

The job shop concept flourished because of its simplicity. Machineswere simply added as demand called for more production capacity. Modern day versions of the job shop exist today in almost the same type of functional layout. It is estimated that 50%-60% of all manufactured products in the world are by job shops, which produce finished goods in lot sizes between 50 and 200 pieces. Processes are still functionally grouped together. Similar types of machines are usually grouped in common areas separated by walls or in separate buildings. It is usually difficult to track and locate work in process without modern data collection systems, and

almost all job shops are characterized by having large amounts of work-in-process using an order release system that release orders to the shop floor as they are received.

1.7.3 Controlling a Job shop

Throughout the late 19th century and well into the 20th century, there was little or nosophisticated tracking and control systems for typical job shop operations. Engineering drawings, shop floor travelers and manufacturing specifications were introduced but most of themanufacturing control was by individual, highly skilled workers. Machines and machine tools were introduced which was single cycle, repeat cycle automatics. Some were integrated into a single platform such as the Bessemer steel making process. Products moved in unit lot sizes (custom products) or in small batches through the job shop. Then as are today many products being unique and one-of-a-kind. Littletheory or scientific management principles existed, and little integrated thought was given on how to change capacity or processes within a factory.

This all changed in the late 1890s and early 1900s when a mechanical engineer named Frederick W. Taylor developed the concepts of *Motion and Time Study*. His pioneering work in improving worker productivity also gave rise to the concepts of *Scientific Management*. Taylor developed "preferred practices" and "standard times" for all sorts of manufacturing activities. Taylor also discovered and invented high-speed steel which doubled cutting speeds.

A colleague of Taylor named Henry Gantt developed charts and procedures for scheduling and controlling work in a factory; he is famous for the *Gantt chart* which bears his name. Almost at the same time, Carl Barth developed the *slide rule*, which was the first modern computational device for complex calculations. Post, Keffel, and Esser, both produced slide rules which soon hung on the belt of every engineer. In this same era, Frank and Lillian Gilbreth began scientific studies based upon *micro motion analysis*. Lillian Gilbreth was formally trained in psychology and sociology, and along with her husband Frank soon developed theories directed to the motivation and attitudes of the workforce. Frank and Lillian more than any other investigator(s) pioneered the *best way* to do a particular task.

The pioneering work of Frederick W. Taylor, Henry Gantt, and Frank and Lillian Gilbreth was the innovative and scientific spark which spawned an entirely new engineering discipline called ***Industrial Engineering*** *which grew up in the era of mass production.* Modern Industrial and Systems Engineering provides the foundation for the new and modern Lean Engineer introduced in this textbook.

1.8 The Second Factory Revolution

The second industrial revolution was development of the *flow shop*. This factory revolution was fueled by the genius of Henry Ford. Henry Ford defined the concepts of *economy of scale* and *work specialization* in his Model T (1908-1927) and Model A (1927-1931) Ford automobile production systems. His innovative approach to large scale production was built upon division of labor, the moving assembly line and the principles of time and motion study. Ford was the first to attempt to minimize work-in- process on a large scale, and he was the first to fully integrate the *supply chain* with the production system. He exploited standard setup and tooling procedures used with interchangeable parts and product standardization. His most famous saying might have been: *You can have one of my Model A motor cars in any color as long as it is black.*Ford and his manufacturing engineers, led by Charles Sorensen, were responsible for many product design innovations and large-scale manufacturing strategies. For example, Ford developed a streamlined process for manufacturing engine blocks. The basic Model A design was a single cast iron engine block instead of pieces bolted together. The use of this engine block led to more power and decreased the weight of each car.

However, almost without question the single most enabling technology Ford pioneered was that of interchangeable parts based upon rigid standards of measurement. Ford demanded that every car produced met standard specifications. He pioneered *gage calibration* to reduce variance. In concept, the second factory revolution was founded upon the principles of the first, but the second factory revolution was rooted in standardization of work and specialization of labor.

What we now call *flow manufacturing* began in the early 1900s and matured with the first Ford motor Company moving assembly line around 1913. By 1915 Henry Ford and Charles E. Sorenson had achieved the first integrated manufacturing system design. All elements in the Ford Manufacturing System….raw materials, machines, machine tools, people and inventory had been integrated into a single, remarkable system. This new factory design and production system attracted people from around the world; all came to see how this new system of manufacturing worked. So, what was the problem?

Henry Ford was obsessed with people productivity and cycle time reduction at each workstation. He implemented specialists who worked only one job. But how would you like to *install shock absorbers* 40 hours a week every week, every year? Ford soon had critical labor problems. He paid his workers an unheard of $5.00 an hour and still could not maintain a critical, trained labor force. But this was not his most serious problem. Obsessed with efficiency, production rate and minimizing costs, his philosophy of standardization soon caught up with him. How would you like to have the option of buying an automobile with nothing but a black car with black seats? Enter Alfred P. Sloan.

Ford refused to change his system and create more variety. His customers wanted other colors, other interiors and other options. Variety did not resonate with either Henry Ford or his Ford production system. Alfred P Sloan recognized the superiority of Henry Ford's manufacturing system, but he adopted the basic concepts of supplying variety to the customer. In retrospect, Henry Ford made a cardinal error…In his obsession to standardize and reduce costs, he failed to recognize customer satisfaction. Sloan reconfigured General Motors Corporation. Not only did he produce a wide variety of specific automobiles, he did so by adopting the basic mass production concepts of Henry Ford. Exploiting Ford's pioneering work; he formed production lines for Chevrolets, Buicks, GMC vehicles and others, adopting flow line and mass production concepts within product lines (car and truck) or car families. The strategy worked, and by the mid 1930's General Motors had passed Ford Motor Company in both volume of cars sold and corporate profits.

General Motors Corporation designed a hybrid manufacturing system, based upon the best manufacturing strategies of job shops, flow shops and mass production. Subassembly lines were designed to feed into main assembly lines. Subcontractors were developed using the same production strategies. This integrated system design enabled General Motors Corporation to produce a wide variety of cars and trucks, while maintaining standardization within product families at lower unit costs. As technology innovations grew, individual machines and processes became highly automated and autonomous. Subassemblies and assemblies were produced in a fixed cycle time on paced, moving assembly lines. The principles of division of labor were still used but unskilled laborers with the proper training replaced the traditional craftsman.

The mass production system pioneered by Henry Ford reached its zenith in the war years between 1928-1944. America was at war and war machines (airplanes, tanks, guns, clothing, etc.) had to be manufactured in great quantities and delivery in a minimum amount of time. During the war, all kinds of new methods, machinery, jigs, fixtures and tooling were developed to support mass production. During this era, the modern *transfer line* evolved. The transfer line is a modern version of the flow line for products like car engines produced in large volumes (200,000-400,000 engines per year for example).The enabling technology is automated, using repeat cycle, mechanized machines. The system demanded interchangeable and standardized parts based upon precise standards of measurement. The transfer line is designed for large volumes of identical parts. These specialized flow lines are very expensive and inflexible.

Work-in-process, which feed the transfer line, usually come from a separate flow shop in large lots; held in inventory for long periods of time; and then moved to the line in large quantities where they are automatically loaded. This type of mass production line was pioneered and implemented during WWII to produce large quantities of fighter planes, tanks and jeeps.
A new product could typically only be made by going through a long changeover and retooling cycle.

After WWII, this form of production line continued in the form of manufactured automobiles such as Ford, Dodge and Chevrolets. Just as it appeared that this new system would dominate automobile production for years to come, a new manufacturing system emerged, which would eclipse all known production systems at that time. This new manufacturing paradigm was called the *Toyota Production System* (TPS), and it would revolutionize the way all manufacturing systems were perceived, designed and implemented. The emergence of TPS ushered in the Third Factory Revolution.

1.9 The Third Factory Revolution

The third industrial revolution was pioneered by the Japanese Toyota Motor Car Company and was led by a genius called *Taiichi Ohno.* The evolution of TPS was generated by the desire of Toyota Motor Company to produce small quantities of Toyota automobiles, at an affordable price, for the emerging Japanese market; and later to overtake Ford Motor Company, Chrysler and General Motors as the leading car manufacturer in the world. Both of these goals were accomplished. The first step that Ohno took was to redesign the job shop into *manufacturing cells*. These cells were configured to hold minimum work-in-progress and produce at an output rate determined by consumer demand.The capacity and output rate of these cells were determined by the *number of laborers* in the cell and not by the *number of machines*. This was a radical design concept. Machines were upgraded to single-cycle, automatics, so operators could run many machines. Final

Taiichi Ohno

assembly was accomplished by merging assembly lines for different models of cars into one final assembly line synchronized by leveling production across all assembly. Toyota called this smoothing of production and it is now called *Mixed Model* Assembly (Black). This mixing of products (any model, any color) requires leveling the demand for subassemblies on the supply chain. Workers on final assembly were empowered to stop production at any time using pull cards if a quality or operational problem was detected. Next, restructure subassembly lines into U-shaped cells with balanced workloads achieved by using roving workers with fixed boundaries and operation assignments to meet output requirements. Methodologies were developed to achieve flexibility, including controlling output as a function of the number of workers in the cell (Chapters 2 and 3).

These innovations were soon followed by a *Kanban system* to control production and reduce inventory, WIP. *Production* and *move* Kanban cards were configured to control the movement of work-in-process and to trigger production runs. On the supply side, external suppliers were integrated into the production system by implementing just-*in-time* rules. Internally, each *downstream* production cell was viewed as an internal customer of the *upstream* production cell. Material handling and delivery was controlled by sending empty carts to trigger upstream production and full carts to satisfy downstream demand. The invention of *autonomation*by Shigeo Shingo replaced pure automation. Automation separated workers from machines through mechanisms that automatically detected production abnormalities and automatically shut down the machine. Many traditional, complex control and operational functions were redesigned to reflect *human reasoning*. This became known as autonomation; automation with a human touch. These efforts to fully exploit human reasoning and actions into a production system did not come easy. Shingo said later that there are twenty-three stages between purely manual and fully *automated* work. To be fully automated, machines must be able to detect *and* correct their own operating problems, which were, at that time, too complex and not cost-effective. However, Toyota attests that ninety percent of the benefits of full automation can be gained by autonomation. The concepts of Just-in-time were fully addressed and exploited by Toyota. For Just-in-Time (JIT) systems to work, it is absolutely vital to produce zero defects. Defects and rework always disrupt the production processes, create variance in the system, and disrupt the orderly flow of work. Taiichi Ohnoconsidered JIT&Automation important pillars upon which TPS is built. The Japanese word for autonomation is *Jidoka, the decision to stop and fix problems as they occur rather than pushing them down the line to be resolved later*; is a large part of the difference between the effectiveness of Toyota and other companies who have tried to adopt Lean Manufacturing. Therefore, Jidoka is an important key element in successfully implementing Lean Manufacturing. If Autonomation and JIT are the pillars upon which a lean manufacturing system is built, the *foundation* upon which these two pillars rest is the aggressive and relentless pursuit of *waste elimination*. In fact, waste elimination has been the single most concepts identified with *Lean Thinking* (Womack and Jones). Lean thinking demands that direct labor utilization be maximized; waste of time and waste of materials be minimized; and machine utilization adjusted to accommodate fluctuations in demand. These goals are achieved by implementing U-shaped manufacturing cells, one-piece flow and the systematic removal of inventory and WIP by carefully designing a *Kanban Control System*. We call this new design a ***L**inked **C**ell **M**anufacturing **S**ystem* (L-CMS).

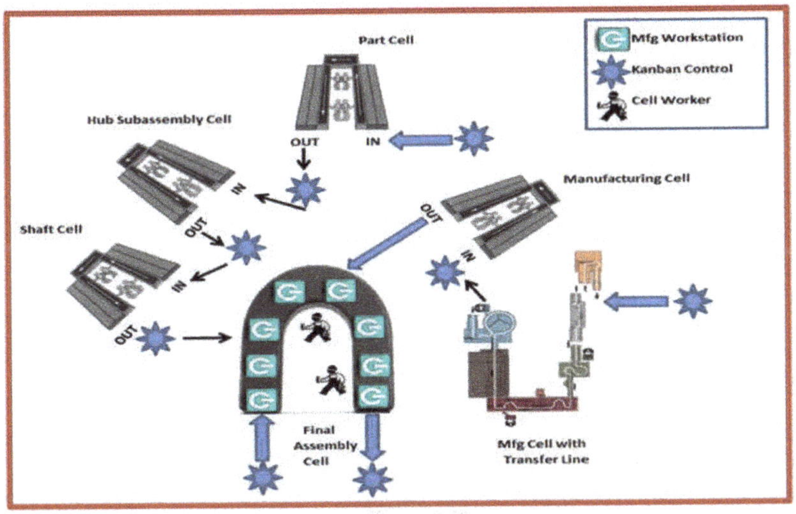

Figure 1.9
A Linked-Cell Manufacturing System Scheduled by Kanban Cards

Part Description				Part Number	
Smoke-shifter, left handed.				14613	
Qty	20	Lead Time	1 week	Order Date	9/3
Supplier	Acme Smoke-Shifter, LLC			Due Date	9/10
Planner	John R.	Card 1 of 2			
		Location		Rack 183	

A Typical Kanban Card

Finally, a key element in the TPS was the ability to produce final product *only when needed, in the right quantities at the right time*. This goal was accomplished by realizing that scheduling production based upon a *push philosophy*, where systems abased entirely upon forecasted customer demand such as MRP II or ERPcould only result in excess WIP, long lead times, long throughput times and inflated finished goods inventory. This gave rise to a new production control system called a *pull system.* Pull systems *ideally* only produce goods and services when the customer demands goods and services, only in the quantities ordered.

The procedure is often misunderstood. Suppose that a production line consists of three workstations, WS1, WS2 and WS3. The entire system is idle. A customer asks for delivery of 100 units of finished product in lots of 10. When the order is received, it is immediately passed to WS3, who is authorized to produce 100 units. Of course, WS2 must send 100 units of work-in-progress to WS3 before the final production operation can be started.

A Kanban (card) signal is then sent to WS2 to supply 100 units of in-process work to WS3. WS2 repeats this procedure and sends a Kanban signal to WS1 to produce 100 units as soon as possible. Suppose that the line is balanced, and each workstation takes 1 hour per part. Units are moved between stations on pallets in lots of 10. It is clear that this pure pull system has a major problem. WS 1 can begin work immediately, but WS2 is idle for 10 hours waiting on a pallet of 10 items. WS3 cannot begin for 20 hours. The first part will not be produced until 30 hours have passed from when the order was received, and the first order will be filled in 31 hours. Of course, one could immediately say, put the stations side by side and movement of parts in unit (1) lot size. Now the first part will come off the line in only 3 hours. The optimal procedure is to immediately start production at all workstations as soon as the order is received. After 1 hour the first unit from WS2 will reach WS3. Meanwhile, WS2 will have immediately begun work at time zero and so would WS1. The entire order will be filled in only 100 hours. However, WS3 must have 1 unit of in-process work finished by WS2 to immediately begin; and WS2 must have 1 unit of work completed by WS1. If raw materials are available when the order is received, WS1 can immediately begin work (assuming zero delivery time and no setup time for all three Workstations): It is obvious that the line would never be idle. To make sure things go smoothly,

1 unit of raw material might be stored for WS1; Note that the system will be restored to its original state after 100 units are produced by Workstations 1, 2 and 3. Of course, the price to pay is at least 2 units of WIP at each workstation at all times during the manufacturing cycle, and 1 unit at all other times. So much for the *myth* of zero WIP! In fact, this is an example of a *production line law* which we will prove in Chapter 9: In an N station serial, balanced production line, the optimal system WIP is N units when the line is operating, which is exactly what we determined in this example. This is called producing *Make-to-Stock inventory* to minimize non-value added (idle) production time. The work-in-process needed to immediately start the line is stored in what is referred to as a *Supermarket*. These concepts for both single product and Mixed Model Production will be further discussed in subsequent Chapters.

1.10 Lean Transformation, Kaizen Events and Continuous Improvement

The relentless reduction of waste and WIP is the primary goal of any Lean Production System. Driving both waste and WIP to lower levels will *always* reduce unit cost and improve customer satisfaction. It also usually results in better quality. Toyota properly recognized that if the entire production system is to be optimized (or run as efficiently as possible), every subsystem must be optimized, but only within the context of every other system component. The entire manufacturing enterprise should (ideally) run like a fine Swiss watch. Smoothing output and supporting balanced production is facilitated by one-piece flow with minimum set up and changeover times Of course, initially switching to a Lean Production system cannot be accomplished by instantly transforming the entire factory unless it is a Greenfield Factory; it must be done in pieces. It must be stated in the strongest terms that one *CANNOT* just pick areas to transform without knowing how the entire system will be affected. This requires system modeling techniques to choose the correct sequence of Kaizen events, such as a Systems Simulation analysis or Queuing Network Analysis (Phillips and Black or Curry and Feldman). Piecewise improvements are accomplished by what is called *Kaizen events*.

Finally, and possibly most important, is that the system is staffed and operated by workers empowered to make continuous improvements and understands how the entire system works. It should be clear that by reducing all setup and changeover times to as small as possible (SMED) eliminates wasted time, creates great flexibility and facilitates smaller lot sizes, balanced production lines and faster throughput times.

It has been almost 50 years since the TPS was introduced, and we are just now beginning to develop a scientific approach to Lean transformation. We have learned that the transformation to Lean is not based strictly upon computer hardware or sophisticated software, but the implementation of a set of Lean Transformational Activities. These set of principles and activities are collected into a body of knowledge which we call ***Lean Engineering***. The presentation and explanation of these principles is the subject of this book.

1.11 The Lean Engineer

The new engineering discipline that we call a *Lean Engineer* is closely aligned with Industrial and System Engineering (ISEN). ISEN has always been tied to manufacturing and factory design. One might conjecture that Henry Ford was the first Industrial Engineer by practice, building upon the work of Frederick Taylor, Henry Gantt, Frank and Lillian Gilbreth and other pioneers. Over the past 100 years, Industrial Engineering has grown into an engineering discipline which embraces Operations Research, Production Planning & Control, Sequencing and Scheduling, Inventory Control, Human Factors & Ergonomics, Methods and Time Studies and Total Quality Assurance. The modern industrial engineer is well schooled in Probability and Statistics to address complex system modeling requirements using Queuing Theory and Systems Simulation. In addition, over the past 20 years many Industrial Engineering departments have broadened their scope of application to *Systems Engineering* and a discipline called *System of Systems Engineering*. The core tools of Industrial and Systems engineering are ideally suited to adopt the principles and Lean Engineering methodologies in this book and become a new breed of Industrial Engineer called a *Lean Engineer*. Although not thoroughly covered in this book, we must stress that Industrial Engineering is also ideally suited and trained to use the modern tools of *Six Sigma Engineering*. Six Sigma Engineering forms a core discipline to execute *variance reduction*…the prime enemy of Lean transformation. We would be negligent and short sighted to not discuss another emerging cornerstone of Lean Engineering: *Environmental and Green Engineering*.

1.12 Environmental and Green Engineering

While all forms of service and manufacturing industries are becoming more efficient, saving time and cost by embracing Lean concepts, there is an emerging if not mandated need to integrate green and environmental issues into Lean transformation. The growth of the human population, which is now proceeding on an exponential rate, has resulted in more and more scarce resources being consumed. In direct correlation to these phenomena has been the need to consider both product disposal and environmental impact. Both issues are beginning to change the face of factory design, mass manufacturing and Lean engineering. New factory designs, product design and manufacturing methodologies are needed for recycling, reuse, recovery of scarce resources and zero waste. It will not be long before a new *Energy Engineer* will also enter the scene…if not already. The need for higher gas mileage, lighter and more disposable cars and zero pollution is already changing the way automobiles are designed, manufactured and disposed. The move has already been initiated to develop new and innovative ways to both supply and reduce energy consumption. The computer controlled; electric vehicle is no longer a dream but a reality. Who will lead this new wave of engineers to accomplish pollution free goals? Again, the *Lean Engineer* is better suited than any other engineer to embrace these needs because of their systems engineering training and unique perspectives.

Green Engineering focuses on the design, manufacture and commercialization of processes and products that can be produced quickly and economically with a minimum impact on *renewable resources*. The term renewable resource is somewhat misleading, since almost any resource consumed at a faster rate than it can be reclaimed or renewed can become a scarce resource, and eventually non-renewable. Feasible and economical solutions to green engineering problems are logically the only set of solutions that will be adopted by any industry or service sector. We will offer only 3 alternative groups of potential benefits, although there are many more.

- Eliminate waste through more efficient and wasteful methodologies and designs. This is clearly within the scope of Lean Engineering and intersects one of its main platforms (waste elimination). This usually means going beyond regulatory mandates and embraces green principles at the source.

- Engineer out costs associated with environmental, energy and green compliance.
- Exploit marketing and sales advantages associated with this new breed of products

Since about 1995, the automobile and processing industries have initiated several efforts in this area…some because of legislative mandates and some to reduce Life Cycle Costs. Concurrently, lean transformation activities and new management structures have been implemented to support these new initiatives.

Toyota has attempted to embrace these principles into TPS and has made significant impact. Toyota's 5R program was initiated to reduce pollution, reduce energy consumption and address *environmental impact*. Tools such as *Green Stream* and *Energy Mapping* have been developed, following the success and principles of Value Stream Mapping. Currently, green methodologies, principles and applications are surfacing all over the world. In 2003, a summit meeting on green engineering produced nine fundamental green principles from a group of 65 participants. Applications and extensions of these nine principles are now beginning to surface in the technical literature, but actual factory transformations are slow to develop. This is an exciting new field of research for the emerging Lean Engineer.

Both a wide variety of Environmental Management Systems (EMS) and ISO standards have been developed to guide this transformation. Sadly, no one engineering discipline has elected to champion Lean, Green, Environmental and Energy efficient transformations. We contend that this new breed of engineer should be called a Lean Engineer and should be spawned out of Industrial and Systems Engineering; just as Industrial Engineering emerged out of Mechanical Engineering during the First Industrial Revolution. This book will introduce the emerging concepts of a *zero-waste, green and environmentally sensitive factory*. Although the focus of this book is on manufacturing, it will become obvious that the principles of Lean Engineering can be applied to service industries, agriculture, banking and other industrial sectors including oil and gas refining operations. Another fundamental goal of zero waste is to recycle all finished products back into some renewable resource or back into the marketplace as a derivative product; just as a caterpillar morphs into a butterfly. At this time, a unified engineering discipline, its fundamental tools and education required, and its professional identification is waiting to be defined and exploited. The best we can hope for in this book is to energize and stimulate such an engineering discipline. We call this ***Lean Engineering***.

Review Questions

1. Define a Manufacturing System.
2. What do you understand by internal customers?
3. List five basis manufacturing system designs.
4. Explain Job Shop Design.
5. How does Flow shop different from Job Shop?
6. List the differences between continuous processing and lean manufacturing.
7. What is Kaizen Event? What is a Kaizen Blitz?
8. What is a Green Engineer?
9. What is an environmental engineer?
10. What is a Lean Engineer?
11. Who was Henry Gantt?Who was Taiichi Ohno? Who was Shigeo Shingo?
12. Who were Frank and Lillian Gilbreth?
13. Who was Frederick Taylor?
    What role did Henry Ford play in the evolution of Manufacturing System?

Chapter 2
The Anatomy of a Lean Manufacturing System

Perhaps we should stop looking for the sharpest needle in the haystack, and find one that we can sew with March Siman, Organizations

2.1 Introduction

In any factory, *manufacturing processes* are assembled together to form a *manufacturing system* (MS) to produce a desired set of goods. The manufacturing system takes a set of orders from customers, translates these orders into resources and raw materials to support production, adds value through a series of manufacturing processes and transforms these raw materials into final goods. It is important to distinguish between the manufacturing system and the *enterprise system* which is comprised of marketing, sales, production planning, supply chain management, finished goods inventory, product delivery and a myriad of other support functions. Figure 2.1 provides a composite view of the enterprise system. All Manufacturing Enterprise components function for only one reason: Manufacture goods and products for the consumer and generate a profit which will satisfy both the owner(s) and stockholder(s), if any.

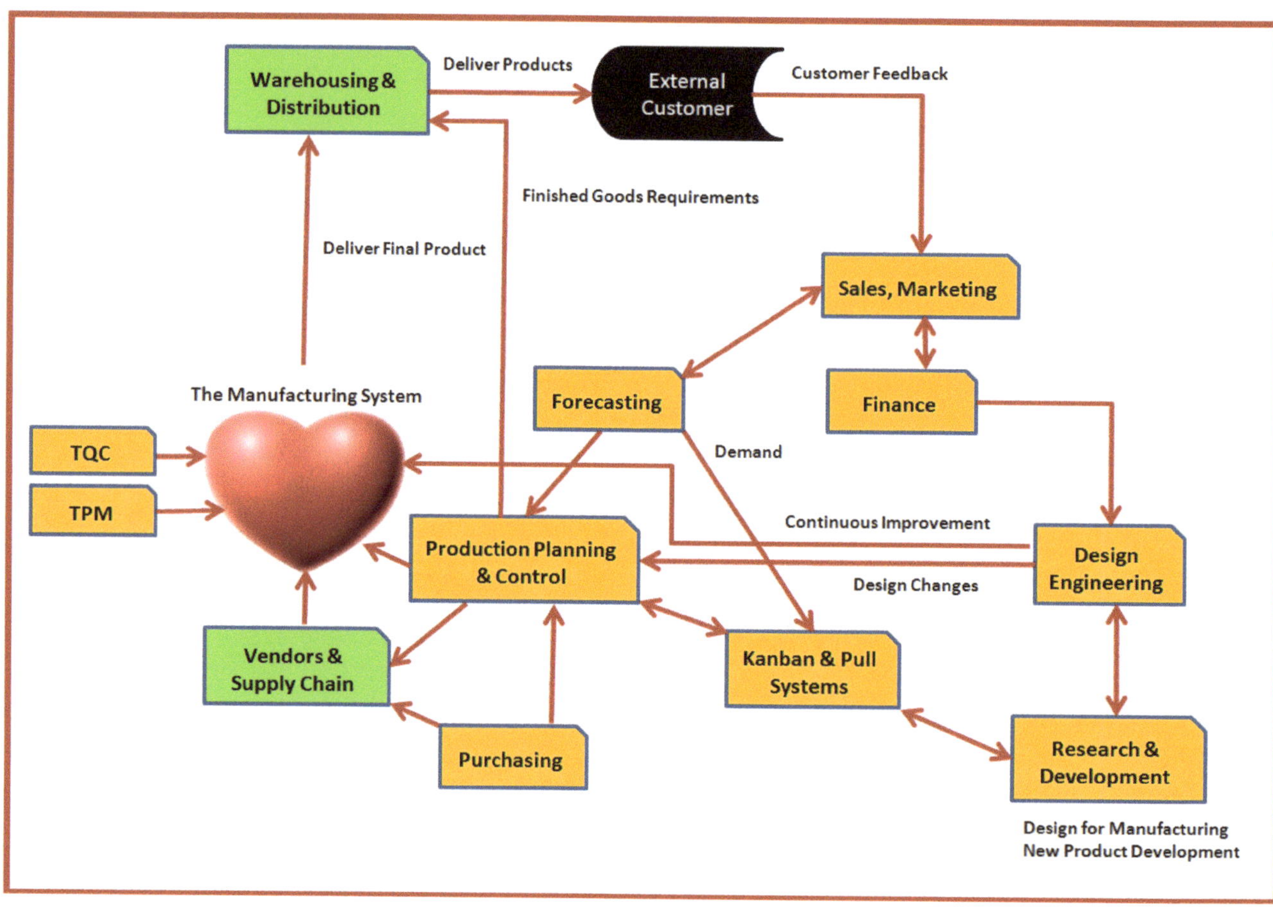

Figure 2.1
Functions of the Production Systems

The heart of any Enterprise is the Manufacturing System. All corporate activities exist to nourish and support the manufacture of goods and services. The enterprise system exists to support the manufacturing system, which is where raw materials are transformed into finished products. The enterprise system is composed of people and information used in product design, analysis and production control functions. These functional areas exist for only one reason…to make money. The entire enterprise system contains all aspects of the business, including design engineering, manufacturing engineering, sales, advertising, production control, inventory control and most importantly the actual manufacturing system.

2.2 The Manufacturing System

The total number of people and processes that actually produce a desired finished product or provide goods, services and raw materials to produce a finished product is called the *Manufacturing System.* The total manufacturing system is a *complex arrangement of physical elements which are guided by measureable parameters* (Black, 1991). Physical raw materials are transformed by a series of *process steps or workstations* which add value to the work in process. The degree to which these measurable parameters dictate both the individual process steps and the entire sequence of operations largely determine how well the system can be managed and controlled. The entire manufacturing system must be continuously monitored to enable the movement of materials, the role of people and the management of value added processes.

The goal of modern scientific management is to design an enterprise system that operates harmoniously as one living organism, and include all the manufacturing system and subsystems as functional components. In nature, there is a law called the *constructional law (*Bejan and Zane) which identifies two basic properties of anything that flows through the system. For example, a river will form through the input of many small tributaries. The main river is a composite structure which is defined by its component parts. This is a natural structural evolution that supports the movement of surface water. In a properly constructed manufacturing system, the same type of principles will exist. For a system to operate effectively and efficiently over time, its configuration (design) must evolve in such a manner to fully support the harmonious movement of parts, resource subassemblies and raw materials through the system.

This paradigm exactly describes the design, control and operation of a manufacturing system constructed around *lean principles*. Just as tributaries to a river constantly erode and change to support water flow, a lean system will continuously change to support cost effective and efficient production of goods and services. The constant search to remove problems enhances product flow and decreases the product throughput time while increasing output.

The main control functions for a manufacturing system are Production Control, product Quality Control, Inventory Control, RAM (Reliability-Availability-Maintainability), Supply Chain Management…etc.

In a manufacturing system, there are two main components that drive many performance indicators: (1) Work-in-Process, including make-to stock and finished goods inventory and (2) Variance. Work-in-Process (WIP) is manifested in three main categories.

(1) Finished goods inventory (product waiting to be shipped to the customer)
(2) Raw materials inventory (raw materials waiting to be used)
(3) Parts or goods flowing through the system waiting (queued) either to be moved or to be processed); Partially finished goods stored to buffer imbalances in production and demand (Make-to-stock inventory)

Major portions of this book will be devoted to how WIP can be reduced or eliminated. Many previous studies across a wide range of manufacturing systems has confirmed that parts which flow through a manufacturing system spend about 95%. Only about 5? of the total time spent in a manufacturing system is in actual value added transformation. The two performance measures that are most affected by WIP levels are the *production Rate* (PR) and the Time in the system or *throughput time* (TPT) per unit. One of the few invariant laws of manufacturing systems was derived by John D.C. Little and is called *Little's Law*. Little proved the following relationship exists for *any* manufacturing system.

$$WIP=PR*TPT$$

Simply put, Little's law is an equation showing that the average number of entities in a general service system (WIP) is equal to their average arrival rate (PR) multiplied by the average amount of time each part spends in the system. . For example, it has been observed that a waiting line of WIP has averaged 21 parts for a production process designed to produce 6 parts per minute. Little's Law predicts that each part will spend TPT = WIP / PR = 21/6 = 3.5 minutes in the system. We will fully exploit and use this law in later Chapters of this book in a variety of applications. For now simply note that the production rate is directly related to and dictated by customer demand. If it is fixed, then as WIP increases for any reason whatsoever, the throughput time also increases. We will see in subsequent Chapters, that there is an optimal level of WIP which exists for a lean manufacturing cell.

The second, and perhaps the most important factor in poor system performance, is *variance*. Using the mechanics of queuing theory and the principles of *Factory Physics* we will analyze various components of variance which manifest themselves in manufacturing systems and analyze their causes and effects. It will be shown that increasing variance in *any way* will *always* increase WIP and hence increase throughput time. A major goal in Lean Thinking is to systematically and aggressively reduce variance in any way possible at every step in the manufacturing processes.

2.3 Typical Manufacturing System Production Facilities

As previously discussed, there are three primary production configurations which dominate current manufacturing philosophies: (1) The job shop (2) The flow shop and the (3) High volume mass production line. The job shop production system has existed since the 1800's, and is still a necessary component of both stand alone and integrated manufacturing system designs. Approximately 60%-70% of and goods and services manufactured are in small lots made by skilled labor. This manufacturing system design is based upon flexibility and functionality. *Flexibility* is traditionally achieved at the expense of high in-process inventories and long throughput times. A part is scheduled through multiple processing steps which usually involve multi-purpose machines which are often manually operated. *Production control* has the responsibility for order release to the shop floor and a material handling function (often manual) is responsible for moving work-in-process from one value added operation to another. The traditional *flow shop* evolves as product demand increases, and involves *machining centers* where different types of machines are clustered together (Lathe, Mill, Grinder, Drill, etc.) The *process layout* is more efficient than typical job shop layouts, but might still result in spaghetti-type flow with multiple set up operations required. The high speed, *mass production line* evolved as demand for a single part or part family increase to the point that dedicated manufacturing cells or production lines can be configured to support higher production rates. Such manufacturing lines are used in the automobile industry and in individual part production such as facemasks .The modern automobile assembly/production line is the prime example of mass production. Automobile production is divided into hundreds of stations sequenced to systematically build a car or truck. There are usually main assembly lines and subassembly lines coordinated and balanced to sustain smooth, uninterrupted flow. This strategy is referred to as *Coordinated one piece flow*. Figure 2.2 shows a mass production line which integrates all three manufacturing system designs into one production system.

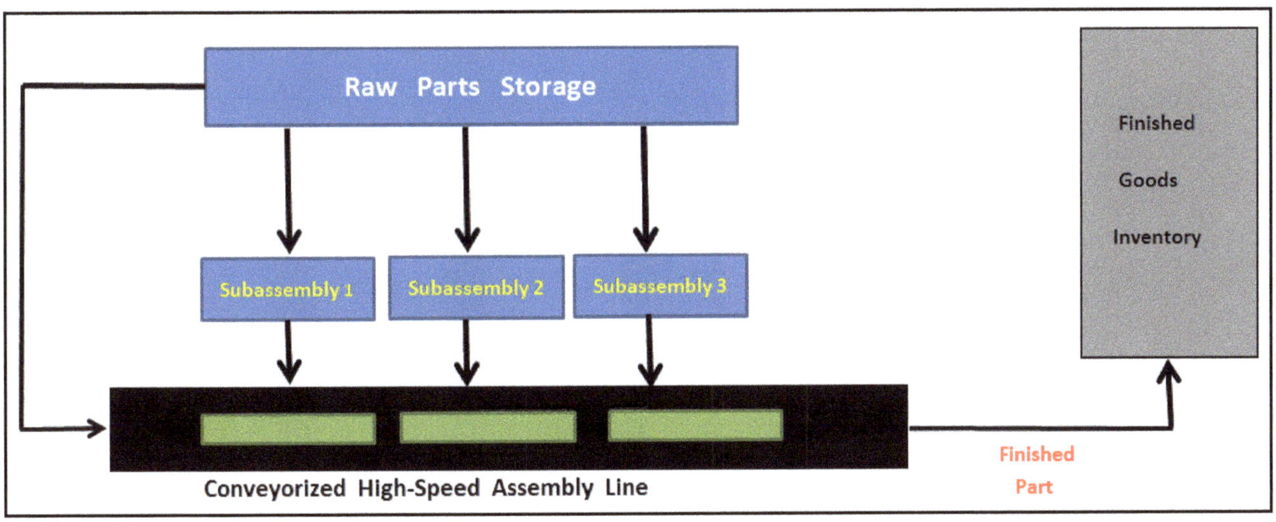

Figure 2.2
The Mass-Production System With Three Subassembly Cells Feeding Final Assembly

Note that from a Manufacturing Enterprise viewpoint; Raw material storage, the 3 subassembly work cells, the main production line and the final goods storage are viewed as one system, but in reality they are not. Each system component will see itself as a separate independent entity with unique operating structures. It is the job of Production Control, Materials handling, raw material supply/ storage and Order Release to manage the system such that every work area seamlessly and harmoniously behaves as a single Manufacturing System. Failure to properly coordinate and link each subsystem properly is a recipe for *waste* and *delays* of all sorts. Lean Manufacturing is a system design such that every support and operational component is directly linked to one another and operates as one, harmonious system. This is accomplished by the design and implementation of U-Shaped manufacturing Cells (Chapter 4).

2.4 The Toyota Production System (TPS)

After World War II, the Toyota Motor Company, led by the genius of a manufacturing engineer called Taiichi Ohno, developed a new manufacturing system design and a new manufacturing philosophy simply known as the *Toyota Production System* (TPS). In the 1980s this system was referred to as Just-in-Time, World Class Manufacturing, or the Pull System to name a few. In 1990, this system was given a name that would become universal called the *Lean Production System* (Womack and Jones). Lean production focuses on waste production and is often characterized by the *two pillars of TPS* shown in Figure 2.3. These two pillars are Just-in-time (JIT) and Jidoka. JIT implies that the system is designed and operated to *produce only what is needed, in the right quantities at the right time*. Jidoka provides the means to achieve these goals and is multifaceted in nature. The primary mission of Jidoka is to produce product with zero defectives and with no in-process quality drop-outs. These objectives are partially achieved by: (1) Enforcing quality-at-the source (2) Empowering workers to stop production when poor quality develops and (3) Training each worker to do the jobs assigned to him/her. Equipment is

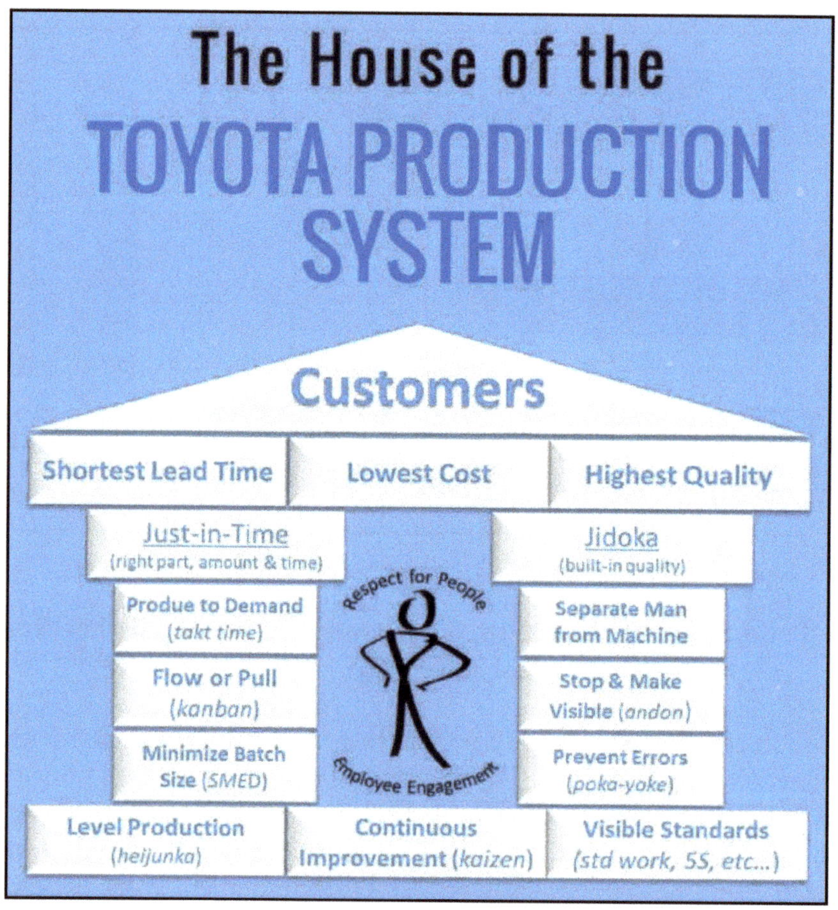

Figure 2.3
The Toyota Production System is based upon the Twin Pillars of Just-in-Time and Jidoka

redesigned to execute self-diagnosis, detect anomalies when they occur and to automatically stop when a problem arises. Operators are also given the power and authority to shut down a line or process if necessary on their own. Stopping a production machine instantly raises *red flags* and highlights a problem immediately. *Jidoka* is called *human engineering* because it empowers a worker to gain complete control over a process or a machine and not vice-versa. In a real sense, every operator is also a quality inspector. These things are an integral part of Lean manufacturing, but: *What is truly different about Lean production when contrasted to traditional mass production or flow shop designs?*

The conversion of a traditional manufacturing system to a lean production system occurs at three levels. *First*, in a lean production system the final assembly area is converted to mixed model final assembly. In automobile production (for example), this means that different models of cars within a car family come down the production line to final assembly in any order. Two door sedans, four door sedans and convertibles all get final assembled on the same line in any order. Really good lean companies can actually run different models (Civics and Accords) on the same time. This is achieved by creating level (not constant) demand for assemblies, subassemblies and main components. Final assembly is run at a mixed production rate based upon customer demand (parts/unit time) and this levels raw material and in-process requirements. The reciprocal of required production rate (Parts / Unit time is called *Takt time* (Time/ Part) and is dictated by customer demand.

Second, as shown in Figure 2.4, linear subassembly conveyor-paced assembly lines are converted into *Lean, Self-balancing subassembly Cells*; often eliminating the need for complex material handling and conveyors. The production rate of each Lean cell is determined by the number of workers in each cell and not strictly the number of machines (Chapters 3 & 4). Each Lean production cell is balanced and produces to down-stream demand that are required to achieve final assembly. There is virtually no sequencing and scheduling in these cells, and balance is achieved by allocating different workers to different sets of tasks. This is similar but not identical to line balancing schemes usually employed in non-lean production facilities.

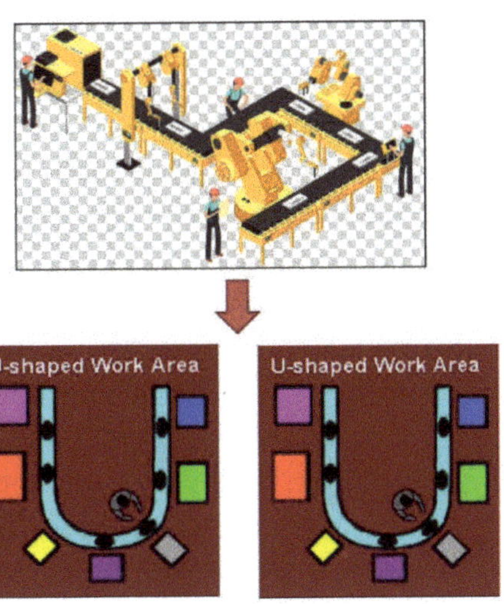

Figure 2.4
Subassembly Flow Line with Conveyor redesigned into two Lean Cells

Third, most support job shops and part manufacturing are also converted into U-shaped manufacturing cells. The redesign of job shops into manufacturing cells requires that an engineer trained in Lean thinking (*The Lean Engineer*) convert all machine tools into *single-cycle automatics* with near Zero Setup and Changeover Time.

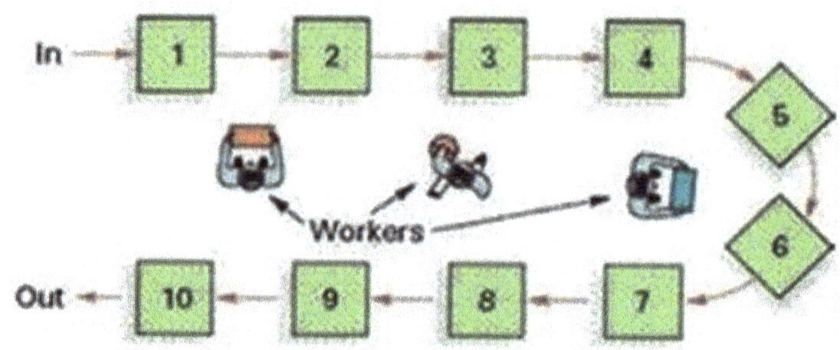

Figure 2.5
Converting to U-Shaped Lean Cells

It is critical that Lean cells minimize set up and change over times, utilize machine tools with jigs and fixtures that can be rapidly reconfigured to accommodate frequent changes in product mix or customer demand. Between each machine in a cell is an area called a *decoupler*, which is used to store one unit of in-process-work (WIP). WIP in this area should be subject to automated Quality Assurance inspections. Devices called *Poka- yoke* are designed and installed to monitor performance, produce zero defects and automatically shut a machine down if it detects a quality problem. This is a radical change from the traditional Quality Control functions of a remote quality control function. There are master Q/C Engineers, but Quality Control is largely a goal of each operator and operating machine. This is called *Quality Control at the Source.*

All *Lean Manufacturing Cells* are designed to provide a *WIP-Cap* within the boundary of each cell. A WIP-cap ensures that work-in-process will not exceed a predetermined level. This WIP-cap is usually no more that than twice the number of machines in the cell, and is strictly enforced (Chapter 6). All component, subassembly, make-to stock WIP locations and final assemblies are linked together by a system called *Kanban* which pulls material through the system. A Kanban is nothing more than a card which is attached to every part, subassembly or lot that moves through the system. A *Kanban Card* is dynamic and it contains both instructions for manufacturing an item and authorization to move an item, part or lot from one location to another. More importantly, in is an authorization to move apart from one location to another. One worker said: "I don't move a part anywhere, at any time, in any quantity unless the Kanban card says that I can". The Kanban card(s) control all material movement, Work-in-progress and production lot size for the entire system. Ideally, these inherent characteristics of Kanban result in a smooth flowing system with minimum WIP and desired output rates characterized by superior quality products and on-time delivery to the customer(s). This new system design can and should operate in ways that the old mass production or job shop system cannot. Each manufacturing cell in the system operates using one-piece flow, just as in the final assembly lines. All manufacturing components are designed for mixed model production. The system is more efficient and totally superior to traditional manufacturing control systems. It was developed and implemented in Japan when production requirements were fairly stable and repeatable. However, we wish to debunk and dismantle the concept that Lean Manufacturing strategies cannot be applied to other systems, such as low volume, high product mix systems. Manufacturers and service organizations around the world are implementing Lean Engineering principles in every type of manufacturing system with spectacular success. The design, implementation and use of Kanban cards will be discussed in Chapter 6.

It should be repeated that perhaps the most important characteristic of a properly designed and operated modern Lean system is *flexibility*. Flexibility and the ability to produce a wider variety of basic products is the key to higher profits and deeper market penetration. Systems designed using the Lean engineering principles in this book allow a company to rapidly adapt to changes in product demand and product mix, and provides a framework by which changes in product design requiring new processes can be implemented. Notice also that a Lean manufacturing system integrates all of the quality assurance, production planning, materials handling and machining /processing requirements into one cohesive, coherent operating framework. The Lean production system is in every way designed to produce superior quality products. Toyota believed in *Total Quality Management and Quality Assurance*, and enforced this philosophy at every level of the company from president to truck drivers. Toyota was transformed from a company which had a reputation for poor quality into one with superior quality products. It is no surprise that the Toyota line of automobiles is generally recognized today as the most reliable, defect free automobile in the world; and it leads the world in sales. This was made possible by enforcing quality upon both the raw material supply chain and every step in the manufacturing system.

We cannot overemphasize enough that the key to both superior quality products and efficient system performance is to aggressively and systematically reduce all sources of *variance*. This is also the key to *continuous improvement*. The kinds of variance that must be considered are:

- Variance in raw material characteristics
- Variance in methods and process activities
- Variance in processing times
- Variance in cutting tools
- Variance in customer demand
- Variance in material delivery times

These and a myriad of other sources of variance can be systematically reduced by applying good, solid Industrial Engineering tools.

2.5 The Quest for Zero Defects

Toyota has made the necessity of *Zero Defects* standard knowledge for all of its workers. Doug Montgomery...one of the best Quality Engineers in the world... has identified and promoted the use of 7 basic tools. He calls these the *Magnificent 7* and they are listed in Figure 2.6. Of course, there is nothing magical or magnificent about any one of them; they simply represent a set of tools that when properly applied, can be used to eliminate variance and improve quality. The interested reader is referred to his book on Statistical Quality Control. It must be emphasized once again that Quality Assurance is not just a goal for Lean Manufacturing....It is a necessity. There can be no rejected parts or subassemblies in any Lean cell. A great deal of time and effort must be expended in seeking the goal of *zero defects*.

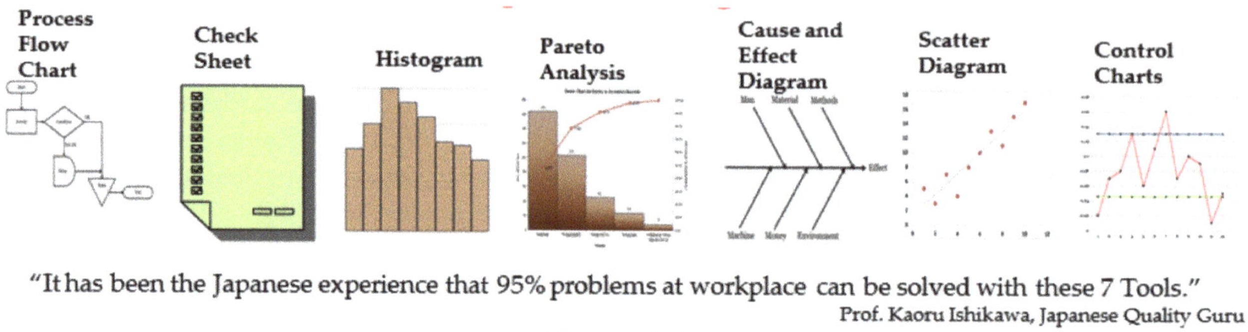

Figure 2.6
Seven Tools of Quality Control (Montgomery)

Quality Control is an area which has been taught in Industrial Engineering for many years, and almost every large company maintains a Q/C division. It is time to recognize that Quality control as it is applied today is largely an outdated and ineffective method of preventing defects or rejected parts in both interim manufacturing and in final assembly. Traditional Quality Control has focused on two main areas: (1) Acceptance Sampling techniques and (2) Quality Control Charts. Both have played a key role in recognizing the impact and cost of poor quality. However, both are *reactive* solutions to Quality Problems and not *proactive* solutions. Both Quality Control Charts and Acceptance Sampling Techniques identify quality problems after the fact. In Lean Manufacturing the goal is not to identify poor quality after it has occurred, but before it rears its ugly head. Lean Manufacturing focuses on *Quality at the Source*. The old saying that "Quality is every ones problem" is absolutely true, and a great deal of effort must be expended in developing real-time monitoring tools which *continuously* inspect work-in-progress both while it is being machined and after each step is finished in Lean Cell decouplers. The Japanese call this *Poka-Yoke*. Poka-Yoke is any mechanism in Lean Manufacturing which can help detect an emerging quality control problem, either by the worker or the machine being used. A shop floor worker can actually shut the entire production line or any cell if a quality assurance problem is discovered. This will instantly trigger a Lean Engineer to "come running" to the source of the problem. A key difference in American traditional quality control and that developed for Lean Manufacturing is that the primary responsibility for product quality lies with the shop floor worker(s), and not in some "ivory tower " office. In fact, the Lean skilled worker is empowered to stop all work until the problem is solved. This would be heresy in a traditional American manufacturing plant.

A major departure from reactive quality control techniques and traditional *Cost reduction* strategies is achieved by not only improving product quality, but by *aggressively reducing any form of waste*, including the efficient use of space, materials, machines, people and storage. This is why the system is called *Lean*. The unique application of Lean thinking is achieved through the creativity and work of the Lean Engineer, who is constantly looking for ways to reduce waste, minimize WIP, improve processes, reduce set up and changeover times, and redesign the workplace. In restructuring traditional manufacturing systems into Lean manufacturing systems, the final assembly line and all subassembly cells are redesigned to support one-piece flow. Workers are assigned to unique, independent tasks and workload balance is enforced (See Black

& Hunter, 2003). In redesigning manufacturing lines into *manufacturing cells*, the axiomatic principles of work cell operation (See Chapter 4) are used to *decouple* or separate one task from another. This also tends to confine quality control problems to where they actually occur.

2.6 Push or Pull ?

The concept of *Push* versus *Pull* material management strategies is fundamental to Lean Manufacturing. This will be discussed later, but in general a *Push system* is one where production orders are released to the manufacturing system based upon forecasts or levels of finished goods inventory. Push Systems fundamentally ignore system characteristics. machine status and WIP levels. Orders for production are released and pushed through the system as quickly as possible. In many system, these production orders seek advice from a Capacity Planning support function, but direct coupling of these two order release components are rarely accomplished. A *Pull System* is an entirely different strategy. Pull Systems will not release an order to the manufacturing system unless an order is in hand and the system can absorb the extra work load by increasing capacity by a unique reconfiguration of Lean Cells by changing labor assignments or number of workers....NEVER by subcontracting or by working overtime which can dramatically increase unit cost.

The WIP level in *each cell* is held constant throughout the entire production period. The output rate is controlled by *worker* utilizations and not machine capacity. Cell workers are *multifunctional* and cross trained on each machine. Every cell operator can perform every operation and also execute routine machine maintenance duties. Lean cell manufacturing strategies completely eliminate the concept of one worker per one machine, except in rare capacity-limited situations. Since each cell operates autonomously of every other cell, the Lean Cell is ideally suited to hosting mixed model production from a common family of parts. Part families can be identified by applying *group technology* or other well-known clustering algorithms. Proper identification of part families help to reduce setup and changeover times. An example of Cell Design and worker task assignments will be given in Section

2.7 Four Key Terms in Lean Engineering

A Lean production system must be continuously refined and improved to achieve success. This refinement must involve four key aspects of Lean thinking: (1) Leveling and Smoothing (2) Balancing (3) Sequencing and (4) Synchronization. Knowledge of these four terms is critical to the success of Lean transformation.

2.7.1 Leveling Production

Leveling production can be simply stated as: *Controlling distribution of both the type of product produced and the volume of product produced over time*. Leveling is the process of planning and executing a production schedule which will not vary drastically over successive time periods. *Leveling production* for a single product assigned to a single line amounts to *leveling the product demand to be* manufactured over a fixed period of time: usually by a week or by a month. For example, suppose that Product A is to be produced according to the following demand schedule over the next 5 weeks. The varying weekly production schedule has likely been created by a *Push* order release system.

Week 1	Week 2	Week 3	Week 4	Week 5
180 units	110 units	205 units	150 units	135 units

An alternate strategy would be to set production at 180 units per week, but at the end of the first week one would observe that weekly demand has significantly decreased (180 to 110), so production would be reduced accordingly. At the end of week 3, suddenly demand almost doubled, so production is doubled (overtime, additional shifts, etc.). The bullwhip effect is still very evident. Suppose that the following production schedule is used.

Week 1	Week 2	Week 3	Week 4	Week 5
156 units	156 units	156 units	156 units	156 units

Clearly, this schedule would be easier to manage. Note that to accommodate this schedule, a certain amount of WIP will be required; which seems to violate one of the main objectives of Lean Thinking. However, note the following.

- Customer demand is fully met on time
- Predictable work schedules
- Fixed production requirements
- Make-to-stock inventory need be accomplished only one time

Another way of creating a non-uniform production schedule is to *produce strictly to demand.* In this case, the weekly production schedule would vary from 180 units to 110 units. If this is a final assembly schedule, all of the upstream processes are going up and down in the same manner. This is called *chasing demand* and the violent, frequent changes in production requirements is called the *Bullwhip Effect.* Clearly, the amount of time devoted to this product and the operation of a production cell would vary from week to week. Inventory and raw material requirements would change from week to week and cost overruns might occur.
The principle behind load leveling is simply to regulate production in all subassembly and final assembly such that constant production rates are achieved. This avoids *spikes* in production schedules which occur when production follows demand. *Final assembly* should not pull production from upstream cells or subassemblies in erratic production requirements as demand changes. Once again, this might result in a fluxuating, changing final goods inventory...but this is in the final stage of customer delivery and is much better than upstream disruptions. Constantly fluxuating customer demand is simply business, but it is the mortal enemy of smooth production: *Variance.* Hopp and Spearman call this the *Corrupting Influence of Variance*.

In almost all Manufacturing companies, production and assembly lines must be designed and implemented to accommodate the manufacture of multiple products in the same facility. This is called *mixed model* production. Mixed model Lean production requires application of the same Lean principles as single product production, implemented in multi-dimensions. The key to equivalent mixed model production is to reduce set-up and changeover times to an absolute minimum. This is called SMED (Single Minute Exchange of Dies) and enforce a zero defects production program. A myth in changing over to Lean production with leveling is that integrated forecasting systems such as those contained in MRP II shop floor schedulers can be abandoned. Forecasted demand is still needed, but must be smoothed to achieve one piece flow. Traditionally, fluxuating demand rates were historically accommodated by over-scheduling normal production, causing congestion, more material handling, WIP levels and excess inventories.

Suppose the demand pattern is as shown in Figure 2.8-a. There are 1000 units which need to be made over a 4 week planning horizon. The schedule calls for 100 units to be produced in weeks 1 & 2; 200 units in week 3 and 600 units in week 4. This calls for a typical rush to produce at the end of the month. To level this schedule, two-2 week schedules are created to divide a typical month into two parts. Next, the two week periods are divided into one week periods, and finally one week periods into daily production requirements. This is to create a leveled production schedule which will greatly influence and improve upstream system behavior. Daily production requirements are set at 50 units/day; working 5 days per week. Of course, if the shipping schedule corresponds to the original manufacturing schedule; this will require an average WIP of 360/units per day in make-to order inventory. If finished goods are shipped every day, this WIP becomes only 25 units/day. The correct policy must be determined by balancing holding costs against shipping costs.

Advantages and Disadvantages

Leveling production has the following *advantages*.

- Raw material and part suppliers can build and ship on a smoother schedule
- Highs and lows in production system demands are smoothed out
- WIP is reduced if shipping is also leveled
- Leveling production can facilitate automated , with standardized containers
- Overall productivity usually increases and waste is decreased
- Direct labor hours may be decreased
- Quality is greatly improved

Leveling production may also have the following *disadvantages*.

- Material handling requirements will increase
- Setup and changeover times must be greatly decreased or eliminated in Mixed model production
- Workers should be trained to be Quality inspectors and maintenance people (these are also advantages!)

Workers must be thoroughly indoctrinated in the principles, techniques and reasons for Lean production and leveling. Their approach to manufacturing must include continuously seeking ways to improve production, reduce costs and eliminate waste (*Muda*). Successful leveling strategies must be fully supported by every factory worker, or the results will be resistance, misunderstanding and mass confusion.

2.7.2 Mixed Model Cells

Almost all manufacturing companies manufacture and sell products of several different types. A company can produce products of different type (General Motors: Chevrolet, Buick, Etc.) or variations of individual products (F250 Ford Truck: Diesel, 4 wheel drive, 2 or 4 door, Etc.).

Consider the following monthly demand profile for products A, B, C and D. Assume there are firm orders which can be billed as soon as they are delivered.

Product A	6000 units/mo.
Product B	4000 units/mo.
Product C	2000 units/mo.
Product D	8000 units/mo.

It is not uncommon to observe that many companies would traditionally run this monthly production profile in units of A=6000, B=4000, C=2000 and D=8000. The sequence of production might be chosen to minimize setup and changeover times. The reason for this line of thinking is that to minimize the time lost to setups and changeovers this is the "thing to do". If in fact, these times are very large; this might indeed be the right solution. Early in the formation of the Toyota Production System, a major emphasis was placed upon reducing setup and changeover times to as near zero as possible. This initiative became known as SMED, which stood for Single Minute Exchange of Dies. Indeed, after much time and effort these times were reduced from hours to minutes. SMED is absolutely necessary to implement Lean Manufacturing and Just-in-time strategies.

Continuing with the example, assume 4 weeks per month and 5 working days per week. This would result in production requirements of A=300, B=200, C=100 and D=400 units per day. This is still not a level production schedule. Maximum level production is ideally achieved by using *one piece flow*: Produce one unit of A, then one unit of B, etc. Any sequence of producing multiple products is called a *Mixed Model Production Schedule*. As a general rule, first calculate the largest integer divisor that would divide evenly into all products. Next, determine how many of each type of product will be included in a partial production sequence by dividing each product demand in the lowest time bucket (called a *Planning Horizon*) selected by that integer divisor. Third, determine a *basic sequence* which would produce that set of products. Finally, determine how many times that sequence would need to be repeated over the Planning Horizon to satisfy the time bucket over which total production is planned; called the *Planning Period.* For this example:

- The planning period is *monthly:* Production Requirements are*:* A=600, B=400, C=200 D=800
- The planning horizon is *daily:* Production Requirements are*:* A=30, B=20, C=10 and D=40
- The largest integer divisor is 10
- The Basic Sequence is: A=3, B=2, C=1 and D=4.
- A possible production sequence is ABDDABCDDA

The production sequence would need to be repeated 10 times per day to achieve monthly requirements. There are many other possible sequences, including one that would result in maximum leveling: DABDABDCBD. Here we must now determine the maximum amount of available time per day that would likely be required to produce the desired sequence of products. Assume that the manufacturing facility is working two shifts per day 7 hours per shift. This will give 840 minutes of available production time per day. Using the sequence of A=3, B=2, C=1 and D=4 if we will assume that the following production times per unit are required: A=240 sec; B=120 sec; C=400 sec; and D=180 sec. Each sequence will require 140 seconds, repeated 10 times or 1400 seconds per day or 23.33 min/day.

Migrating to Lean production also requires maximum flexibility in final assembly lines so that production can be scheduled around firm customer orders rather than chasing demand. So basically: *How is flexibility achieved and how is final assembly effected?*

2.7.3 Mixed Model Final Assembly

Lean manufacturing companies have yearly production plans that combine orders-in hand with a forecast of how many items need to be produced over the next planning period to cover new orders or unforeseen delays. This is why an MRP II system should not be scrapped. For example, a yearly production plan might employ a running two month plan. Products are scheduled into production and a detailed routing sheet is constructed built around a one month planning horizon and the amount to be produced on a daily or weekly basis is placed on the order release schedule. The production plan should ideally support minimum time in finished goods inventory and just-in-time delivery. The final product will be the construction of Kanban Cards to control production and authorize material handling moves. Leveling production is an important piece of the actual production schedule.

For example, suppose that a final assembly line is required to produce a total of 10,000 vehicles per month according to the plan shown in Table 2.1 for a mixture of 4-door sedans, 2-door sedans and mini-vans (Phillips, Black 2014)

Table 2.1
Mixed-Model Final Assembly: Sequencing to Smooth the System

Assembly Sequence
4DS, 2DS,4DS, MV, 4DS, 2DS, 4DS, MV, 4DS, 2DS

Model	*Monthly Demand*	*Daily Demand**	*Takt Time*
4-door sedan (4DS)	*5,000*	*250*	*1.92 minutes*
2-door sedan (2DS	*2,500*	*125*	*3.84 minutes*
Minivan (MV	*2,500*	*125*	*3.84 minutes*
			500 = Daily Demand

$$Takt = \frac{480 \text{ min. } x \text{ } 2 \text{ } shifts}{500} = 1.92 \text{ min. } per \text{ } vehicle$$

**Assuming 20 working days in the month*

Using a 20 day work month, this would require a daily production of 500 vehicles; 250 4-door sedans, 125 2-door sedans and 125 minivans to satisfy demand. The factory is scheduled to work two 8-hour shifts per day. Each shift would contain 480 working minutes. In addition to the three basic types of vehicles, suppose that due to options there are 300 different variations to the basic vehicles. The production of these automobiles is a good example of *Mixed Model Manufacturing*. For convenience in this example, there will be no set up or changeover times between models or variations. Inclusion of these times is not hard to accommodate. Once the production sequences are established, all of the setup and changeover times are simply deducted from the available working time.

The assembly line receives a production schedule at the end of every month. The average daily production time required to meet the schedule is then calculated. The process step(s) are then examined to see if sufficient capacity exists for the schedule: If not, capacity is adjusted. If a Lean, U-shaped manufacturing cell is being used, then capacity is easily adjusted by assigning the appropriate number of workers to the appropriate number of machines to match required Takt time (Chapter 3). All operator tasks must be standardized and all workers cross trained to operate any machine in the cell. Overtime on another shift can be used temporarily when demand exceeds capacity, but this should be a next-to-last resort. The last resort is *subcontracting*. If this condition persists, then parallel machines could be purchased to increase capacity.

If demand decreases, the reverse strategy is used. Workers take on more machines in their assigned work cycle. and machine utilizations decrease. Excess workers are moved to other cells or other duties until needed. A goal of Lean manufacturing is to produce goods and services with a minimum of skilled laborers. Conversely, it is not necessary or desired to keep machines running all of the time or configure a cell with the minimum number of processes. For this example, the Takt time for this cell is to produce a 4-door sedan every 1.92 minutes; a 2-door sedan every 3.84 minutes; and a Minivan every 3.84 minutes.

2.8 Mixed-model Final Assembly

To perform leveled production in a factory that manufactures several types of final products, Mixed-model *final* assembly lines must be designed and implemented. Mixed model *subassembly* lines may or may not be needed. Every U-shaped Lean manufacturing cell might be dedicated to a particular type of subassembly, but some may need to be Mixed Model cells. The decision depends largely upon the production quantities that need to be produced over a fixed planning horizon. Here we will constrain our thoughts to Mixed Model final assembly, although the same principles apply to both Subassembly cells and component manufacturing cells.

- Eliminate or bring to a minimum all setup and changeover times.
- Standardize all material handling as much as possible.
- Determine the required Takt time(s)
- Determine the sequencing and dependency of all final assembly operations
- Level the Mixed model production schedule by properly sequencing and grouping parts
- Design workstations to balance line workstation times

2.9 Determining Takt Time(s)

In final assembly, the required cycle times between part releases are called *Takt* times. Takt time is mathematically the inverse of production rate . Takt time is determined by dividing the operating minutes per day (scheduled time-breaks minus allowances, etc.) by the daily demand. This sets the pace of manufacturing at final assembly, which in turn sets the pace for every operation and subassembly cell that precedes final assembly.

Let: TT=Takt time (Minutes/Part)

H=Number of operating minutes per day
DD = daily demand (Parts)
OM= Required output/month
M= Number of operating days/month
TPT=Throughput time per part
PR= 1/TT=Production Rate

Then: TT = H/DD in minutes/part
And: DD = OM/M in parts/day

As usual, Little's Law can be written as:
WIP = TPT *(1/TT) = TPT* PR

In Lean production systems, final assembly Takt time is held constant over the planning/production period. By Little's Law, any reduction of WIP will result in a corresponding reduction of throughput times. This is a major goal of Lean manufacturing.

Contrary to popular belief, a Lean system does not strictly react only to customer demands. We have already pointed out the fallacy of this policy. The required output from final assembly comes from a combination of actual orders-in-hand and a sales forecast. Here we kill another widespread misconception of Lean....... Sales forecasting is not a necessary part of Lean manufacturing: This is not true. *So how is this done*? A common belief is that Lean thinking and Lean production will replace MRP, ERP and other shop floor control systems. While this is partially true, these legacy systems can still play a major role in production planning. A major contribution can be sales forecasting.

Once the required *cycle time* is determined, the Lean Engineer develops standard procedures and best practices for every worker in both *preceding subassembly cells* and on the *final assembly line*. Both subassembly cells and final assembly line has been designed for flexibility and reconfiguration as demand/product mix changes. Likewise, workers must be cross trained to execute many process steps. Standard procedures and best practice is the best arrangement of men, materials and machines coupled with the best methods used to produce a defect-free final product. It is Takt time which determines equipment capacity, work breakdown structure, work balance, leveling and product mix.

2.10 Determining the Proper Sequence of Operations

The sequence of operations in final assembly is easily determined by a process routing sheet. This simply identifies which operation must precede another. Closely associated with sequencing is the granularity of processing steps. In general, each processing step should be as short as possible. This will make it easier to balance production on the line. It is worth noting once again that no product can move from any one Lean Manufacturing Cell to another unless such a move has been either requested or planned by the down-stream process (consumer) and is authorized by a Kanban card. Kanban Cards augment and largely replace an expensive, computerized Scheduling System and drives the entire Manufacturing system.. Except for subcontracts or specialty items, scheduling a Lean Manufacturing System starts with Final Assembly. Final assembly is the ultimate fulfillment of customer orders and periodic make-to-stock manufacturing to buffer customer orders. A common misconception is that a Lean Manufacturing System will not need integrated computer support. Legacy systems can be interfaced to Lean Manufacturing to execute forecasting, generate work orders, build the necessary Kanban cards and execute accounting functions.

2.10.1 Determining the Mixed model Sequence of Finished Product

The sequencing problem at final assembly is the quantity, type and mixture of each finished product that needs to be introduced to smooth and balance the work at each workstation. Sequencing involves two main goals (Chapter 8).

- *Balance* the time spent at each step of final assembly
- *Level* the amount of product produced over time at different workstations
- Keep the rate at which each part is produced constant. This is called *leveling demand.*

In Mixed model assembly, not all products have to exactly meet Takt time requirements. Some may actually be lower or higher than Takt times (Just as in traditional Line Balancing). Often one is forced to average individual product Takt times. For this reason, products introduced into the line with assembly times which exceed Takt time(s) should be followed by products which are less than Takt time(s). Workers are often trained to assist across station boundaries, thereby achieving workload balance. When Mixed Model production is in place, inventory (WIP) levels are also affected. Generally, WIP levels in Mixed Model assembly lines are larger than individual product lines. WIP levels are required to prevent starvation and to accommodate temporary overproduction.

2.10.2 Scheduling

Scheduling a Mixed model, final assembly line to achieve good line balance and leveling is a key component in designing the entire production system. If a station on the final assembly line starves or is blocked, the entire system might be starved or blocked. All upstream cells and subassemblies must be tuned to final assembly. Great flexibility is required to seamlessly move between different types of products with different cycle times. Lean Mixed model final assemblies usually result in higher on time delivery, lower finished goods inventory and customer satisfaction.

After setting weekly and daily production schedules, determination of the daily sequence is critical. The sequence of products dictates the order of final assembly, the just-in-time resources required and the balance. It is important to realize that if three products are being made (A, B & C), and if the final assembly Mixed model sequence is 1000 replications of AAABBCC, this sequence is important not only to final assembly, but also to the upstream processes or cells. In order to synchronize upstream cellular manufacturing with final assembly, upstream production must mirror the final assembly schedule. There is only one way to avoid this: it is to store upstream production in an interim *Supermarket*, which in turn will ship to final assembly. A supermarket is nothing more than parts or subassemblies which are strategically located on the production floor to avoid either starvation or bottlenecking. In a properly designed and smooth running Lean system, this should never happen. While this will work, it will create excess WIP, longer throughput time and violate lean manufacturing principles. This is the most important and fundamental aspect of linked-cell, Lean production system, and it distinguishes Lean Mixed model final assembly from traditional Mass manufacturing systems. In general, upstream component and subassembly lines receive only monthly or weekly production schedules. Workers are assigned to cells and processes based upon this schedule. In the Lean linked-cell system, Mixed Model final assembly builds a final product (s) by pulling components and subassemblies directly from upstream production cells with only a small buffer to keep from starving final assembly. This is accomplished by issuing a Kanban withdrawal signal to the upstream subassembly cells. This in turn causes a production Kanban signal to upstream cells to deliver subassemblies/components to final assembly. Hence, upstream production does not need short term or detailed production schedules. Kanban signals provide all of the scheduling orders to trigger upstream production as parts are pulled toward final assembly. In general, Taiichi Ohno identified 8 categories of waste which must be aggressively attacked and minimized or eliminated. One of these is inventory of all kinds...In-progress-inventory (supermarkets), finished -goods-inventory, Cell inventory and just-in-case inventory. These eight categories of waste (called *Muda*) are shown in the following graphic.

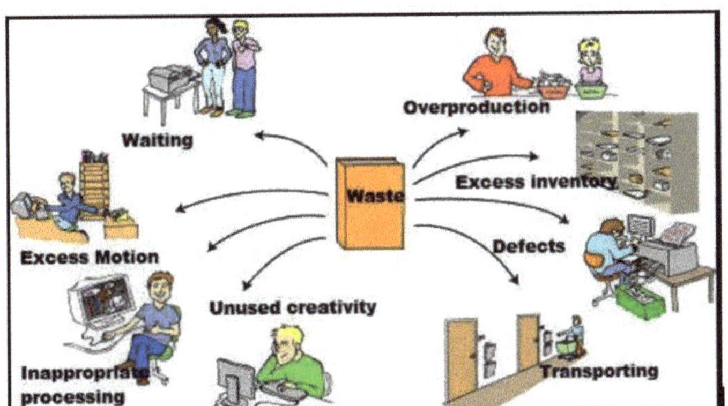

Graphic Courtesy of https://tulip.co/resources/5s-guide

Eliminating waste is a key function of every Manufacturing employee. If the problem is severe, it is investigated and solved by what is called a *Kaizen Event*. A Kaizen event is a focused, intense investigation by a team which should be led by a Lean Engineer and include shop-floor workers.

Another distinction between final assembly and upstream suppliers is that workers both on the final assembly line and in upstream manufacturing cells must know both the quantity and identification of what final product they are building next. This information is usually generated from a central computer via a local computer terminal. In most cases, detailed assembly instructions are also available, but after a short period of time experienced workers need little or no help. We will warn that the best final assembly sequence for Mixed Model production can be difficult to optimize and implement for smooth production. The goal is, of course, to minimize variation in both the processes and in the required output rates. Consumption rates, subassembly deliveries and processing times must be as consistent and constant as possible. This requires great ingenuity, much work and technological innovation as product demand and variety increases.

2.10.3 Long Range Forecasting

The Toyota Motor Company advertises an outstanding 70 or more inventory turns per year and his there is only about 3 days of inventory in the system at any one time. Toyota calls this their *instant delivery system*. Toyota spends a great deal of time and effort to support forecasting methods to predict short and long term demands for their automobiles. *How does Toyota do this? Do they know something about forecasting methods that the rest of the world does not know?* To answer these questions requires a deeper understanding of how the TPS actually works.

Toyota claims that they can produce customized vehicles in two days or less, but the production *throughput time* is much longer. *How can two day delivery be accomplished?* The answer lies in how they schedule production, manage subassemblies and make-to-stock certain key components. The body, frame and other key parts are manufactured to a relatively fixed production plan, and each has been standardized for a variety of vehicles. The real customization of populating the interior, painting and other custom add-ons must be accomplished in two days. There is no real magic here, only good design for manufacturing and good manufacturing planning.

Toyota determines to a large degree its basic production plan by constantly performing sales forecasts and tracking monthly trend; and surveying over 60,000 potential customers twice a year. As a result, Toyota establishes fixed monthly production plans for key parts and then breaks those plans into daily manufacturing orders. Precise daily production plans are produced to level production, producing the same amount every day until circumstances dictate otherwise. This daily schedule is only communicated to Mixed Model final assembly. Specific requests for all components, parts and subassemblies are then passed *upstream* to precedence activities. The final assembly requirements are manifested in *Production Kanban Cards* and *Move Kanban cards* as needed. For example, the subassembly line that puts together custom dashes with gages and trim only has 2 hours to respond to an assembly line order. In contrast, the typical American mass production system is a combination of the functional job shop and flow lines that build-to-stock for final assembly, creating large banks of in-process inventories and longer throughput times.

Many American companies, recognizing the superiority and advantages of the TPS, have launched efforts to become Lean; but they have failed to convert job shops and flow shops into cellular manufacturing. This is a sure recipe for chaos. Small lot production with minimum WIP requires that the traditional job shop be eliminated (possibly not totally) and replaced by a linked-cell system in which zero defects are produced, setup and changeover times drastically reduced or eliminated, and final assembly is converted to Mixed model production. All of this is achieved by driving the system with accurate forecasts of potential sales, actual sales, and the versatility to correct any planning errors. The ultimate goal is to produce all parts one-at-a-time, or at least using minimum lot sizes. Coupled with new design for manufacturing innovations, this new system can quickly and easily respond to fluxuations in demand and support shorter forecasting windows.

2.11 Line Balance

Line balancing is making sure that each worker in final assembly works the same amount of time on the processes assigned to him/her. Task times assigned to each workstation must, of course, be less than the *Takt* time required by final assembly. The traditional line balancing problem is an old Industrial engineering problem that has now been mostly solved using computer methods. The goal is to meet Takt time while minimizing idle time and imbalance between workstations. When the final assembly line is Mixed Model, both Takt time and line balance are subject to constant change. This characteristic is a problem at final assembly that must be solved by *line flexibility*.

2.11.1 Balancing Subassembly Cells to Final Assembly

As the level of inventory in the factory shrinks and processes become more tightly linked, production rates for each component and subassembly cell become closely aligned to those of final assembly. Ideally, the cycle time of each component, part, and the subassembly would also be close to the final assembly line's Takt time(s). Balancing the output of the subassembly cells with final assembly carries over to balancing labor and processes. In a traditional factory line, balancing entails shifting people and tasks along the assembly line. This only creates a temporary balanced flow line. In a lean production factory, workers are seldom moved and balancing extends upstream to subassembly cells and manufacturing cells for the component parts.

As a production job shop converts to a lean shop, different types of machines are rearranged into manufacturing cells. The machines in these cells are placed in close proximity to one another and move parts between them one piece at a time. Excessive WIP and inventory between machines disappears when only one part flows between machines, which lead some people to call this a *Zero Inventory System*. Those who use this term do not call raw material being transformed at each machine WIP. Close attention to machine setup time, quality (no defects) and total preventive maintenance (TPM) is essential for unit in-process-inventory between adjacent machines.

It should be reiterated that the critical design or redesign of a traditional manufacturing system into a Lean Manufacturing system involves the transformation of traditional part manufacturing and subassembly lines into U-shaped Lean cells (Chapter 4). Operations and processes are placed near one another, and arranged in a U-shaped configuration. The necessity for a U-shaped manufacturing cell will be explained later. Subassembly cells consist of small, simple pieces of equipment. Workers walk and work at individual machines in loops following a precise time schedule. If possible, the cells are designed so that their cycle times are slightly less than the Takt times of final assembly. Balancing each cell means that it can meet the daily demand Takt time even if the cycle time is not exactly equal to Takt time. Output is balanced to support all part manufacture or subassembly cell(s) downstream. The Kanban link between cycle and Takt times absorbs any mismatching of cell production and usage rates by subassemblies or final assemblies. The fewer mismatches, the smaller cell output buffer WIP is required.

Suppose a cell is making two products: A and B, which is subsequently used downstream in equal quantities. While it is making Product A, there must be a sufficient amount of Product B in output WIP between cells to meet downstream demand for Product B. Of course, if setup and changeover times have been reduced to a minimum, the feeder cell might make one unit of A, then one unit of B in repeating patterns. If SMED cannot minimize or virtually eliminate setup/changeover times; it will be necessary to maintain a small amount of WIP using a Kanban controlled link between the upstream and downstream manufacturing operations.

In cells dominated by manual *Machining Workstations*, different parts produced in the same cell usually require different processing times, but parts made in one cell should strictly adhere to labor times that are dictated by *Standard Times* for each part produced in the cell. This does not mean that total machining time between two parts must be balanced or equal because machining times are decoupled from the cycle time. The cycle time depends on the number of machines a worker visits on a trip around the cell (Chapter 4). Hence, cell production rates can be increased or decreased by adding or subtracting workers...not machines.

Adding or subtracting workers changes the output rate from cells. During any fixed order release, the number of workers within a cell should remain constant. However, sometimes workers cross cell boundaries when they have highly automated machines and occasional idle time. When the daily cell demand changes, personnel and personnel loop assignments might both change. Occasionally a cell is designed to add or subtract machines. This may be acceptable if a cell is reconfigured and staffed to support long production runs, but if machine capacity or modification is required frequently, it will be better to let some machines sit idle for a particular production run. In any event, a change in either demand or product mix (*Mixed model* cells) will usually cause the work pattern in the cell to change. These changes are necessary to match cell cycle times to those required by a new downstream final assembly schedule (FAS).

Workers who have performed only a single function or operated a single machine may find that Lean manufacturing requirements are difficult to accept. Cells are designed with simple, single-cycle machines and equipment modified for short set-up time and maximum flexibility. A properly trained operator should be able to operate and set up every machine in a cell. Higher management, finance and accounting functions may also resist Lean philosophies which often result in lower utilization of machines.

Supervisors are responsible for managing cell performance. They will usually determine how many workers are required for different cycle times in a cell, but in mature cells experienced workers may want to do this themselves. This results in an attitude of ownership for worker assignments. A computerized routine which can assign the required number of workers to different machines in the cell was developed by Curry and Phillips, and is given in Chapter xxx. A trained Lean Engineer can assist in solving complex balancing and assignment problems, although

Balancing a cell is easier when the required waking time of a cell worker between assigned machines exceeds the cell Takt time. However, when this occurs replication of an existing cell should be considered. If there are two different parts being made in a Mixed Model cell, each cell should be dedicated to a single part. This requires twice as much equipment. However, since machines are simple, single-cycle automatics this may not be a large capital investment.

2.11.2 Balancing the Entire Factory

The steps for balancing an entire factory are:

- Balancing the rate of parts production to match the final assembly-line rate (overall cycle times are the inverse of production rates).
- Adjusting work content and cycle times at each cell or station until times match system cycle times are nearly as possible; and

- Trying to off-load work content of selected stations until selected stations are no longer needed.

As previously mentioned, the final assembly line Takt time requirements might dictate some WIP storage or raw material storage adjacent to selected final assembly operations. These storage locations are often called *supermarkets*. Ideally, a supermarket might be managed and supplied by an outside vendor but they could also hold subassembly units or individual parts made upstream from final assembly, particularly in some Mixed Model production. The decision to manufacture either parts or subassemblies in-house is largely an economic decision. A *Cost* or *Industrial Engineer* should run a cost analysis to determine whether make-to-stock in house or subcontractor make to demand is best. If subassemblies or parts are subcontracted, many companies have a working agreement with preferred subcontractors to both supply, store and manage their usage. There are tremendous advantages to such an arrangement. This is a special form of *Just-in-Time* material supply. The commonly held and proclaimed notion that there are no WIP storage areas on a Lean factory floor is incorrect. However, the idea is to minimize material in these storage areas. The best place to minimize storage areas are near points of use and close enough to other areas so operators have visual signs of part usage.

In the traditional mass production system, line balance refers to balancing the amount of work or labor or at each station, regardless of the Manufacturing System cycle time. There is almost never any attempt to achieve overall factory balance. In lean manufacturing, balancing the line or plant also refers to balancing material flow. The pull system of material control will also balance material flow. Material balance, labor balance and part production balance are all multiple faces of the same problem and they are dependent upon each other. The primary indicator that labor is out of balance may in fact be a periodic excess or shortage of material, particularly if required material is made upstream and there is no Supermarket. The amount of work-in-progress between cells can be drastically effected by unbalanced cycle times. The objective is to set cycle times as required as close as possible to the final production schedule and then shift tasks and operations to achieve that goal. Remembered that systems are designed to minimize labor content as well as improve labor efficiency.

Plant balancing is a dynamic, ongoing process. Normally, a plant must be rebalanced whenever the production rate changes. The fear of losing line balance is the major reason for reluctance to stop a balanced assembly line once it is up and running. This is also true of an entire factory. Properly designed Lean Factories are not balanced as a complete system. Balance is automatically achieved by forcing every operation to obey required output Takt time and properly designing and operating Lean Cells to meet those requirements. There are Lean principles that support this goal, including but not limited to the following.

- Visible signaling systems, called *andon*, allow a system to respond to temporary variation in part-usage rates and to changes in product mix.
- Flexibility built into cells permits a system to quickly adapt to requests for increased or decreased production rates or to changes in product mix.
- Less-than-full-capacity scheduling means keeping a little slack time, perhaps 20-30 minutes per shift at each cell, so a system can respond to variations from the planned schedule .This adds flexibility to a system.

Since a plant's schedule changes periodically, the operations may also have to periodically change. Lean Manufacturing cells and subassembly cells can usually continue working without rebalancing if the cycle times required do not vary more than about 10% from one case to another. Cycle-time variations beyond 10% usually require rebalancing. Detailed planning for production by a Lean engineer must also take this into account.

2.12 Standard Operations (Also called Standard Work)

Standard operations are designed to allow manufacturing cells and subassembly cells to use a minimum number of workers. A *Standard operations chart* is a form used to define standard work in a manufacturing cell. It shows the sequence of operations, the manual times for the operator, the walking times and the machining cycle times. Workers are expected to never violate the standard work/machine sequence or standard time spent at each machine. This may seem strange to traditional supervisors who prefer and often offer bonuses to those workers who produce high quality products in less than standard time. This is a mistake. Workers should never exceed or work faster than a time standard. Standard times and standardized routing helps ensure that new workers are performing the correct steps in the correct sequence in the correct time.

Standard operations try to achieve a balance among operations so the cycle time is in compliance with the final assembly cycle time. Producing the daily demand supports the final assembly Takt time. Therefore, the cycle time concept should be incorporated into standard operations.

With the use of cams, micro-switches, and similar mechanical devices, different types of machine tools can be designed to have a processing time less than the necessary cycle for the cell. At the completion of each machining cycle, the machine is stopped automatically and its components and attachments are returned to the start position. There is usually a WIP holding area between machines which can only hold one unit. Hence, the amount of WIP in any one Cell with M machines is (M-1). The total amount of work-in-progress in a Lean Cell is usually bounded by 2M-1. The unit WIP between machines serves to decouple each machine from every other machine. If there is space to hold a finished work piece on the output side of a process (the downstream decoupler), the work piece can be automatically ejected from each machine to the next machine in sequence. If an empty space is not available, this is a clear and visible sign that something is wrong. The line should be shut down and a Lean Engineer called to fix the problem. An empty decoupler provides a signal to the upstream machine to release and then make another part. If an unprocessed work piece is available in an upstream decoupler, it may be automatically fed to the machine using guides to position it without human assistance. Once the unprocessed work piece is located and clamped into the downstream machine, the next machining cycle is started.

Leveling quantities and synchronizing processes can significantly reduce delays, thus greatly reducing manufacturing throughput time. Using one-piece flow to eliminate lot delays for two serial processes reduces throughput time. It has been reported that the multiplier effect of eliminating process delays and lot delays with one-piece part movement in cells improved throughput time for a traditional manufacturing sequence by almost 100% If 10 or more processes are involved. Since any one Lean Manufacturing cell with n machining stations will only hold (n-1) units of WIP, this cycle time reduction is highly predictable by applying Little's Law. Cycle time reduction is directly proportional to WIP reduction.

With the use of jigs, fixtures, micro-switches, and similar mechanical devices different types of machine tools can be set-up to have a machine processing time less than the necessary cell cycle time. Upon completion of each machining cycle, the machine is stopped automatically and its components and attachments are returned to the start position. If there is space to hold a finished work piece on the output side of a process (the downstream decoupler), the work piece can be automatically ejected from machine to the decoupler. If an empty space is not available, the machine must wait until the downstream operation is completed and the machine reloaded. An empty decoupler can provide a signal to the upstream machine to make another part. If an unprocessed work piece is available in an upstream decoupler, it may be automatically fed to the machine using guides and rails to position it without human assistance .Once the unprocessed work piece is located and clamped into the machine, the next machining cycle can be started.

Leveling, balancing and synchronizing processes can significantly reduce delays, thus greatly reducing manufacturing throughput time. For example, using one-piece flow to eliminate lot delays for two serial processes reduces throughput time. It has been reported that the multiplier effect of eliminating process delays and lot delays with one-piece part movement in cells has improve throughput time by 98% if 10 or more processes are involved.

By designing and implementing specialized jigs, fixtures, micro-switches, and similar mechanical devices.... different types of machine tools can be set-up to have a machine processing time less than the necessary cell cycle time. Upon completion of each machining cycle, the machine is stopped automatically and its

2.13 Other Key Features

An important principle of lean manufacturing is that the *internal customer* is just as important s an *external customer*. To emphasize this, some Lean companies have created separate cost centers for each manufacturing cell, and created competition between system components to reduce costs

In the lean production factory, processes produce only the amount of product required by downstream pull. Lean design concentrates more on worker utilization than machine utilization. This has sometimes driven the Western world accountants crazy. Traditional Manufacturing has used large, expensive multi-function machines to reduce labor costs. Lean engineering uses low cost, ordinary machines when possible and designs low utilization into their usage. Overproduction and shortages cannot be tolerated in a Lean Factory. Using processes to make more than is required violates the basic principles of lean production. Overproduction means that eventually inventory will increase, and increased inventory in WIP, intermediate storage or final storage increase product cost. Excess must be stored, tracked, and retrieved – all costly and wasteful operations. Frequently, these wasteful operations lead to the purchase of automatic storage-and-retrieval systems that require a maintenance person to keep the system on-line, and a computer system to track materials, another worker to program the automatic storage, retrieval-system, the computer: etc, etc, etc. This scenario is a total waste of valuable resources and it lowers an organization's ability to compete.

Lean production uses a minimum number of workers to achieve daily demands. Locating Lean manufacturing Cells next to each other minimizes both part movement time and move quantities between cells.. Cells close to other cells sometimes also makes it easy to balance tasks among workers. In mature Lean Factories, it is not uncommon to see a worker in one cell help a worker in another cell to resolve a temporary problem.

What can be done at full capacity if the production rate must be increased and the cycle time minimized? One approach would be to develop two cells with half the number of workers. This change will cut the per unit cycle time in half while producing the necessary number of parts in the same amount of time. In some cases, a little overtime would permit workers in cells to meet system-level requirements. In a critical situation, some of the required parts or subassemblies can be subcontracted, but this creates a host of operational problems. Notice the objective of *improvement* is not to reduce cycle time. Changing cycle time only is necessary for meeting schedule changes and meet Takt time requirements dictated by final assembly. Changes resulting in *any* excessive inventory stored in the manufacturing system are deceptive and really do not improve productivity.

2.14 Inventory Revisited: Not Zero but Minimized

Two historical names for lean manufacturing have been *stockless production* and *zero inventory*. We must emphasize again that nothing could be farther from the truth. Every manufacturing system has inventory. It is the life blood of the factory. No factory can run with zero inventories or have a stockless production system. A Lean Manufacturing System can drastically reduce both work-in progress inventory and finished goods inventory but cannot completely drive either to zero. In a Lean Manufacturing System (LMS), the inventory is held between Cells and is tightly controlled by Kanbans. The material in each cell is not called WIP (it really is WIP) but is called *stock-on-hand*, and it is not considered inventory. Hence, the term zero inventory has some validity. In the LMS, the inventory is minimized and controlled. It is routinely reduced in order to expose problems in the system. Taiichi Ohno's most famous schematic is called *Rocks in the River*, where the river represents the WIP inventory in the system (flowing or stagnant) and the rocks represent problems. Lowering the level of the river (the level of inventory) exposes the rocks (problems) in the system. These can then be resolved and the "rocks' are no longer seen. WIP in the system at different areas are again reduced, and the smooth river is now littered with "rocks". The process is repeated until no further reduction in WIP is possible or can be justified.

2.15 Synchronization Revisited

Synchronization refers to the process of timing movement of material between portions of the assembly line and the major subassemblies such that raw material and/or parts/subassemblies is at the right place at the right time. Even when material quantities have been leveled and balanced, unnecessary storage of in-process material can occur between synchronized operations. However, once operations are leveled, synchronization is just a matter of efficient, integrated scheduling. Leveling must precede synchronization because it helps eliminate process delays that make synchronization difficult.

Subassemblies that are synchronized for final assembly in the automobile industry are large items such as seats, panels, headliners, cockpits, doors, and engines. These subassemblies are all specific to certain vehicle models. They are made in sequence with the final assembly and delivered to designated workstations on the final assembly line at the same time as the vehicle needs that subassembly.

Only the best lean manufacturers can accomplish synchronization, since any subassembly-line failure can also stop a final assembly line. Conversely, if final assembly is stopped, then synchronized subassembly cells and lines also must stop so elements stay in sync. Production processes that have stopped must restart together.

2.15.1 Yo-i-don : Synchronization

Yo-i-don in Japanese means: *ready, set, go*. It is the name given to synchronizing Lean Manufacturing Processes. Suppose a final assembly line is producing one finished part per minute. Each individual operation at every step of the subassembly, processing, and main assembly areas must be completed in one minute or less. Since each car needs a body, the car-body welding area must also produce one body every minute. Workers and spot welding robots in each area must complete tasks and pass welded parts to subsequent workstations at the end of prescribed one-minute-cycle times. After completing assigned tasks and passing the welded part downstream, each operator presses a job completion button. This is called Yo-i-don by the Japanese.

2.15.2 Andon Signals

When a part is completed a button is pushed which illuminates a green light on a visible board called an *andon*, indicating that a part has been completed. The andon is a signal board with a colored lights, one for each workstation on the line. The andon is located where it can be easily seen. In this case, a *yellow* light illuminates if there are any incomplete tasks. This can mean that synchronization has been lost. When this happens, a Supervisor and/or a Lean Engineer should immediately investigate. If synchronization cannot be resolved in a short time, a *red light* illuminates and the entire line comes to a halt. Once the problem is resolved the red light is turned off...a *green light* is turned on... and the line or cell starts again with all processes beginning at the same time. This synchronizes operations by getting them all to start together, otherwise there will be *starvation* or *excess WIP* at each workstation. The andon board is called a *bingo board* in some factories. If there is an external subcontractor feeding parts or subassemblies to a production or final assembly line, they must also be synchronized. The part or subassembly must arrive at the right time, in the right place in the right quantity to the correct station *just-in-time* to be installed on the final assembly item.

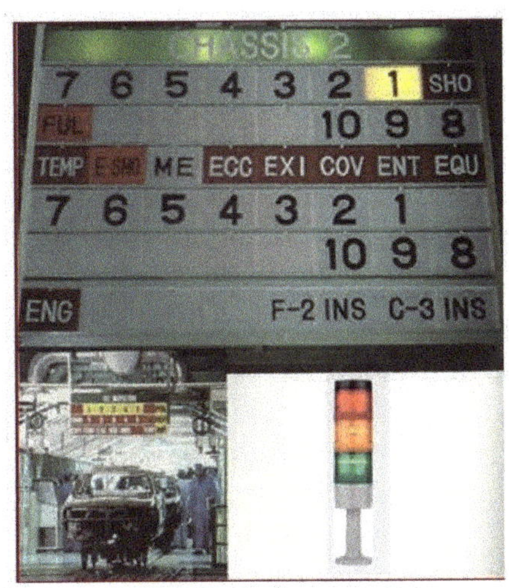

Figure 2.7
Andon Signals

2.16 A Lean Manufacturing Example

An example of a U-shaped Lean manufacturing cell that has seven machines and input/output shipping carts is shown in Figure 2.8.

There is waiting space for only one in-process part between each machine (Decoupler). The cell is arranged in a U-shape to minimize walk time between machines and to provide full line visibility to any one worker. A part is produced in this cell by a single male worker who will walk in a loop beginning at Machine 1 (a **S**haping operation) and ending at Station 8 which is final **I**nspection. He will either automatically or manually accomplish the following tasks at each machine: (1) inspect the part (2) unload the part (either manually or automatically to the next decoupler location (3) Reset the machine (4) Load the machine with the next part (5) Start the machine again. Each machine is usually automated and once the worker has completed his assigned tasks he will hit a "Walk-away switch" which will start the next machine cycle. He will then walk to the next machine in sequence and begin work. The machine which was just started will automatically cut off when it finishes. Note that at the top of this cell there is a **M**illing machine which can produce a part every 20 minutes. Assume for simplicity that the time to walk from 1 one machine to the another is 2 minutes and that worker task at each machine can be accomplished in 5 minutes per machine. The following performance measures can be verified.

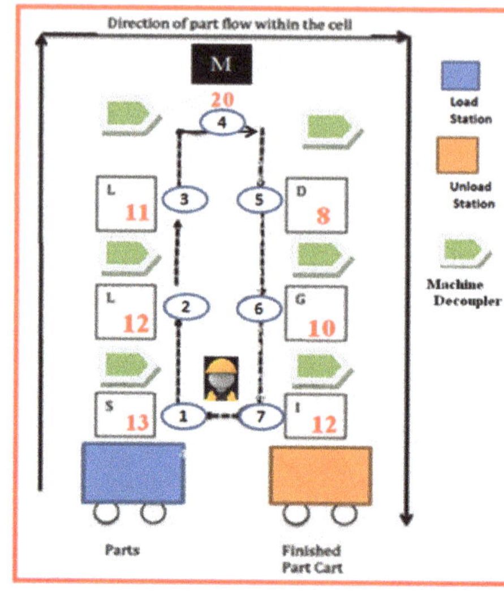

Figure 2.8
A U-Shaped Manufacturing Cell

- The *Actual Machining Time* per part is 86 minutes (Value added machining time)
- The *Worker Cycle Time....* time to walk around his work loop is 49 minutes (Walk time = 14 minutes and work assignment time = 35 minutes)
- *Cell Cycle Time Per Part* = 343 minutes (Time a part spends in the cell)
- *Cell Production Rate* = 1.22 parts/hr   [60 min/hr] / [49 min/part] = 1.22 parts/hr

This Manufacturing Cell with one worker is capable of producing 1.22 parts per hour or 10 parts/8-hour day. Now suppose that Sales calls and says that they need to have 15 parts per day....a 50% increase in production. The typical manufacturer would say....Oh Boy! We can easily do this by buying faster machines, or adding parallel machines or working a second 8-hour shift each day. The Lean Engineer would say...No Problem. We can do that starting tomorrow! How can this "miracle" take place. The answer is to simply add one more cross-trained worker to the cell and redistribute the work. The diagram on the right illustrates this solution. Several comments will clarify the diagram.

- No new machines or costly automation is needed
- The first worker is assigned a loop which contains 4 machines (machines 1-2-6-7)
- The second worker is assigned a loop which contains 3 machines (machines 3-4-5)

The worker responsible for machines 3-4-5 can complete his loop in 6 minutes walking time plus 15 minutes of task time = 21 minutes. The worker responsible for machines 1-2-6-7 can complete his loop in 8 minutes walking time plus 20 minutes assigned task time = 28 minutes.

Hence, a part can be released from the cell only every: Max { 21, 28} = 28 minutes. The upper loop worker has 7 minutes idle time every cycle. This is not perfect worker balance, but it is not bad. Checking machine capacity, each machine can complete its work in less than 28 seconds.

- At a production time of 28 seconds per part the Production Rate = 2.14 parts/hr = 17 parts/8-hour day. This will satisfy the requested production rate of 15 parts per 8-hour day.

This clearly demonstrates the power and versatility of Lean Manufacturing Cells. *What is the maximum production rate which could be achieved by adding more workers*? By applying Lean engineering logic, maximum production can be achieved by assigning one worker to each machine. In this case the *Bottleneck Operation* is **M**illing and the maximum production rate is one part every 20 minutes or 3 parts/per hour or 24 parts per day.

> *Comment*: It is actually possible to achieve this output with less than 7 workers. Can you see how?

Once the Manufacturing Cell reaches maximum output capacity, more production can only be achieved by adding parallel machines, working overtime or subcontracting work.

The cell operators are critical to cell operation. He/she is handling every part and checking the finished part at every process. Each worker must be cross-trained on all machines in the cell and know how to immediately correct minor quality problems and execute limited machine maintenance. It should be obvious that rejected parts, broken tools and poor quality product CANNOT be tolerated. This is why *In-Line* or *At-the-Source Quality Control* is so important. In Lean manufacturing cells, the operators are considered to be the company's most important (fixed) asset, a point of view not traditionally held in U.S. factories. People are viewed as more important than machines, and much more flexible than computers. Computers are viewed only as a useful tool in the cell. This example was presented to summarize the fundamentals of Lean Work cell Design and Operation.

2.17 Linked Work Cells

Lean Manufacturing work cells can be complete, independent cost and work centers. This will occur when a Work cell is making a complete part within its boundary. Even so, these types of cells are usually making complete parts and subassemblies for other downstream cells. Hence, they function as a supplier for other cell(s) and operate exactly as a production cell would operate which is directly making consumer goods. The demand for such cells is dictated by downstream, internal demand...which might be regularly fluxuating.....A good reason for using Lean Production Cells. Linked, U-shaped Lean production cells are quite common in a Lean Manufacturing System. Linked cell production systems will be discussed in some detail in later chapters. Figure 2.9 illustrates such a system.

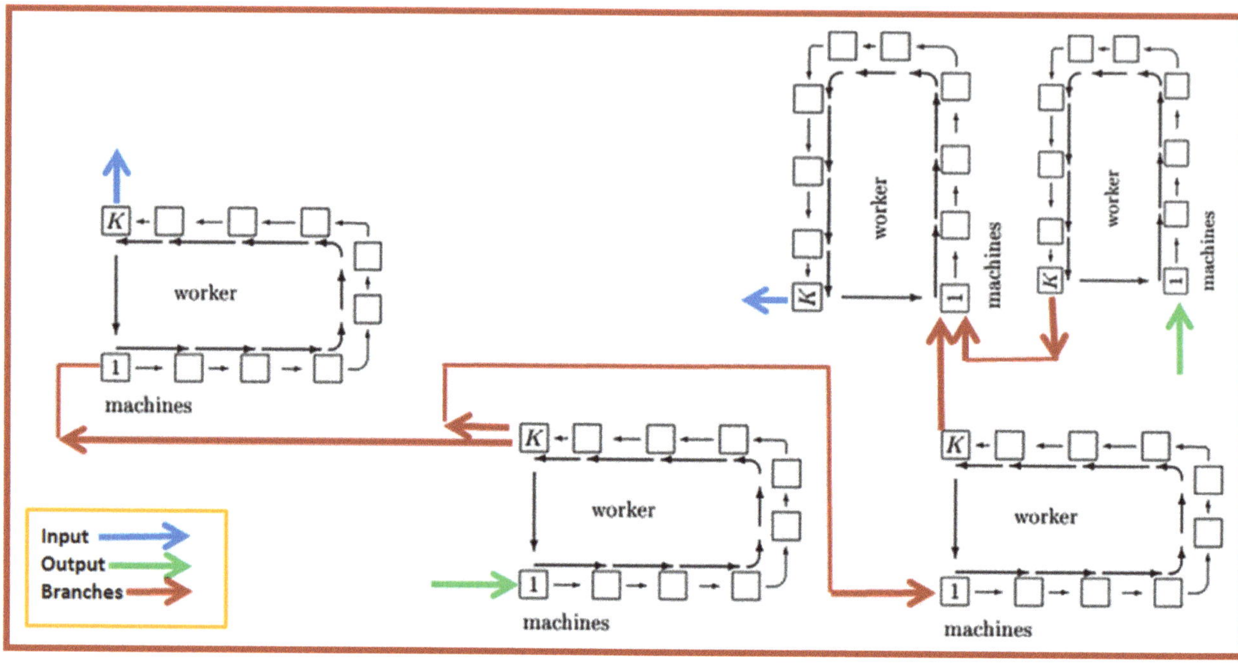

Figure 2.9
A Linked Cell Manufacturing System

The demand for any one work cell is completely determined by all downstream consumers (except in the case where a cell is producing parts under subcontract). The production rate of each cell must be Balanced and Leveled to achieve integrated product flow. For example if a table is being assembled in cell K and uses 4 legs for each table, the Lean Work cell that produces table legs must produce each leg in 1/4 the time required to produce a table. This is called *Takt Time* and is determined for each cell. The inverse of Takt time is the required production time per unit at each cell. Cells CANNOT produce at any other rate than that set by 1/Takt time. Overproduction will simply create WIP and underproduction will starve a downstream operation. A Lean Manufacturing System is a living, changing organism that is designed to operate smoothly at minimum cost-per-unit.

2.18 Cost Considerations

Capital investment costs of specialized machines and tools are very high, as are the risks. If design changes are necessary, highly automated and dedicated machines are often not flexible enough to make the transition. Many productions skills are transferred from operators to machines resulting in lower levels of manual labor skills than in a production job shop. Items flow through a sequence of operations by material-handling devices (conveyors, moving belts, transfer devices, etc.). Items usually move through operations one at a time.

A line can be dedicated to a single family of products or designed for Mixed Model production. We will discuss this in Chapter 4. The line advances the work (transfers it) forward one station at a time, with each station performing some operation(s), all done to satisfy the required Takt time. The automated-transfer machine that assembles automobile engines (at a rate of 100 per hour) is an example of a super machine for mass production. Transfer machines are specialized, expensive to design and build, and usually not capable of making another product. These machines must be operated for long periods of time to spread the cost of the initial investment – typically $20-40 million over many units. Although highly efficient, they usually make products in large volume.

Desired design changes in a product must be avoided, because it would be too costly to scrap the dedicated and specialized machines. Such systems are usually not flexible enough for product or process-design changes. Smaller versions of transfer machines developed for smaller-sized products made in large volumes range in cost from $2-$3 million.

Since the advent of the *mass production system*, various approaches and techniques have led to the development of automated machine tools that that are highly effective in large-scale manufacturing. Their effectiveness is closely related to product-design standardization and the length of time required between design changes. A machine producing a part with a minimal amount of skilled labor can be developed if the part can be highly standardized. The part can then be manufactured in large quantities. An automatic screw machine (a complex lathe) is a good example of a machine for the manufacture of small parts.

2.19 Automate or Not?

The development of the *Computer Numerical-Control Machine Tool* in the late 1950s and early 1960s permitted programmable control of the position of cutting tools in relationship to the work piece. By the late 1960s, automatic tool changers had been added to the numerical control machine, launching the birth of a workerless machining center. Computers and material handling devices (pallets) were added to support fully automated production. Today, the computer-numerical-control machine tool is readily available to all manufacturers. The machine shown in Figure 2.10 is an example of a CNC machining center.

Figure 2.10
CNC Machining Center

We will now turn our attention to the detailed design and operation of Lean Manufacturing Systems.

Chapter 3
Operational Rules for Lean Manufacturing Systems

Build a better mousetrap and the world will beat a path to your door *Ralph W. Emerson*

3.1 Introduction

The Toyota Motor Company has risen to lead the world in automobile manufacturing. In a relatively short period of about 35 years, Toyota has moved from last place to first place in both sales and profits. How was this amazing feat accomplished? Under the leadership of Taiichi Ohno and Shigeo Shingo, they redesigned traditional Mass manufacturing systems and implemented a new paradigm of manufacturing called the Toyota Production System (TPS). TPS is now known worldwide as *Lean Manufacturing*. In implementing TPS, they changed all final assemblies and key subassemblies into Mixed Model production lines; they balanced and leveled production; converted linear flow and mass production lines into U-shaped manufacturing cells; and almost totally eliminated Job shop manufacturing. They called WIP the *Root of all Evil*, reduced Work-in-Process (WIP) to a minimum and drove production with Kanban cards, not by complex computer scheduling. They developed pull strategies instead of push, and produced according to a measure called Takt time. Starting at final assembly, every component of production is designed to respond only to only Final Assembly demand. It other words, order release is only to final assembly which is driven by demand. Final assembly is configured to have an output rate only slightly higher than demand (Takt time) and to operate using one piece flow. Single cycle machines are designed to operate with flexible jigs and fixtures and to automatically check for quality and defective items (Poka-yoke). In all Lean production Cells a single item of WIP is maintained between adjacent machines (decouplers) designed to render each machine independent of any other machine in the cell. When the Toyota Motor Company invented and implemented TPS they introduced an entirely new way of thinking about how manufacturing systems are designed and operated. This new way of thinking was called *Lean Thinking* (Womack and Jones) and focused on only one paramount and overriding goal: *Eliminate all forms of waste*.

Figure 3.1
The 8 Wastes

The details of TPS are well documented by early proponents of TPS (Ohno, 1988; Sugimori, 1977; Hall, 1982 & 1983; Shingo.1985; Mondon, 1983 and Shoenberger, 1982). These authors visited Toyota and described how they changed their final assembly lines into Mixed-model production lines and leveled demand from the manufacturing shop floor through their supply chains. Other authors focused on the redesign and operation of traditional Mass production lines into U-shaped manufacturing cells (Black, 1986; Black, 1991; Black and Cochran, 2000), this new design Black (1986) called a *Linked Cell Manufacturing System* (L-CMS). The L-CMS is a radical departure from traditional Mass production lines. The output rates of L-CMS are controlled by allocating workers to machines in a self- balancing paradigm. L-CMS cells are flexible by design, they execute one-piece flow, they can easily accommodate changes in product mix or customer demand, and they are easier to accept changes in technological innovation and product designs. Figure 3.2 shows a typical U-shaped cell manned by two workers.

Each worker completes the tasks (machines) assigned to them, and repeats the assigned sequence of operations. The cell output is the maximum cycle time of either worker. Ideally, this cycle time will match Takt time production requirements and every worker will work exactly the same amount of time per cycle (Balanced production). How is this *magic* accomplished?

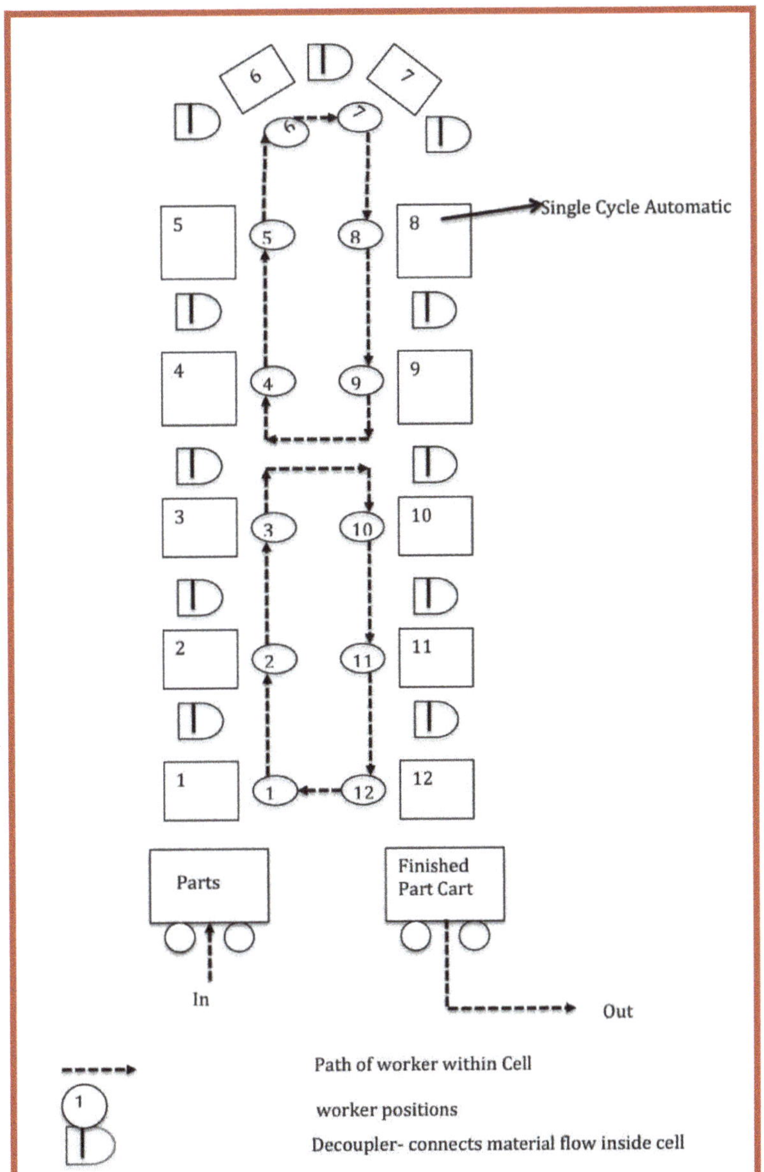

Figure 3.2
Example of Lean Manufacturing Cell Operated by 2 Workers

3.2 Design Rules for the L-CMS

The design and operation of a L-CMS is based upon a set of simple axiomatic design rules. Axiomatic design rules are based upon *tautology*; they are operationally necessary and true but they are based more upon operational knowledge rather than mathematical proofs. The force of gravity is a tautology: Everyone knows that gravity exists and holds things in place, but the theory of gravity is extremely difficult. Tautologies usually evolve out of practice and observation. A minimum set of design rules for a L-CMS are given below.

- ***Axiom 1***
  - All functional requirements should be independent of one another
  - An optimal design will always maintain independence of functional requirements

This Axiom states that each machine in a U-shaped cell operates independently of any other machine. This does not imply that there is not a precedence relationship between machine centers. It does imply that the work assigned to each individual processing step is a complete and uninterrupted operation.

- ***Axiom 2***
  - Minimize any needed information
  - Provide any information needed in the simplest form possible

The best cell design is one that has the minimum set of specifications and instructions.
A design that satisfies Axiom 1 by maintaining functional independence is called an *uncoupled design.* Any design that does not satisfy Axiom 1 is called a *coupled design*. Any coupled design can always be decoupled by separating the coupled functional requirement(s).

- ***Axiom 3***

Cell cycle time is dictated by the immediate downstream Takt timerequirements. The output rates (Takt time) from all cells upstream from Mixed Model Final Assembly are synchronized by the Takt time of final assembly, and whether or not there is a one-to-one coupling relationship ora many-to-one coupling relationship (four table legs are shipped to final assembly to be bolted on to a single table top).

The third axiom is critical to regulating and bounding cell output. The output(s) of a Lean Manufacturing System are dictated and set by the final assembly line(s), which produce at a rate determined by customer-driven Takt time. Takt time is based upon daily demand (DD), which is a smoothed weekly and monthly demand. Contrary to popular belief, Takt time for final assembly does not *chase* demand but is based upon actual sales and forecasted demand. Leveling the actual demand enables all final assemblies, subassemblies and component production to produce only what is needed, when it is needed.

- ***Axiom 4***

Balance the amount of time required by each worker in their assigned loops.
We will shortly demonstrate that the output of Lean manufacturing cells is controlled by creating enough capacity to meet demand, and then controlling the output rate by appropriately assigning machining tasks to one or more workers in the cell. Each cell requires at least one worker, and no more than one worker at each machine. Assignment of processes to workers should be such that all workers work the same amount of time. This is called *intercell balance*.

- ***Axiom 5***

Worker loop times should be as close as possible to cell Takt time. A Cell It should be obvious that if each worker assigned to a U-shaped cell works the same amount of time, this time should be as close as possible to Takt time. It is usually impossible to satisfy this axiom exactly. In most real-world cell designs, the individual worker's cycle time should be as close to one another as possible, with cell Takt time based upon the maximum cycle time of any worker in the cell.

- ***Axiom 6***

All equipment in the cell should be arranged in a U-shape in such a way as to guarantee that precedence constraints are satisfied and to minimize walking time.
Just as in a serial production line, machines or sequences of machines in a U-shaped cell, should be ordered from first to last reflecting precedence relationships.

Each machine in a U-shaped cell is operated independently of every other machine in the cell(Axiom 1). The sequencing of work at each machine should be the same at all machines. (1) Unload the machine (2) Inspect the part. (3) place the part in the decoupler (4) Load the next waiting part to the machine (5) Start the machine with a simple walk-away-switch, which will execute the processing step (6) Walk to the next machine in his assigned loop and begin this sequence again. We will soon see that this same basic sequence of steps is followed by both *assembly* and *machining* operations/cells; except in an assembly cell the operator manually executes an assembly operation, provided it is not totally automated Any cell can contain both machining and assembly operations in any order.

- ***Axiom 7***

Minimize as much as possible all set-up, changeover, inspection, load and unload times.
It should be obvious that since all setup, changeover, mount and dismount times at each machine (if manually executed) can be a large part of the cell cycle time. These times should be minimized as much as possible (SMED) through autonomous machine automation. This is particularly important in Mixed-model production cells.

- ***Axiom 8***

Enforce a Zero defects policy and execute Total Preventative Maintenance at each process step (machine)It will shortly be made obvious that the utilization of every machine in a Lean cell is usually considerably less than that one on a corresponding, traditional Mass production line. While each machine can absorb small amounts of unscheduled downtime, Total Preventative Maintenance (TPM) should be introduced and set the goal of 100% availability. Each machine must finish its assigned manufacturing operation, and shut itself down before the worker assigned to that processing step returns. Steady, uninterrupted flow of parts through a Lean cell cannot tolerate any rejected or scrapped part in the production cycle. A policy of *zero defects* must be strictly enforced. The worker serves as both a Quality Control Inspector and Preventative Maintenance Engineer.

- ***Axiom 9***

Redesign as many machines as possible to be single-cycle automatics. Each machine in a U-shaped Lean cell should be retrofitted with a simple single cycle, start switch. This is called a *walk-away switch.*

- ***Axiom 10***

Introduce no parts into the Manufacturing cell unless they are authorized and waiting in WIP before the first workstation in the sequence of workstations. All parts are removed from the cell to an output station and are not moved downstream except when authorized. Authorization to start a piece or small originates from another downstream cell and is triggered by a Kanban card.

One of the most important characteristics of a U-shaped Lean cell is that the amount of stock on hand allowed in the cell is strictly regulated and enforced. A properly designed cell for one worker will have one part residing on each machine and one unit of WIP waiting to be loaded in the decoupler. In an N-Machine cell, there should be no more that 2*N pieces of WIP being worked on in each cell. In some companies, the N parts being worked at each machine are not called WIP. It is necessary to make the work cycle of multiple workers assigned to a cell independent. For example, if there are three workers assigned to a single cell, their work patterns define three mutually exclusive and separate *loops*. Each loop is sometimes made independent of every other loop by holding one part in a special decoupler at the *loop interfaces*. This is to insure that a worker with a faster loop time is never starved by a slower loop time. A part is never introduced into the cell unless a part has finished all cell operations and has been placed into shipping. Individual Cells are buffered (made independent) from every other cell by placing a limited amount of WIP before the cell loading station. This WIP storage is usually only one unit waiting in a place called a *Decoupler*. Parts are moved to a cell loading area in small lots where they wait to be introduced into the first workstation. They are moved into the cell by a Kanban or some other *pull* mechanism. In a similar manner, the output station will hold a limited amount of parts (WIP) in a conveyance device (pallet, box). This final cell WIP can never be moved without Kanban authorization or a similar light pull system (Traffic Cop) triggered by downstream requirements. If the output device is full and cannot be moved, all cell production ceases. This prevents excess WIP and system congestion. This is a major departure from how parts are moved in a traditional manufacturing plant.

- ***Axiom 11***:

Decouple all operations/machines contained in the cell with a single item of stock between machines.A single item of stock-on-hand between before each workstation*decouples the* machining operations. It can also help to buffer any irregularities or variance in the processing cycle. Decouplers can and should also perform quality control checks on the most recently completed part.

- ***Axiom 12***:

All workers should be cross trained to operate any machine in the cell.
The key to responding to different demand patterns or product mix is the ability to add workers or to change worker assignments. This is only possible by cross training each worker to run any machine in the cell. One worker is often assigned to cover the input and output processes. This also educates a worker concerning what each value-added operation means to final assembly.

- ***Axiom 13***

Any individual machining time in the cell must be *strictly less* than the Takt time which is driving cell output.

This is one of the most important design principles in U-shaped Lean cell design. Since the cell cycle time is based on the required Takt time(min/part), every machine in cell k must be able to meet the following rule:

MT_{kij} <Takt time MT_{kij} is the processing time on machine i for part j in cell K

This Axiom insures that cell production will meet the Takt time requirement. Once it is started, a machine will usually automatically complete the machining time assigned to that process step. Upon completion, it will automatically cut off. Meanwhile, the worker that started that machine is working his/her way through all other machines in his/her assigned group of machines (his/her work *loop*). The *machine cycle* must be completely finished before the worker returns, or the machine will cause a *bottleneck*and cause the worker to wait.

Bottlenecks can occur because: (1) MT_{kij} > Takt time(2) Quality Control problem is detected and the machine cuts off or (3) A problem requiring Maintenance has occurred. Quality must be strictly enforced and defects eliminated. Any Preventative Maintenance must be conducted on an off-shift.

- ***Axiom 14***

In a cell with multiple workers and/or workers operating multiple machines/operations, his or her walking pattern in the assigned loop should not cross that of any other worker.

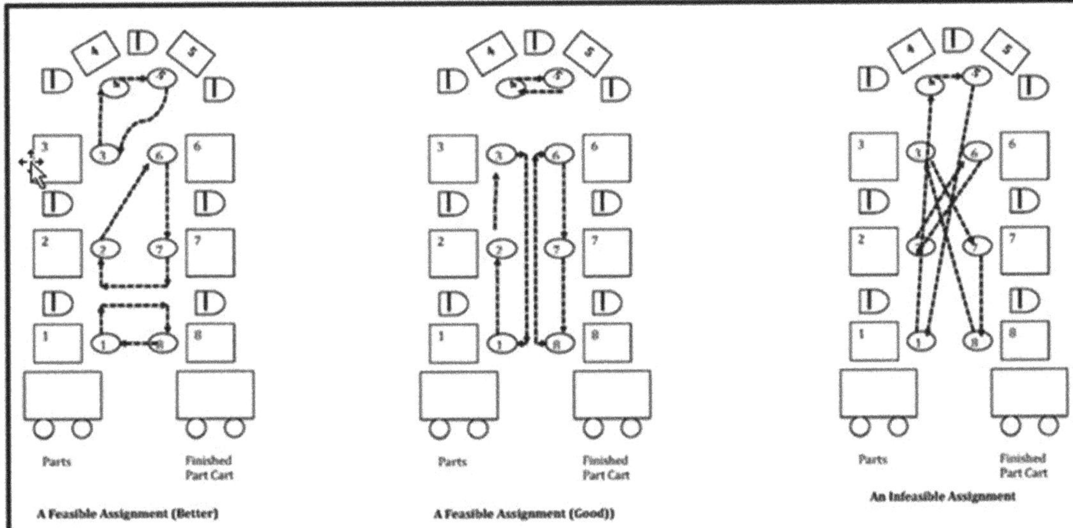

Feasible and Infeasible Worker Assignments

This axiom is given to prevent any interference between workers. It is a *rule of convenience* and may or may not be invoked. When it is applied, it will constrain the maximum number of total worker assignments which can be made between N workers and M machines. The operating principle is not an absolutely hard and fast rule, but the idea is that as each worker walks through his/her assigned route, that route should not cross another worker's route.

Using these 14 Axiomatic design rules, a U-shaped Lean manufacturing cell can be designed and implemented. Beneath each rule are a lot of innovative and difficult design activities. Naturally, the cells will be operated by people who will use and improve *autonomation* principles. Everyone who works in the system must understand and buy into Lean thinking. The result of following these design rules will result in a U-shaped Lean cell which is robust, flexible and simple to operate with a minimum of complex calculations and operating data.

3.3 The Evolution of Lean Factory Design

The Lean linked-cell factory design (L-CMS) is an outgrowth of the job shop and the flow shop, but it is neither. The Job shop as a manufacturing system emerged in the early 1800's. Job shops replaced craft or cottage manufacturing when it became necessary to replicate individual part types and produce them in larger quantities. This functional design was initially made possible when new power sources were developed to automatically run many machines simultaneously. The sources of power was*waterdriven turbines* which used waterfrom streams and rivers. The second source of power changed the entire face of automation in wide variety of applications: This was the invention of *steam powered turbines*. The father of modern steam powered turbines was James Watt, a Scott in 1784. It was not until William Avery built the 1st truly successful steam power plant in 1837 that steam power became a source of virtually unlimited energy.

In the early 1800s, the first vestiges of the flow shop began to emerge with flow lines designed for small part manufacturing. The flow line reached maturity at the Ford Motor Company around 1913. The world came to see how this new system was designed and operated. Over the next 20 years, hybrid shops with Job shop and Flow shop components emerged using the pioneering work of Henry Ford and Frederick Taylor. Taylor was the first to define and implement the principles of *scientific management*. New system designs enabled factories to produce large volumes of identical products at low unit cost.

Subassembly lines were designed for make-to-stock inventories creating large in-process-inventories but sustained continuous production of the main flow lines. These systems helped America win World War II and become the world leader in manufacturing. After WWII, the Toyota Motor Company, led by an engineering genius called Taiichi Ohno, pioneered a new manufacturing system design called the Toyota Production System (TPS). In 1991 it was given a name that promoted its universal application to any form of service or manufacturing system: The *Lean Production System* (Womack, Jones and Roos).

3.4 Defining a Manufacturing System

The Manufacturing System is the beating heart of any Enterprise system.

The manufacturing system is defined as a complex arrangement of physical entities which are characterized by measureable economic and time parameters. The physical entities are machines, tooling, material handling resources, raw materials and people. Manufacturing systems systematically transform raw materials into finished goods which are sold for profit or built for special use (Aircraft Carriers). Companies and individuals who buy finished goods are *external customers*. The workers performing the processing steps and machining operations which add value to the finished product are *internal customers.* In designing a manufacturing system, this is a key concept. Any manufacturing system must satisfy the needs of *both* external and internal customers. Both types of customers respond to measurable parameters that define how well a company can respond to customer needs . There are many measurable parameters which must be carefully controlled. Profits, costs, cycle time, throughput time, the amount of work-in-process, scrap and rework rates and customer satisfaction to name a few.

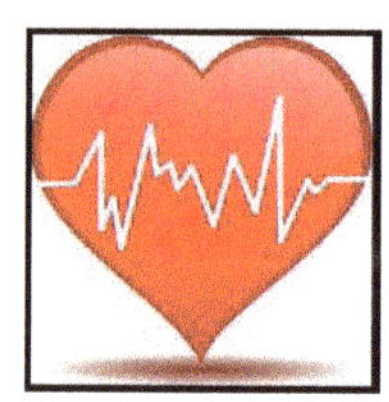

Different system designs can greatly influence the response of measureable parameters. Time is one of the most critical drivers, measured by throughput rates, inventory turns, cycle times and customer response. Lean systems has as one of its main goals the reduction of *throughput time*. Lean systems are continually seeking ways to reduce WIP levels, improve quality by eliminating rework and scrap rates and eliminating non-value added activities. Using Little's Law, if the desired output rate is fixed by Takt time, then systematic reduction of WIP will directly reduce throughput time. This is one of the main reasons to design and operate U-shaped cells. We have already pointed out that cells exhibit a rigid WIP Cap. Work in process held between cells is minimized by invoking Kanbancontrol. Inter-cell inventories are determined by the needs of *internal customers*. Final assembly is driven by *Customer Demand.* Kanban control is the key to maintaining minimum inventory levels in the Manufacturing System.Final goods inventory is primarily used to support on-time delivery and to smooth production requirements caused by varying demand. The entire Lean, linked manufacturing system constantly tells all manufacturing components *what* to make, *how much* to make and *when* to make or store product. Parts or subassemblies being transformed at machines within cells or are often called*stock-on-hand* (SOH). SOH is not considered to be WIP inventory, but part of *standard work*. The design of the factory and the manufacturing cells are tightly coupled to product design (work sequencing) and required production rates (upper bounds). System design includes close attention to Ergonomic and human actors issues. The Lean Engineer should be well schooled in these principles, and ideally derived from Industrial Engineering.

Unless there is a high-volume, final assembly line dictated to a single finished product,final assembly is converted to a *mixed model*assembly line and dictates how all upstream subassembly and component cells operate. Leveling and smoothing of all production components follow customer order generated Takt time specifications. All subassembly and component production is accomplished in relatively independent U-shaped Lean cells, with Takt time requirements linking all downstream cells to upstream cells. Cells operate on a one piece flow basis just like final assembly. All subassembly and component cells are linked to one another and to final assembly by Kanban and pull philosophies. This usually eliminates the need for paced conveyor lines.

One thing often misunderstood is the fact that a linked-cell Lean manufacturing system has inherent, integrated control functions built into the system design and operation. This type of built-in production control is quite different from traditional manufacturing systems that operate by pushing orders into the manufacturing system. Traditional manufacturing systems require a constant dialog between order release systems like MRP II and production control systems with Computer Aided Manufacturing (CAM) modules.

Every Lean system is designed and operated built upon a *zero defects* manufacturing philosophy. Toyota believed in company-wide Total Quality Control (TQC), and taught zero defects operational procedures to everyone in the company from the president to the shop floor worker. The enabling technology to support TQC was to redesign the manufacturing system to support one-piece flow, where work-in-process is inspected after each processing step. Every worker in the factory became a quality control inspector.

A second equally important concept was to aggressively and continuously work on *variance reduction.* Variance reduction is the key to consistent, steady production and is a key component of continuous improvement, waste (*Muda*) elimination, and the efficient use of people, machines, on-time delivery, and supply chain integration. Components of variance to be eliminated include variation in quality, variation in hourly/daily output rates, variation in processing time and variation in standard practices. We will continually stress the importance of eliminating variance in any form.

A Lean Manufacturing System is designed to be versatile and flexible to respond to customer demands. It is the responsibility of every shop floor worker to develop and implement manufacturing processes, reduce setup and changeover times, improve quality, and continuously redesign the workplace (methods improvement. Many innovative solutions become company secrets, closely guarded to gain competitive cost and time improvements. The result is that innovative and creative breakthroughs in Lean Manufacturing are not achieved overnight, but is an on-going process. A Lean system operates in ways that the old mass production system cannot. It is responsive and easily reconfigured to accept design changes, product mix and changing demand patterns. *So what steps need to be taken to begin a Lean transformation?*

3.5 Steps Necessary to Implementing Lean Manufacturing Strategies

Changing traditional manufacturing system design and integrating the control functions into a new Lean system design requires commitment from top level management to the lowest factory worker. Every employee must be fully committed to system change. Total employee involvement is absolutely necessary. If the manufacturing enterprise is unionized, the union(s) must be fully briefed and involved in the transformation. In general, it is not the labor union or the shop floor workers who raise barriers to Lean production, but it is usually middle management (MM) and complex corporate control systems. Middle management and computer controlled manufacturing will have the most to lose in implementing Lean manufacturing strategies. Many of these functions get integrated into the operational strategy and design of a Lean manufacturing system. We have identified some basic steps which must be taken to implement Lean production.

1. All levels of the plant from the president to hourly workers must be fully educated in Lean production philosophy and concepts. There must be a clear understanding of how a Lean manufacturing system differs from traditional Mass manufacturing systems.
2. Top down support and solid commitment is critical. The entire company must be involved and committed to change, particularly the existing design and manufacturing engineers.
3. Top management must realize two fundamental truths. The first is that once the company begins to embrace and switch over to a lean system design, things will get worse before they get better. Initially, it may appear that Lean principles simply do not work. There will be a desire to move back to how the company was operating prior to change, then a steady improvement will occur in system performance measures, cost and customer satisfaction. This cycle might be repeated several times as technological and operational changes take place. There will be resistance to change: equipment utilization will be lower in (by plan), workers roles and responsibilities will be

broadened, and corporate costing strategies will need to be changed to abandon the traditional cost per labor hour.

4. The selection and monitoring of system performance measures to reflect change is critical. Everyone in the company must be aware that Lean thinking is directed to waste elimination, WIP reduction, quality improvement and throughput time reduction. Cost and not profit margins determine profits.
5. All engineers and shop floor workers must be trained to find waste, and to eliminate all forms of waste as a part of their job description. There must be a relentless pursuit of eliminating waste and continuous improvement. New and more comprehensive educational programs are essential. Lean cell operators must understand why change is necessary and how change takes place. Operators need to be involved in cell design; cross trained to operate all machines in a cell; be proactive in methods improvement; and function as Quality control inspectors empowered to shut a system down if problems arise. A critical function of all cell operators is to continuously find ways to reduce setup and changeover times.
6. Shop floor personnel and cell workers should be rewarded for reducing waste and improving system performance. Innovation awards and bonus payments are commonly used to reward innovative workers. NOT PIECE RATE
7. Middle management and corporate executives should take an active role in Lean transformation and avoid passive participation.
8. Large production control systems such as MRP II should be used and utilized to *support* production...not *drive* production.

Table 3.1 on the next page provides an overview of a successful implementation strategy. The suggestions in Table 3.1 were extracted from several American companies who have successfully implemented Lean Engineering principles. These implementation strategies are discussed in more detail in subsequent Chapters.

The remainder of this Chapter will provide a high level presentation of some important Lean design rules. The existing manufacturing system must be redesigned, simplified and integrated before any computerized systems are put into place to support production. *Autonomation*, which is automation with a human touch, becomes a large part of final transformation after system integration has been achieved. *Concurrent engineering* becomes the last step in the effort to restructure the entire manufacturing enterprise. A word of caution: piecemeal and isolated implementation of Lean manufacturing will never realize the full potential of Lean transformation., and may even turn management against the entire concept. Isolated pilot projects may, in fact, prove to be counterproductive. The order in which Lean transformation occurs is also important. Many companies have attempted WIP reduction byfirst implementing Kanban control and pull versus push methodologies. These companies have, in fact, experienced immediate WIP reduction even before U-shaped cells were implemented. However, because this is only a part of Lean production, many such approaches have resulted in failure. Even worse, many companies have resisted continuing to run the company with large production control software systems such as MRP and ERP. Such an approach fails to realize the full potential of total Lean transformation.

Table 3.1
Successful Implementation Strategies

Strategy	Actions
1. **Level and Balance the manufacturing system: Smooth the material flow (Monden, 1983)** Leveling involves the design of Mixed-model U-shaped lean cells and the implementation of Mixed-Model final assembly. Balancing is making sure that every set of processing steps are equalized in terms of time required to execute each step.	• Calculate Takt times and daily demands • Implement Mixed-Model final assembly • Balance output from each supplier • Develop one-piece flow • Sequence products and subassemblies with other final assembly
2. **Design and reconfigure the Manufacturing system** Design and implement component, subassembly and final assembly cells. System design must consider the manufacturing specifications of each part and product, and the requirements of both internal and external customers	• Set standard work for operators in cells • Implement U-shared lean cells • Implement self-balancing subassembly cells
3. **Initiate SMED to reduce or eliminate all setup and changeover times** All setup and changeover times are non-value added and increase throughput time. By Little's Law, this also creates WIP, Redesign to accommodate small lot sizes.	• Teach everyone SMED • Develop one touch setup procedures • Train operators to decrease these times and to perform all setups and changeovers.

Finally, it should be stressed that the main manufacturing facility cannot be isolated and transformed by itself. The Supply Chain and the final distribution system must also be changed. The principles and methodologies used to change logistics and the total supply chain are just now beginning to emerge. This book introduces the Kanban system and sole – sourcing supply chains while supply chain management is the subject of the book by Ravindran.

3.6 Takt Time Design Rules

The subassembly and component manufacturing cells are linked to Mixed Model final assembly by a Kanban control system. Each manufacturing cell is designed to produce product as near as possible to customer demand; which coupled to downstream requirements defines Takt time for each cell. Within cells, a serial set of processes is put into place to both balance the cell and to meet the Takt time requirement. Takt is a German term that historically refers to a symphony conductor, who is responsible for keeping every member of a large orchestra in time and sequence. Takt time (TT) is available production time divided by demand.

$$\text{Takt Time} = \frac{\text{MinutesavailableforProduction/Day}}{\text{DailyDemand (DD)}}$$

Takt time is the inverse of Production Rate (Parts/day) that must be produced to satisfy customer demands over a specified period of time (Lead Time). Clock time available for actual production must not include start-up, shut-down, lunch breaks and other non-productive times.

Takt time for multiple products is based upon a Weighted Average of individual product times if the production line is a Mixed Model Production Line. This will later be covered in detail.

Takt Time is the *heartbeat* of a Lean production system. Takt time calculation is easy to understand, but requires a lot of Lean Engineering to properly execute. Final assembly (FA) lines must be converted to a Mixed-Model (MM) final assembly that can make multiple products in any order. Takt Time is simply the required cycle time for final assembly. The inverse of Takt time is *production rate* for final assembly. While the required production rate is mathematically the inverse of Takt time, there is a subtle difference in these two measures. A typical production rate might be 5 parts per hour. This is the rate of production that must be met to satisfy customer sales. The reciprocal of production rate is one part produced every 12 minutes. In Lean manufacturing, the production rate is a long term average of output per unit time, but the *timeper part* is a goal which must be met. Takt time sets the pace for the entire factory. Mixed-model final assembly sets the requirements for balanced and level production for the rest of the factory and for any Kanban-linked supply chain. Each Cell is Producing to final assembly; Takt time sets the functional requirements of producing the right quantity, at the right time in the right place. All operations in final assembly are designed to be rapidly changed to Mixed Model requirements, so that finished products can be made with great variety in small batches.

3.7 Subassembly Cells vs. Manufacturing Cells

In the Linked Cell Manufacturing System (L-CMS) there are multiple subassembly and U-shaped cells, all aligned to the Takt time of the Final Assembly Schedule (FAS). Some assembly lines are synchronized with final assembly, and stop and start when necessary. In manufacturing cells with automatic machine tools, each operator only remains at each machine through the unloading/loading cycle. Once the machine is restarted, the operator moves on to the next machine in sequence. Cells are manned by multifunctional, cross trained workers who move in circular patterns between machines arranged in a U-shape to minimize walking time.*So how do Lean subassembly cells operate differently from flow lines and manufacturing cells?*Figure 3.3 shows a typical linear flow line.

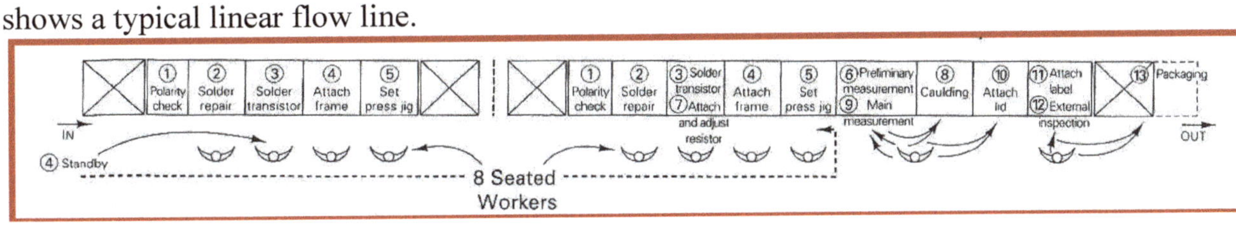

Figure 3.3

A Typical Subassembly Line for the Mass System has Conveyors and Seated Workers.

This type of linear flow is representative of most mass production lines. Multiple operations are grouped according to precedence constraints and are synchronized to meet required line production rate. These stations are typically assigned to individual workers. Ideally, every machine or machine grouping consumes the same amount of time. This is called *line balancing*, and is well-known to all Industrial and Manufacturing engineers.

In a Lean, U-shaped subassembly cell, walking workers are assigned to one or more machines in such a way as to achieve the Takt time requirement as discussed in Chapter 2. Multiple workers are responsible for one or more operations, and ideally the worker cycle time of each worker is identical. Task times at each machine can be very different. Different output requirements or changes in product mix are easily accommodated by reassigning tasks and machines to workers, and by increasing or decreasing the number of workers in a cell. There are 3 commonly used variations of this basic scheme: (1) Rabbit chase (2) Sub-celldesign and (3) The Toyota sewing system, also known as *Bucket brigade*. These alternatives will be explained in Chapter xxxx.

In most Lean manufacturing cells, almost all the machines are single cycle automatics (A machine completes its work cycle and automatically stops). The operator unloads, checks, loads and starts the machine, and then either automatically or manually places the part in a *decoupler*.This sequence of work is called **Lo-Co-Moo** (**L**oad one, **C**heck one, **M**ove **O**ne**O**n).

3.8 Decouplers: Revisited

Decouplers are placed between adjacent machines or at the point where the work cycle of any one worker interfaces to the work cycle of another worker. This point of interface is called the *Baton passing zone*. This zone usually contains only a unit of WIP. The purpose of a decoupler is to: (1) Create operational independence between all machines in the cell and(2) To absorb unavoidable differences in the worker cycle times or interruption of serial flow. In this way, variations in smooth, continuous flow are absorbed and each worker operates independently of every other worker. The term decoupler originated by Black based on pioneering work by Nam Suh in axiomatic design. Decouplers can play many critical roles in the operation of U-shaped Lean cells (Black and Schroer).

- Decouplers greatly enhance the flexibility of U-shaped cells by eliminating any co-dependency between machines or worker sub cells.
- Decouplers facilitate flexible worker assignments or reassignments as demand and/or product mix changes
- Decouplers promote continuous, single piece flow by buffering disruptions
- Decouplers provide the opportunity to perform automated quality control inspections without adding to the cell cycle time which is determined by the workers and not the machines. Decouplers can be automated to perform part orientation for machine loading and unloading functions.
- If two machines are located some distance apart, part of the decoupler functionality might be to execute material handling responsibilities.
- If a particular machine is functionally coupled to another feeder machine or cell in an assembly operation, the decoupler can execute orientation and staging capabilities.
- Decouplers can be useful in smoothing production if a cell is a Mixed Model production cell.
- Simple process transformation requirements such as heating, cooling or drying can be part of decoupler functionality. Figure 3.4 shows a paint drying decoupler. The drying decoupler in this example has six different fixtures with different surface orientation. Each fixture has a 2 minute drying cycle, so the total drying time per part is 12 minutes. Of course, the required Takt time for the entire cell cannot exceed 12 minutes.

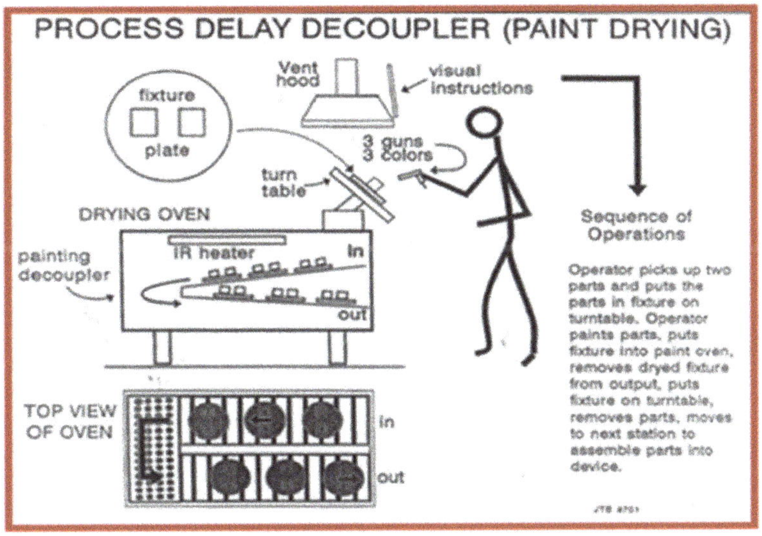

Figure 3.4
A Decoupler to Dry Paint Using a Gravity Chute

- Decouplers control the level of Work-in-Process within a cell, and are part of the cell WIP cap. The capacity of a decoupler should never exceed two units except in rare cases, or in Mixed Model cells.

Any of the machines can quickly switch from one part type to another. A decoupleris between each machine. The number of parts to be made by type is determined from product demand and final assembly, and might change frequently. A particular mixed model sequence will require different changeover and setup times between different part types. These times must be held to an absolute minimum, since every minute of time lost to setup and changeovers is a minute of lost production time. Setup and changeover times must be incorporated into the worker's loop duties. Suppose that the following production sequence is to be repeated 10 times at Machine 4: **AAAABB.** Suppose the 4th **A** has completed work at Machine 4. As the walking worker approaches Machine 4, a simple signal will alert him to reset machine 4 to make a **B**. Some machines of this type might have the next part to be loaded indicated by a colored light to guarantee the correct part sequence. The *Lo-Co-Moo* strategy is modified as follows. The discharged Part A is inspected and approved. Machine 4 is reset to make Part B; Part B is then loaded and the operator hits a walk-away switch. The Part B machining cycle then begins.

AAABB

Figure 3.5
A Mixed Model Sequence That Minimizes Changeover

A *Mixed Model Sequence* requires a modification to the Axiom that a decoupler will hold only one unit. If a cell is designed for Mixed Model production, one unit of each part type must be held in the decoupler. Once the Mixed Model sequence has been determined, there is no sequencing or scheduling or routing sheet needed whatsoever in the cell. As parts are removed from each decoupler, the next part in line is loaded into that same area. It makes no difference at all to a walking, moving cell worker if machines are not in the correct routing sequence of a particular part. Clearly, the cell must be initially loaded with parts, processed one part to one machine at a time, and properly staged in each decoupler. In some factories, all parts are staged and sequenced in a flow-through rack at the input station. Properly staged and sequenced, any mix of parts can be processed without error. As usual, there is one vitally important design principle which must be followed.

The processing time for each part in each machine must be less than the cycle time.

Remember that CT is determined by the maximum amount of time the worker takes to complete his loop around the cell. The operation in a mixed model production cell requires that setup and changeover times are reduced to near zero; every machine produces no defective parts; and TPM practices make the availability of each machine 100%. It makes no difference if any one machine is fully automated or not. Each machine can be an assembly operation or a machining operation.The common Cell/subcell cycle times are determined as usual

3.8.1 Classifying and Defining Decouplers

In examining the various uses and functions of decouplers, a more comprehensive definition of a decoupler might be appropriate.

> *Any functional component in a U-shaped, Lean production cell that holds, delays, orients, inspects or transports parts in a temporary holding area*

In Mixed Model Lean cells, a primary function is to introduce flexibility into cell operation and facilitate smooth flow of parts through the cell. In other words, decouplers can be classified as an *operational control device* and not simply an in-process material storage device. The primary and secondary functions of a decoupler are given in Table 3.2.

Table 3.2
Primary and Secondary Functions of a Decoupler

Primary Functions	***Secondary Functions***
Decouple process steps	Poka-Yoke (Defect prevention)
Control cell WIP levels	Inspection/Quality checks
Stock-on Hand for Mixed Model production	Part transportation
Part orientation	
Part registration	
Process delays for Inspection, Etc.	
Part transformation: Heating, cooling, etc	

Finally, it should be noted that decouplers can often relieve humans of boring, simple repetitive tasks and improve cell efficiency. Historically, the important role and functionality that decouplers provide has been overlooked. In Lean Engineering, they now play extremely important roles in designing and implementing Lean production cells. They are vital to Lean systems with Kanban control and pull philosophies and facilitate real-time Quality assurance.

3.9 Integrating Production Control

The traditional mass manufacturing system is very complex. Large, multiple function software systems such as MRP II and ERP interface to literally hundreds of functions and sometimes thousands of workers. People who execute production control functions sequence and schedule the entire manufacturing facility, which means that they literally determine *where* purchased parts and subassemblies go; *when* they should be scheduled to go there; and *how many* should be moved. Many companies have attempted to completely computerize these functions and eliminate production control personnel, but the general result has been frustration, wasted time and money, disappointment and largely failure. Why has this been so difficult? There are many reasons, but we believe that the main problem rests with the software systems being used. Software packages are basically designed for planning, and not control. Many software systems being used today were developed and installed by consultants and outside people who do not use these systems, and they were designed and configured to be a general use software system across multiple platforms. One of the missing ingredients is that manufacturing control involves people and not strictly things. The other major problem is the inherent incapability of hard rule-based systems to intelligently respond to uncertainty, change, variations and disruptions. This latter problem has recently been addressed by Artificial Intelligent software.

The Lean manufacturing approach greatly simplifies and redesigns traditional system control functions. Many system and subsystem control functions are integrated into the system design using versions of Kanban. *Physically*, production control and synchronization is achieved by creating Lean, U-shaped assembly, subassembly and component cells linked by Kanban control and pull philosophies. *Logically*, nothing is moved or manufactured without some form of signal from the Kanban system. Lean layout and product flow rigidly define the paths that parts can take as they move through the plant. Each manufacturing component is precedent linked to every other component, and Lean Cell; eliminating the basic need for a routing sheet.

3.10 Production Control in a Linked Cell System

Production in upstream U-shaped Lean cells is pulled by downstream demand, and is initiated/authorized by a Kanban card. For discussion purposes we will refer to the diagram in Figure 3.6.

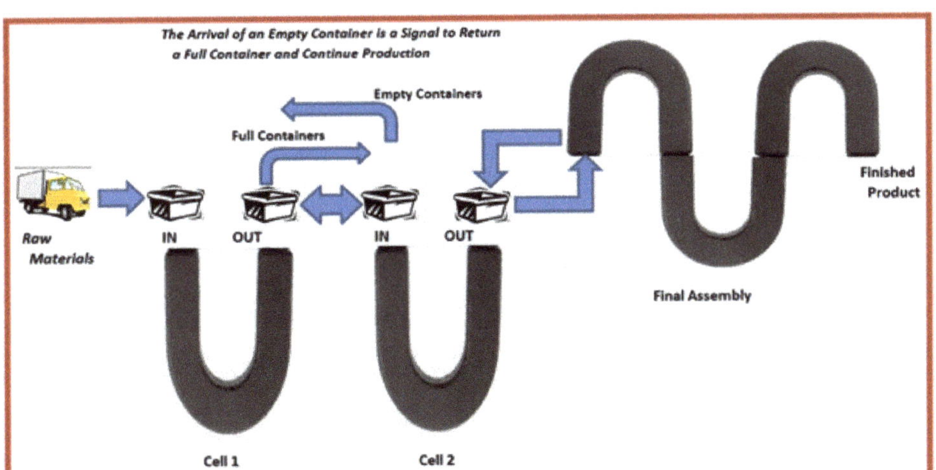

Figure 3.6

Cells are Linked by a No-Card Kanban system. The Design Rule for Determining the Maximum Inventory in a k-link is an Operational Form of Little's Law.

Assume that cell C1 is providing parts to only cell C2. The *output station* of cell C1 unloads parts into a material handling carrier. These parts will eventually be moved to an *input station* in front of the receiving cell (C2). We have previously indicated that the WIP in carts at both stations is tightly controlled by Kanban signals. Further, nothing can be moved from C1 to C2 without authorization. In Chapter 6 we will describe several types of Kanban control systems that can be used. For now, we will describe a simple pull system that does not require any formalKanban controls at all. Let us briefly address the following questions. (1) How many units should be stored in the output station of cell C1? How many units should be stored in the input station of cell C2? What triggers part movement from C1 to C2? How many units should be placed in a full cart?

The system shown in Figure 3.6 is an example of how Lean U-shaped cells can be linked to one another. A design objective is to minimize the total amount of WIP at the output of the supplier cell (C1) andat the input of the consumer cell (C2).

Consider the number of units in Cell1 and Cell 2, particularly the amount of WIP contained at the output of Cell1 and the input of Cell2. For illustrative purposes, assume that there are no variance components in the linked-cell system. If Cell1 contained six workstations and Cell2 eight, and if unit moves are used inside the cells, there would be fourteen total units of WIP interior to both cells plus 14 units of Work-in-progress .At the cell interfaces, there will be an output station at Cell1 and an input station at Cell2. How can the output of Cell1 be synchronized with the input of Cell2? Assume that Cells 1 and 2 are both producing at a Takt time of 30 seconds per unit or yielding an output rate of 2 parts per minute. Hence, the two cells are perfectly balanced. The request/delivery cycle between Cell2 and Cell1 will not be *controlled* by a Kanban card, but one will probably be attached to each container. We will show in this example that this Kanban system will not require any cards or lights. The containers used to move material from Cell1 (full) and Cell 2 (empty) will themselves execute Kanban control. There are two important questions to be answered: (1) How are container moves initiated? and (2) How many parts are to be moved in each container?When Cell 2 removes the last part from a container at the input station, it will request a full container from Cell1. When an empty container arrives at Cell1, it is a signal to produce another full container.

- The roundtrip delivery time is 10 minutes.
- It takes 1 minute to swap a full/empty container with an empty/full container at either Cell1 or Cell2.
- Dedicated material handler containers will be used to move parts or subassemblies between cells.

The roundtrip cycle (load/unload) will take a total of 12 minutes. During these 12 minutes, Cell2 will consume 24 parts. The Kanban control will operate as follows.

(1) Assume only full containers will be moved from Cell1 to Cell2, and empty containers from Cell2 to Cell1. Assume a full container has just been emptied at Cell2.

(2) Material handling moves the empty container from Cell2 to Cell1. This takes 6 minutes.
(3) Upon arrival at Cell1, the empty container is placed at the output of Cell 1 and a full container moved to Cell2. The full container arrives at Cell2 exactly 24 minutes after the empty container was picked up.
(4) It should be easily determined that each full container generated at Cell1 will hold exactly 24 parts.

An empty container at Cell 2 initiates a request for more parts to be delivered to Cell2 and an authorization at Cell1 to produce another full container of parts.Cell1 cannot produce any parts until an empty container arrives.It is likely that 2 containers will be used at each cell (why?)

3.11 Kanban Signals

It should be obvious that more than one container might be moved between two Lean cells. For example, if each part in the previous example was either large or heavy, each container might be able to carry no more that 6 parts. In this case, eight containers would be required: Four empty or being filled containers and four full or being depleted containers. There are many different types of Kanban signals which might be used to trigger amultitude of events, but the basic design rule to determine both how many containers should be used given by:

$$K = \{(L*D)+SS\}/a$$

Where:

L = Lead time or time to traverse the complete inter-cell cycle
D = Demand or Production Rate of the consumer
SS = Safety Stock or the amount of extra parts which should be included to buffer against delays or variance
a = Number of parts/items which can be put into a standard container
K = Number of containers required

Note that if there is no Safety Stock this formula reduces to:

$$K*a = L*D$$

Interestingly, since K*a is the total number of parts to be moved and L*DD is the number of parts consumed during the move cycle, this equation is just another application of *Little's Law*to a Kanban Control system.

3.12 Mixed Model Cells

The same basic Kanban Control rules apply to Mixed Model production systems. The only difference is that when multiple products are made in the same Lean cell, the changeover/set-up times mightreduce the actual production capacity of the cell if the set-up and changeover time for the next product in sequence is greater than the machine idle time. In general, each product will have its own processing and set-up times. The reduction of set-up and changeover times between products requires the aggressive application of SMED principles. The frequency of set-ups and changeovers depends upon the sequencing scheme used to level production. The following example illustrates the impact of product sequencing on set-up and changeover times. Suppose that a company has contracted to make 3 different types of table legs. A possible Mixed Model assembly sequence which meets a particular demand pattern might be ABABCThe following Table shows these total times.

	To		
Type	A	B	C
From A		3	4
B	4		5
C	6	6	

Part Set-Up and Changeover Times (Minutes)

Assume that this sequence needs to be repeated 1000 times to meet demand.This sequence will require 5 set-up and changeover activities per set. Assume that changing from one part to the next is pure lost time. This production run would result in 21 minutes of lost time per set or 21,000 minutes. This is per one cycle of production time (not cell cycle time). This is 350 hours of lost production time. Another possible sequence might be AABBC. This would result in only 233 hours of lost time. This is still a lot of time. The correct solution would be to have Lean Engineering develop quicker change over strategies. If this work is done on a single machine, the solution will require a lot of skill and ingenuity. In any case, the point made is that Mixed Model Manufacturing times can always be improved by choosing the correct sequence of activities. The optimal solution can easily be found by trial and error if the number of different parts is small. Larger part sets can be solved by using Network Analysis Optimization (Phillips and Garcia, *Fundamentals of Network Analysis*).

3.13 More Kanbans?

A popular type of Kanban is called the *dual card* system. In a dual card system there are two basic types of Kanban: *Withdrawal* or conveyance Kanban (WLK) and a *Production Ordering* Kanban (POK). In this type of system, the output (order of production) of each cell is directly controlled by the input of another cell. The output of the upstream cell contains a fixed (one or more) number of material handling carts, and each cart contains the same number of parts. Cart capacity should be as small as is practical or possible to meet downstream demand or Takt time. We will call the upstream cell a *production cell,* and the downstream cell a *consumer cell*. In Each Lean Linked Cell Manufacturing System a single cell is in fact both a production cell and a consumer cell (Why is the 1stand last cell both a production and consumer cell?). Assume that each production cell has full carts sitting on the output side ready to be delivered to the consumer cell. Each of these carts will be filled with parts or subassemblies until it is full. Downstream, the consumer cell is pulling parts from a container sitting on its input side. When a container is emptied, the downstream cell puts a WLK card and a POK card in the empty cart. The empty cart is now moved upstream to the output side of the production cell. The WLK card is attached to a full cart and the POK is now a signal for the upstream cell to begin production to fill another cart. The quantity to be made is on the POK card. Note two important aspects of this simple visual control system: (1) Nothing can be made in the upstream cell unless authorized to do so by a downstream cell. This is a *Pull* mechanism. (2) Nothing can be moved downstream unless authorized to do so. The number of carts in the system multiplied by the number of parts per cart put a *WIP Cap* on both the output of the production cell and the input of the consumer cell. Note further that the ability to respond to changes in demand in both cells are still dependent only upon available (bottleneck) capacity, smoothed production (proper product mix), the number of workers in the cell and the number of machines in the cell. and work done at each station) and the number of workers active in each cell. This same kind of linkage is implemented between all cells.

The logic in specifying Kanban links to execute minimum WIP, synchronized production in a two card system is again just an operational implementation of Little's Law. Recall that Little's Law states that in any type of manufacturing system, there is an invariant mathematical relationship between WIP (Work-In-Process inventory), Throughput time (time spent in any system) anda specified production rate (the inverse of Takt time). Little's Law is as follows.

WIP = TPT*PR

Where: WIP = Work-in-process
TPT = Throughput time or Time in the system
PR= Production rate or demand rate (1/Takt time)

A *system* can be any manufacturing component which has boundaries. A single machine with waiting space, a group of machines with individual or pooled waiting space, a Lean cell or even a total factory. Another useful version of Little's Law is as follows

$$WIP_{Queue} = Time_{In\ queue} * \lambda$$

Where: λ = demand rate or 1/(Takt time)

This form of Little's Law mathematically relates the amount of inventory waiting to be worked on in front of a manufacturing component (WIP_{Queue}) to the amount of time per unit spent in queue ($Time_{In\ queue}$), multiplied by the demand rate. These two forms of Little's Law are coupled by the following relationship's.

$$WIP_{System} = WIP_{In\ queue} + WIP_{Currently\ in\ Manufa\ cturing\ component}$$

$$Time_{In\ System} = Time_{In\ queue} + Time_{In\ Manufacturing\ component}$$

3.14 Characterizing and Defining WIP

It should be recognized that WIP is just another way to characterize inventory in a system. Inventory may be waiting, blocked or delayed parts; parts being worked upon or raw materials being transformed; Make-to-Stock, Make-to-Order, intermediate storage or final finished goods inventory. Parts residing in decouplers are actually WIP also, although they are frequently and erroneously not designated as WIP. All material in a manufacturing system....in-process or final goods.... is considered to be inventory. System inventory of any form should be considered WIP. We will focus on three different types of WIP: (1) Raw materials (2) Work-in-Process and (3) finished goods inventory. WIP has often been analogized as *water in a river*. When water (inventory) is high, any rocks (problems) in the river cannot be seen. If the water level in the river (Manufacturing facility) is reduced, then the rocks (problems) are exposed. If inventory (work-in-Process) is high then every machine and every worker feels the need to produce whether needed or not. This type of traditional manufacturing behavior is typical of manufacturing control systems that *push* work orders into the system with little or no consideration of current WIP levels or machine status. This causes workers and machines to work harder to try and catch up. Overproduction, subcontracting and overtime are then employed to bring the system back under control. This results in high make-to-stock and make-to-order inventory levels. Inventory turns radically decrease and customer delivery times cannot be met. The result is disaster. On the other hand, if WIP (inventory) levels are systematically reduced instead of increased, the problems (rocks) are more easily exposed and can be addressed. Taiichi Ohno and Shigeo Shingo both steadfastly maintained that inventory is the *root of all evil*: lower the inventory levels (the river) and expose problems (rocks). Fix the problems, stabilize production, and lower WIP again. Their analogy is absolutely correct. In Lean manufacturing, any problems which are exposed by lowering WIP level(s) receive immediate attention when exposed. Figure 3.10 illustrates this analogy.

We must point out that many Lean strategists loudly proclaim that one of the main objectives of Lean implementation is *Zero Inventory*, The goal is to reduce all WIP to zero… or as near as possible. We must in the strongest possible way expose this type of thinking as dangerous and incorrect. The river analogy is that if the river goes dry, there is no flow at all. Of course, the same Manufacturing engineer will then claim to be misinterpreted. It will be explained that the goal is to *minimize WIP* outside of any work directly being manufactured. In particular, the material which resides in a Lean cell is called *stock-on-hand* in most of the Lean literature; so that it is not classified as WIP.No matter what it might be called; a rose by any other name is still a rose and any material between entry into the system and exit from the system should be called work-in-process. To placate the zero WIP engineer; we can also claim that there is no WIP in any material handling functions; no WIP to buffer any unexpected delays; no WIP in manufacturing cells and no WIP to smooth and balance production in Mixed-model manufacturing. To be fair, if a manufacturing system could be constructed in which every processing step took exactly the same amount of time; material handling, the supply chain support was just-in-time...just in the right quantity...at just the right place; this philosophy might be worthy of exposition. However, goals are fine but the Lean engineer lives in *Realville, USA*. The real problem as we will repeatedly point out is the corrupting and disrupting effects of *variance*. When variance components of any kind are present in a manufacturing system, the system MUST pay for it in degraded system performance. With all due respect and gratitude, we will quote from the work of Hopp and Spearman in their landmark book *Factory Physics*.

"Any form of variability causes congestion. This is stated as one of the *Laws of Factory Physics* (Hopp and Spearman). Congestion means more inventory, and more inventory in a system with fixed demand causes a corresponding increase in Throughput time by *Little's Law*. This implies that variability reduction is a key for improving system performance. Unfortunately, when proposing policies or solutions to reduce variance it is common to encounter a response such as: *Well, that might work somewhere else, but our problems are unique; variability is just an unfortunate part of our business and we simply cannot control it.*

Sometimes…rarely…this could be true, but is often used as simply an excuse for not attacking a difficult problem; one that they may not be (statistically) equipped to solve". Hopp and Spearman categorically state that solving variability problems might cost time and money, but if problems are ignored they will *always increase WIP* and will cause more problems. Granted, reducing variability anywhere it is found might be expansive and time consuming, but Hopp and

Spearman formulate another *Factory Physics Law* that states what will happen if variability is either tolerated or ignored.

Factory Physics Law: Pay me now, or Pay me later..........

> *If you cannot pay for variability reduction, you will pay for it in one of three ways.*
> *1.0 Long throughput times and higher WIP levels*
> *2.0 Wasted machine capacity and lower utilization of scarce resources*
> *3.0 Lower production or output rates*

Hopp and Spearman continue:

> *"This law does not imply that variability reduction is always the preferred course of action. In a system where capacity is inexpensive, low utilization may be perfectly acceptable. The point to be made is that there is no free lunch. Since we will pay for variability in one way or another, the decision of what to do (or tolerate) should be made with full knowledge of the consequences.* Let the engineer who has never examined variability or its effects be aware...*You will pay in one way or another"*
> *Factory Physics, 2nd Edition (Hopp and Spearman)*

In light of these penetrating observations, it should be obvious that regardless of how it is done, minimizing WIP whenever possible is a *Design Rule* and a *Goal* ofcontinuous improvement. It should also be clear that one of the primary goals in designing and implementing Lean, U-shaped cells is to minimize the stock-on-hand within the cell. Simultaneously, the WIP levels between cells and in final assembly are controlled and set by the Lean Engineer, driven by customer demand and set to smooth production. Here is an example of what Lean Engineers can do.

Suppose that in front of final assembly there are 10 carts and that each cart holds 40 parts. Empty carts signal upstream production to refill and deliver a full cart. The maximum inventory is therefore 400 parts in front of final assembly. Suppose that one cart is purposely removed: the WIP cap is now 360 parts. The foreman now waits to see if and where a problem might arise: (Lower the river to expose rocks). To prevent a negative impact on final assembly, when a problem surfaces it is immediately identified. A *Kaizen Event* is then launched to fix the problem, When the problem(s) are resolved new procedures are implemented; a cart is again removed. If the problem has been properly identified and properly solved, the system might function with no problem, but it is likely that new problems will surface (rocks in the river appear). *Some* improvement in system performance should occur by the Laws of Factory Physics and Little's Law. After allowing the system to stabilize, the foreman will now remove another cart, and resolve new problems. This procedure is repeated several times. If solutions are formulated and are effective, suppose that final assembly is operating with only 5 carts (WIP is now 200). Now something interesting occurs! Over a weekend, the number of carts is restored to 10, but now each cart contains only 20 parts. Note that both upstream and in final assembly, this new policy helps to smooth production....... a primary goal of Lean Engineering. By continuing in this fashion, WIP can be systematically reduced. Note carefully that this necessitates changes in both material handling and in setup time reductions in all areas. Any problems which surface, any material handling problems and any setup time reductions required are solved one by one. Following these procedures, WIP (inventory) between cells is continuously reduced. The effect on system performance is to reduce throughput time,increase flexibility and smooth production.

Minimum levels of required inventory can be achieved in many ways. The *zero defects* and *zero rework* goals must be simultaneously pursued. Unscheduled machine downtimes must be eliminated. Transportation and material handling functions must be redesigned to guarantee *Just-in Time* delivery. Workers must engage in educational programs to eliminate mistakes, become cross trained on many machines and understand their role in the system. Finally, we cannot stress enough that the principles of *six-sigma engineering* must be coupled to Lean Engineering principles to aggressively and continually reduce variance whenever and wherever it occurs. This is how continuous improvement is executed in a Lean factory. Considering the importance of WIP and variance reduction, it must be realized that inventory should not be considered anindependent, uncontrollable factor...... but a dependent, controllable factor.

3.15 Lean Manufacturing Cells Can Replace the Job Shop

Most companies that design their first Lean manufacturing cell use their existing job shop or flow shop equipment. Job shops traditionally are used to produce unique or one-at-a-time finished goods. Lean cells are built around part or product families which produce larger quantities of each type. Rearranging, converting and using existing equipment is a natural precursor to the implementation of mature, Lean cells. This stepwise conversion to Lean manufacturing and U-shaped Lean cells is good, because the company can learn how Lean cells work, how to drastically reduce setup and changeover times, enforce in-line Quality Assurance and Preventative Maintenance, implement MO-CO-MOO machine cycle time procedures and operate U-shaped Lean cells. As conversion to Lean systems mature, lean machine tools can be designed and specialized to Lean practices. The initial use of existing machines is a relatively low-cost, risk-free approach. We will call these initial Lean manufacturing cells *interim cells* to distinguish them from complete, *mature* Lean cells.

Consider how a traditional job shop might be used to manufacture a drive shaft, a gear box and two types of automatic transmissionsfor different pickup models. Possible routing sequences are shown in Figure 3.7.

This is typical of what might be observed in a traditional job shop. These types of routing patterns are often called *spaghetti flow,* which is typical of a traditional job shop operation. In the job Shop, a single skilled operator would usually man each machine, and parts would move from machine to machine in tote boxes. A process plan which gives the *sequence* of operations (machines) to be used in manufacturing a part like the drive shaft is crafted by either a Manufacturing engineer or by production planning. This sheet is sometimes called a *routing sheet* or a *traveler*.

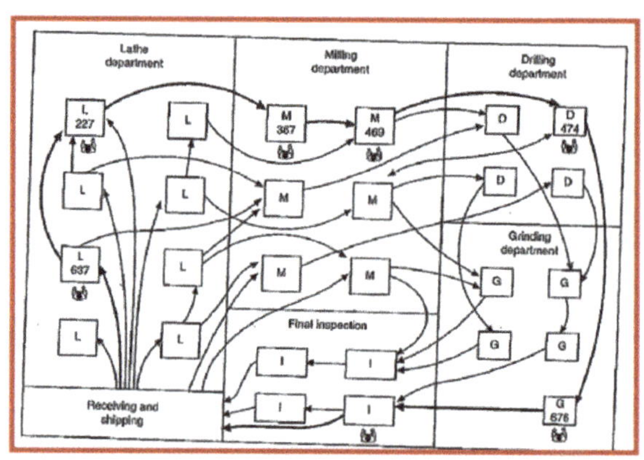

Figure 3.7
Typical Flow Patterns in a Job Shop

An *operations sheet* is prepared for each part flow that contains information about how to set key processing parameters at each step (speeds & feeds), the tooling required, type of raw material, etc. Sometimes standard processing throughput time or machining time is also shown for each operation: These are usually Industrial Engineering stop-watched, sampled times or artificial engineered time standards. These times are also used to calculate process and product cost. The routing sheet is accompanied, preceded or augmented by machining instructions, manufacturing/move lot sizes and set-up instructions. All of this information and much more is used to produce one or more parts in a traditional job shop Production of each part is labor and information intensive.

Figure 3.7 is a typical job shop layout. Assume that this job shop is being used to produce automotive assemblies: (1) drive shafts (2) gear boxes and (3) transmissions. There are many different types of machines, most of which are not used except when demand dictates their use. In this case, another skilled worker is usually hired. The skilled labor pool is constantly changing, creating a dichotomy of fully skilled workers and new hires. This particular company is experiencing worker inconsistencies and quality control problems. Now let us see what happens if we convert this manufacturing system over to a Lean manufacturing system.
All machines in a job shop are arranged into U-shaped configurations such as those shown in Figure 3.8.

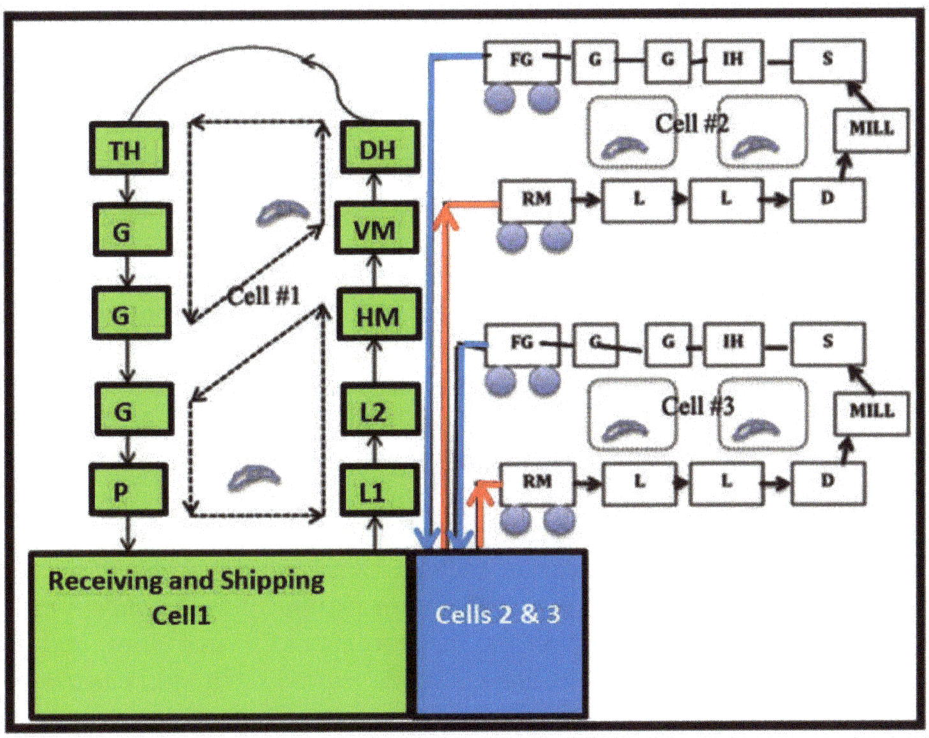

Figure 3.8
Converting a Job Shop to a Lean Shop

Cell 1 makes two types of transmissions. A drive shaft is completely manufactured in Cell 2, and the drive shaft is feeding a drive train subassembly cell. Cell 3 produces a gear box.All three subassemblies are moved to Final Assembly (not shown).

It should be emphatically stated that the conversion to a U-shaped manufacturing is much more than just moving machines around as many group technology gurus and value stream mapping technical papers might suggest.
A lot of work is required by the Lean Engineer to make this transition, involving everyone in the factory. For example, when designing these cells a new sequence of operations might be implemented by adding a horizontal band saw to Cell 1. The saw might be used to cut bar stock into the correct length,depending upon which of the two different transmissions is being made. This modification might eliminate the need to perform facing and cutoff operations at the first two lathe workstations and reduce the required time. Every machine should be modified to execute autonomous, single cycle operations unattended (single cycle automatics). Jigs, fixtures and cutting tools are designed or modified to drastically reduce setup and changeover from one type of drive shaft to another (SMED), and tooptimize cutting parameters. Machines should all be equipped with walk away switches, safety devices, QC checks and automated loading/unloading devices. All machines are arranged according to required precedence relationships in a U-shaped layout with a narrow internal isle. Between each machine, a decoupler is added which holds only one unit to support one-piece flow through the cell. A small WIP area is designed to hold incoming parts before the first workstation, and following the last workstation in the cell, designed using Little's Law.

A complete transformation of the original job shop to the three Lean cells shown in Figure 3.8 is called a *Mature Lean Cell Design*. Any partial conversion would result in what we call an *Interim Lean Cell Design*. These three cells are designed to be flexible and versatile enough to rapidly adjust to demand fluxuations, and even changing product mix. This is done by scheduling cell production based upon the number of workers assigned to each cell and the assigned tasks for each worker.

Once the cell capacity has been reached by assigning a single worker to each machine, the cell operates as a traditional serial production line with output determined by the bottleneck operation. At this point, the cell must again be redesigned. Within each cell, cell output is set to meet Takt time requirements. In steady-state operation, every machine in each cell must obey the *First Axiomatic Design Rule:* The machining cycle time for every machine or parallel machine configuration in the cell must be strictly less that the necessary required *cell cycle time* (Time between part completions). The key step is to determine (1) how many workers are needed in the cell and (2) the number of machines to be assigned to each worker (worker's loop) to meet the required Takt time for the cell. Each worker will spend a certain amount of time at each machine, determined by: (a) whether or not it is a machining or assembly operation (b) the changeover/setup time required (c) any visual inspections or QC activities, and (d) the walking times between machines. The time it takes each worker to complete his/her loop of assigned processes is the sum of all these times in the assigned loop. The cell can often be manned by one or two workers. The maximum output rate is determined by the longest machining processing time. For any number of assigned workers less than the number of machines in the cell, the cell cycle time in terms of minutes/part is the *maximum of any of the assigned worker's loop times*.

For a numerical example, consider Cell 1 in Figure 3.9 which manufactures two types of transmissions, Type A and Type B. This is a simple example of *mixed model*. The numbers shown beside each workstation are the workstation processing times for Parts A and B. Both A

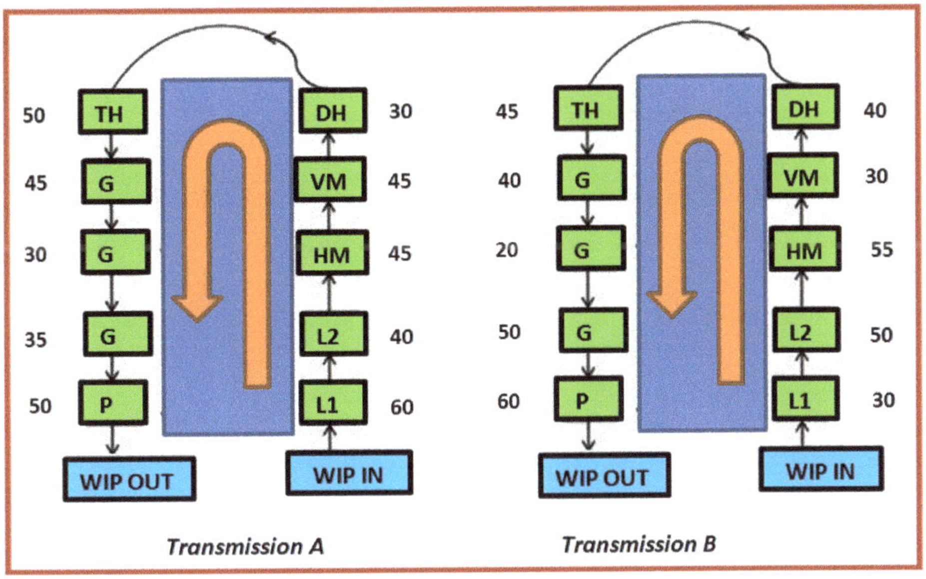

Figure 3.9
A Transmission Cell

and B machine (unloading, checking and loading) and that about 5 seconds is spent walking from one machine to the next: 10 seconds between Machine DH and TH, as well as Machines L1 and P (Packaging).

Cell loading and cell unloading takes 50 seconds. At each machine, the acronym Lo-Co-Moo applies (**LoadO**ne-Check **O**ne-**M**ove **O**ne **O**n, and QC checks are done manually in the decoupler. Every processing step is a single cycle automatic machine except for packaging, which is a manual operation. Hence, the cell is a hybrid machining and assembly cell.

To illustrate cell balancing and how to meet Takt time requirements in *Mixed Model* production, assume that the daily demand for Transmission A is 200 units/day and Transmission B is 300 units/day. If the line runs 8 hours/day there are 489 minutes or 28,800 seconds of production time per day. The daily demand is 500 units per day, so Takt time is 57.6 minutes/part with a production rate of 0.96 parts/minute. It appears that the current system cannot meet the daily demand since the L1 operation for Part A and the packing operation for Part B both take 60 seconds. However, this conclusion is incorrect. The correct procedure in *Mixed Model Manufacturing* is to calculate a *weighted average* processing time per part. For example, at L1 the average processing time per part is a weighted average given by: $APT_{L1} = \{200*60 + 300*30\}/500 = 42$ seconds/unit. Similarly, $APT_P = 56.0$. Figure 3.10 provides the average processing time per part in the composite cell.
The composite cell now adheres to the first axiomatic design principle and can be implemented.
First: *What is the cell Takt time if one worker is assigned to*

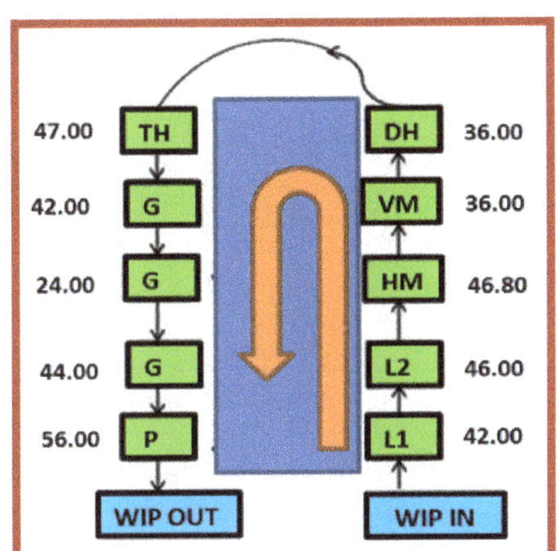

Figure 3.10
Mixed Model Production

the cell? The worker loop time is given by:

$$Loop\ Time_{One\ worker} = (10 * 20)_{Task\ Times} + (2 * 50)_{Load/Unload} + 55_{Walk\ time} = 355 \text{ seconds.}$$

This is clearly infeasible.

Using two workers, assign Worker1 to machines 1-5 and Worker2 to machines 6-10. The two loop times are identical for this example and are given by:

$$Each\ Loop\ Time_{Two\ workers} = (5 * 20)_{Task\ Times} + (50)_{Load/Unload} + 40_{Walk\ time} = 190 \text{ seconds.}$$

This is also infeasible.

Continuing to assign workers, the reader can verify that a feasible solution is given by:

- Assign Worker1 to the Cell loading task
- Assign Worker2 to the Cell unloading task
- Assign Worker3 to Machines 1 & 2
- Assign Worker4 to Machines 3 & 4
- Assign Worker5 to Machines 5 & 6
- Assign Worker3 to Machines 7 & 8
- Assign Worker3 to Machine 10 & Unloading

Align Machines 5 & 6 so that the walking time between these two machines is 5 seconds.

The previous mixed model example was not realistic, because it ignored set-up and changeover times. In doing so, any sequencing of the 200 A's and 300 B's could be used. Let us revisit the example and assume that any changeover will take an extra 30 seconds. The effect of nonzero changeover times is to reduce available production capacity. The magnitude of lost production is totally dependent upon the manufacturing sequence used. For example, if 200 A's are run followed by 300 B's every day, only 1 minute of production time would be lost per day. However, this would result in uneven and "lumpy" delivery of A's and B's to downstream consumers. In general, small production lot sizes create less variance, reduce WIP and *smooth production* downstream. For example, another production sequence would be to run 40 A's and 60 B's; and repeat this sequence 5 times per day. In this case, 20 setups would be required and 10 minutes of production time lost. In this example, the reader will be asked in the problem set for this chapter to verify that if the sequence AAAABBBBBB is run and repeated 50 times, this will require 100 set-ups and lose 50 minutes of production time per day but the current cell configuration can still be staffed to run this pattern and meet required Takt time. Note that since the demand per day is the same, but the available production time per day has been decreased, the cell cycle time must now be shorter to meet customer delivery schedule. Overall, a part must still be passed on to final assembly every 60 seconds.

The cell cycle time and the cell throughput time are based solely upon the number of workers assigned to the cell, the operations in their individual loop(s), the time spent at each machine in their assigned loop and the walking times between machines. Here is the unique aspect of *Lean Manufacturing Cell Design*. The throughput time and production rate of the cell does not depend upon individual machine times but on the number of workers assigned to the cell, as long as there is sufficient capacity at all cell workstations to meet Takt time requirements. The machining times have been *decoupled* from the cell cycle time by the way the cell operates. In other words, *the production schedule has been made independent of the processing times at each machine.* This is the magic and genius of the TPS, linked-cell production system. *The job shop and flow lines cannot be operated in this way.* The output of any cell can be easily changed by assigning more or less workers until cell capacity is reached....then the cell must be redesigned. Generally workers assign themselves to the machines without changing any machining operations (provided capacity exists for expansion).

The number of workers and the machines operated by each worker are determined to meet two critical goals: (1) Balance the workload across all workers and (2) produce to Takt time. Balancing the workload is accomplished by assigning machines to each worker in such a way that the time required for each worker to make one cycle around his/her loop is the same for all workers. In most situations, this goal cannot be accomplished. There is usually a relatively small difference between worker cycle times. The differential can be reduced in a number of ways, including workers sharing one or more machines at the interface of their loops, but since both the number of workers and product demand will frequently change, one of the best ways to balance is to reduce long machine cycle times into multiple short, sequential operations.

Since a cell's cycle time is dictated by operator and machine assignments, and each machine will automatically cut off and wait for the operator's next arrival, the throughput time per part is not determined by the machine processing time, but the maximum loop time of all workers multiplied by the number of workstations. Hence, machine times and machine cycle times can be changed (within limits) without affecting the production rate of the cell. In traditional manufacturing lines, changing machine times would affect both the output rate and the delivery lead time. Note that the number of output cycles that any one part spends in the cell depends not only upon the maximum loop time of any worker, but also upon the amount of stock on hand in the cell and whether each machining operation is an assembly operation or a machining operation. Cell stock-on-hand (cell WIP) is limited to the number of machines in the cell. If an operation is assembly, the operator will need to be at the machine during the assembly cycle time. If the operation is machining, the operator will MO-CO-MOO, hit a walk-away switch and go to the next machine. The machine will automatically shut itself down after the machining operation is completed and wait for the operator to return.

3.16 Summary

Lean Engineering is a discipline which is primarily responsible for designing and implementing a radically different design for a manufacturing system. The fundamental premise of Lean thinking first identified by Womack and Jones is still true: Lean production is fundamentally *Waste elimination*, in which all non-value activities are systematically removed. The most important aspects of Lean thinking are to Minimize WIP by enforcing WIP caps at all levels of the Manufacturing system; design and implement Lean, U-shaped manufacturing cells; aggressively reduce all sources of variation in the system; seek one-piece flow in minimum quantities; smooth and balance production; and implement Kanban /Pull control between all functional components. All subassembly, component and final assembly cells should be converted to Mixed-model production, and changeover/setup times reduced to as small a time as possible. Downstream functions, starting with final assembly, dictate the Takt time requirements for all upstream functions. All parts or subassemblies are*pulled* through the system using Kanban mechanisms. Transformation to a Lean manufacturing system first requires implementation of lean, U-shaped cells and Mixed-model final assembly, before attempting to implement high technology and innovative solutions such as automation, *Autonomation*, Poka-Yoke, computerization and robotization. We present the following observations based upon experience and observation.

- When implementing Lean transformation, *start at the top*. System level changes demand top level support and strong leadership
- Before starting the journey, *educate everyone in the company* on what you are trying to do. *Involve everyone in the company from top to bottom.*
- Stress that the *purpose* of Lean Manufacturing is *not* to eliminate jobs or to fully automate, but to make more profit and improve customer satisfaction.
- Total Quality improvement and defect free manufacturing is not optional, but *mandatory*. Perfection is the goal. Defects can be prevented. Fix problems at the source. If something goes wrong, shut down the process immediately and fix the problem.
- If ask to choose between cost and quality.... TAKE BOTH.
- Total Preventative Maintenance using proactive and not reactive maintenance is mandatory. The squeaky wheel gets the grease. An hour lost at a bottleneck or critical production station can never be fully recovered without increasing cost and time.
- The *First Key* to Lean production is to *reduce work-in-process* to the lowest possible level. AXIOM: The lowest possible level is NOT ZERO in almost all cases. Determine the optimal WIP level(s) and go there.
- The *Second Key* is to aggressively and systematically reduce all forms of *variance*.
- The *Third Key* is to design and implement Lean, linked, U-shaped manufacturing & subassembly cells. *Link all production cells* with Kanban/Pull control.
- Implement a continuous improvement program to continuously change and improve the Lean manufacturing system design.
- Lowering WIP will reduce throughput times. Do not confuse Takt time, a cycle time with throughput time.
- Build a system design that can be understood by everyone involved. *Keep it simple, or at least as simple as possible.*
- Do not forget that the Supply Chain can drastically affect the performance of the Manufacturing system. Drive Lean principles all the way to the bottom of the supply chain. Implement Lean in second and third tier subcontractors.
- Lean engineering system design must precede concurrent engineering efforts. Eventually include all aspects of the Manufacturing Enterprise, including logistics, marketing, purchasing and sales.

Finally, Lean transformation calls for rethinking and redesigning traditional Mass manufacturing systems. This is difficult to accomplish because most corporate managers are risk averters and are resistant to radical change. Remember, things will get worse before they get better. Stay with the plan, but expect initial resistance at all levels. All companies want to raise productivity, lower cost and make more money; but few are willing to pay the price to facilitate long-term change. The conversion of final assembly to Mixed-model assembly fed by synchronized Lean cells is

just the first step in the Lean journey. The entire journey will involve the relentless elimination of all forms of waste, on-time delivery of goods and services; and perfect quality.
The TPS (Lean manufacturing system) is designed so that every person in the system has responsibility to meet goal; understands how the system works; and realizes that people are responsible for making superior products. The L-CMS discussed in this Chapter simplifies the Manufacturing system and seeks continuous improvement. The remainder of this book will mostly address the Lean Engineering design tools and methodologies needed to execute change.

Review Questions

1. Define an enterprise or production system versus a manufacturing system. What are the classic manufacturing system designs (MSD)? How is the mass production system different from the lean production system in terms of their systems design?
2. Review the axioms (Design Rules) for the L-CMS and explain these 4 terms
   Leveling, Balancing, Smoothing, Sequencing, Synchronization
3. In comparing a linear subassembly design with a U-shaped cell design, what measurable parameters were used?
4. What is the main difference between the U-shaped subassembly cells and the U-shaped manufacturing cells?
5. Explain this statement: Cycle times for Lean manufacturing cells do not depend on machining times of the manufacturing process.
6. What is the ramification of this independence [MT<CT] on production control (the scheduling of the system)?
7. What is standard work?
8. W. Edwards Deming said that to know when a problem exists, we must distinguish normal from abnormal. How does standard work make this idea possible?
9. What does it mean for a system design to be robust or versatile?
10. Explain the differences and relationships between cycle time, Takt time, throughput time and production rate.
11. If a machine's processing time exceeds the cycle time in a manufacturing cell, what alternatives might the lean engineer consider to ensure that a cell meets the necessary cycle time?
12. One-piece flow in lean cell design has many benefits. Discuss.
13. What is the key role of the worker(s) in the manufacturing cell?
14. Describe a common error-proofing device (poka-yoke) you are frequently in contact with.
15. Why is mistake proofing important in a cell's operation?
16. Name one product that has had a life of 50 years and is produced in volume over 10,000 per year. How is it made? Is it a candidate for lean manufacturing cells?
17. What functions can decouplers do in manufacturing cells?
18. Why do lean companies work toward having sole suppliers for their subassemblies and components?
19. Explain how a Kanban design rule is an application of Little's Law?

Chapter 4
Manufacturing Cell Design

The superior man, when he sees what is good, moves toward it; and when he sees his errors, he turns from them   *The Book of Changes, China circa 1200 BC*

4.1 Introduction

The Toyota Production System (TPS) is a radical redesign of the mass production system which evolved out of Henry Ford's assembly line. Under the leadership of Taiichi Ohno, Toyota invented and implemented a new type of Manufacturing system which we will call the *Linked-Cell Manufacturing System* (L-CMS). A schematic of L-CMS design is shown in Figure 4.1.

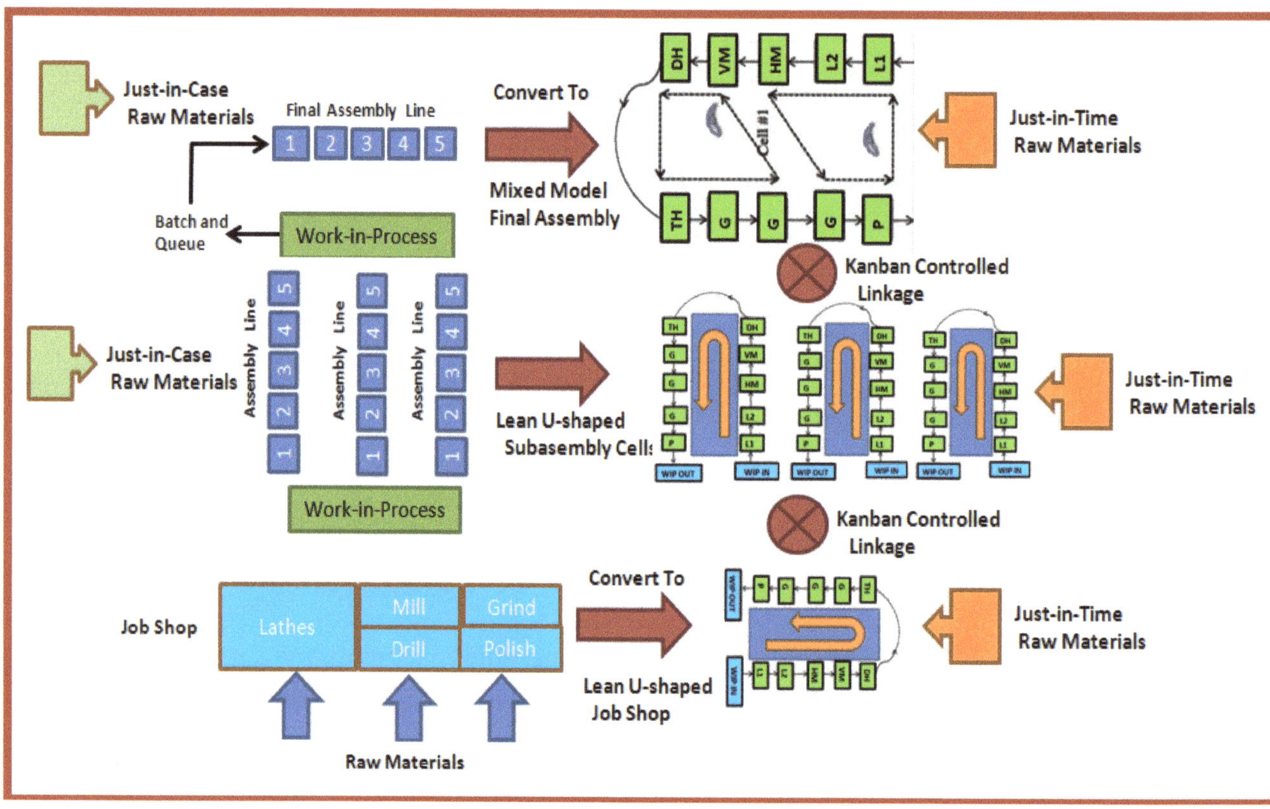

Figure 4.1
Converting Mass Production to Lean Occurs Final Assembly, Subassembly and Component Manufacturing

For final assembly, a traditional manufacturing line is converted from large batch production to a mixed model final assembly (MMFA) plant. All subassembly lines are converted from linear lines into U-shaped cells. The job shop is converted into U-shaped manufacturing cells which can produce components for the subassembly cells as required in whatever mix the subassembly cell needs: Figure 4.2 shows a cell which has been designed to manufacture a beveled gear. This cell might be connected to another subassembly or to a final assembly by Kanban links which strictly control part flow. All cells are equipped with many devices to prevent defects. These devices are called *Poka-yokes*. All machines are modified so that they can be quickly changed from one part in a part family to another, and they are maintained so that they do not fail during an operation (Preventive Maintenance strategies). The goal is to make perfect parts every time using one-piece flow (Zero defects). Workers supporting one piece flow follow a MO-CO-MOO for Make one, Check one and Move one on at each machining workstation.

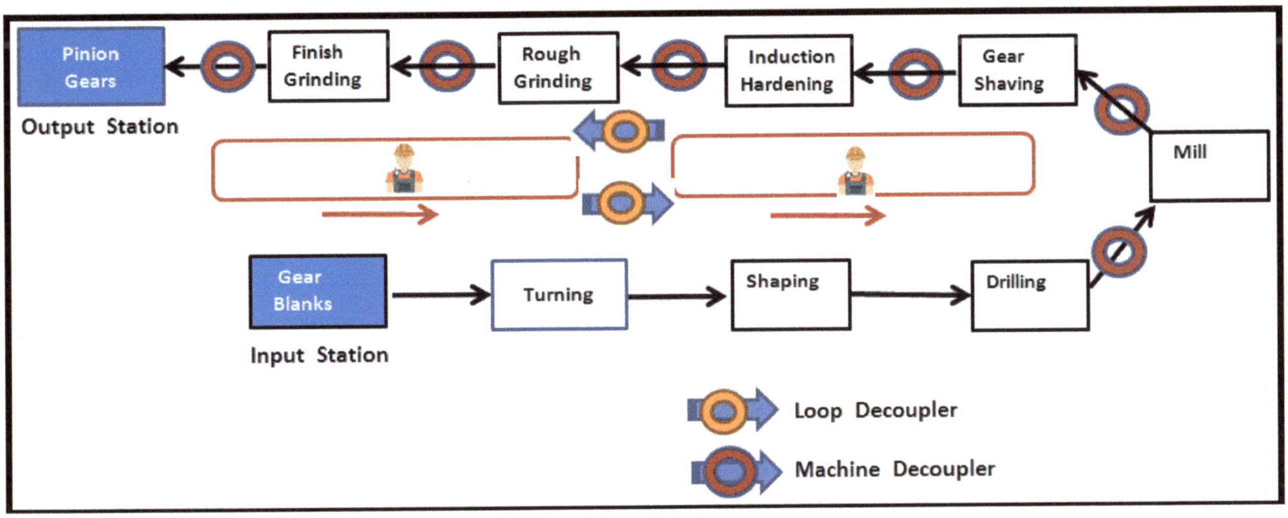

Figure 4.2
A Cell with 2 Operators and 2 Loop Decouplers Supporting One Piece Flow

The symbol "blue arrow with a red circle" is a *machine decoupler*, and a "blue arrow with an orange circle" is a *cell decoupler*. A Cell decoupler serves to make the two workers independent of one another by holding one item of WIP to make sure each worker can begin his work without delay. This configuration can support either assembly or a subassembly cells, with manual or automated machine stations. The cycle time for the cells should be a bit less than the Takt time for the final assembly line to allow for any problems that will occur. The CT for the cell is based upon the Max$\{t_{j1},t_{j2}.....t_{jk}....t_{jN}\}$ Where t_{jk} is the time that the j_{th} worker takes to complete the k_{th} loop. Loop time is the sum of the manual times for the operator(s) for loading, unloading, checking parts and walking from machine to another. This time must be strictly less than the cell Takt time.

Contrary to popular belief, Job Shops can be converted to Lean Cells. In converting from job shops to initial Lean cell configuration, cells are configured and built using machines that were originally designed and built for the job shop. The original machine tools from standard mass production lines are extensively modified for Lean cells. Toyota has developed a stepwise strategy for converting machines designed and built for the job shop so they can be used in a lean shop. Typically the machines operate on a single cycle automatic (SCA) basis (once loaded, you hit the start button and they complete the cycle and then turn off – like your kitchen toaster).

True lean manufacturing cells have machines and processes designed and built specifically for the cell. The machines are designed ergonomically for easy loading and specialized automatic loading/unloading devices. These cells may also be served by *storage* locations which are stocked with nuts, bolts, parts...Etc. which are used at each machine. These items are usually stored adjacent to the point of use and are called *Supermarkets*.

Another interesting aspect of cell design is the manner in which robots can be used with people to make parts and subassemblies. Figure 4.3 is an example of a manufacturing cell for sheet metal parts where sub-cell decouplers separate robotic operations from normal cell activity. The objective of Taiichi Ohno was to build a simple system that could make complex goods. Toyota did not have sophisticated computers, and manufacturing instructions were difficult to convey with written documents, so most of the procedures and controls were visual at each workstation. Graphical displays were extensively used to show the workers what to do.

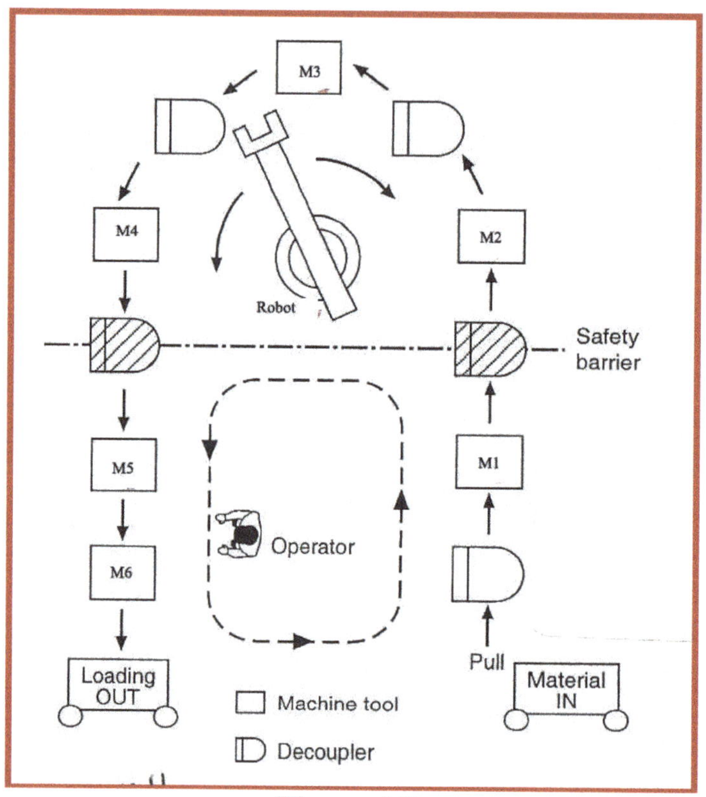

Figure 4.3
A U-shaped Cell Used in Sheet Metal Fabrication is Manned by an Operator and a Robot with Part Flow Connected by 2 Decouplers (Honda).

Because Lean manufacturing system design is simple, the Kanban system for production and inventory control operates on a few simple rules. The most critical and but poorly understood aspect of Lean system design is that the design of manufacturing cells make the throughput times (TPT) for the delivery of the parts independent of machine processing or machining times. Cell cycle time and throughput rate depends only on cell cycle times which depend only on manual and walking times for the operators.... provided capacity exists to achieve Takt time. Thus, TPT, which is set by customer demand, can easily be ramped up or down by adding and subtracting workers, not machines. Changes in demand are easily and quickly handled, making the system very flexible

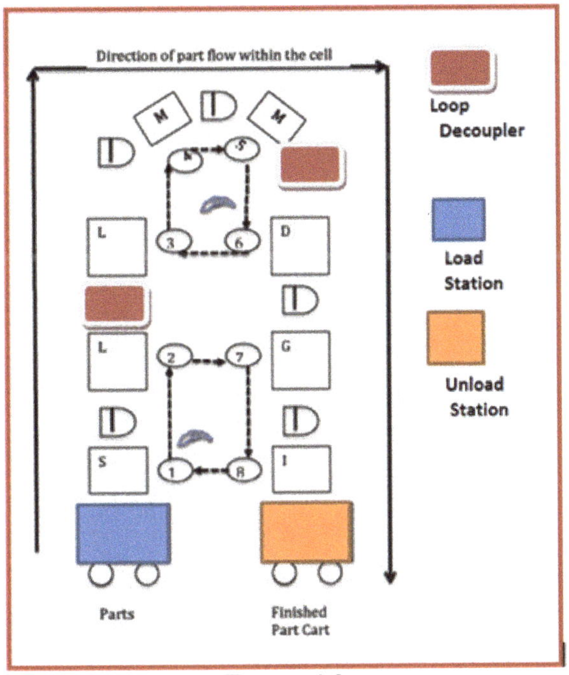

Figure 4.4
A Lean Subassembly Cell Operated by Two Workers

Operation Number	Manual Time (Sec)	Machine Time (Sec)
1	6	20
2	10	35
3	10	30
4	20	80
5	20	80
6	10	40
7	10	30
8	15	0

Manual Task Times and Machining Times

4.2 A Subassembly Cell: Example

A drive pinion gear to be manufactured in a U-shaped Lean manufacturing cell is shown in Figure 4.4. The cell must be capable of producing up to 50 parts/hr. This cell will contain six different types of machines (processing stations) plus a final inspection station. All of the machines are automated. Figure 4.4 shows that a Lean Engineer placed two **M**illing machines in the cell. This is because one **M**illing machine can produce a single part every 100 seconds. This is a production rate of only 36 parts per hour. Since a minimum production rate of 50 parts per hour is desired, a second **M**illing machine was added in parallel to the first. This resulted in an average production rate of 50 seconds per part or 72 parts per hour.

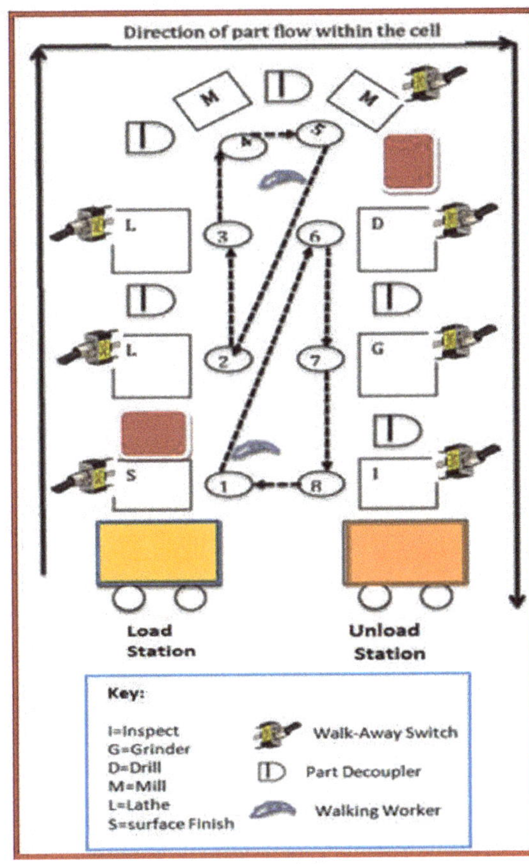

Figure 4.5
A Lean Subassembly Cell

Figures 4.4 and 4.5 shows assignment loops for two workers. Note that the final inspection in this cell takes about 15 seconds. This task is performed by the worker who has Output Station 8 in his/her loop. The machining time for this pinion gear on each machine ranges between 20-80 seconds. Workers spend about 3 seconds walking between machines and from the output storage area to the input station. We will use the configuration shown in Figure 4.5 to demonstrate how to calculate system performance measures.

4.2.1 Cell Cycle Time

Cycle time is calculated as follows.

$$CT_{Each\ Worker} = \sum Walking\ times\ between\ machines + \sum Manual\ tasks\ at\ each\ machine$$

One Worker

Suppose that only one worker is assigned to the cell. The time it takes one worker to complete the cell loop is calculated by adding up the sum of all task times spent at each machine and the walking times. The time required using one cell worker is 86 seconds of task times and 24 seconds of walking times.

CT_{Worker} = 81 sec (task times) + 24 sec (walk time) = 105 sec = 1.75 minutes/part.

Hence, one part will be produced about every 1.75 minutes. One must now check to see if any one machine time is greater than 105 seconds. It is not.....so there will be no blocking/delay as the worker reaches each machine in the loop. This yields the following results.

The output rate of the cell using one worker is approximately 34.3 parts per hour
This shows that one worker cannot achieve an output rate of 50 parts/hr.
The amount of time that each part stays in this cell is:

CT_{Cell} = *Number of cycles part was in the cell (8 loops by 1 worker)* * Time per cycle
TPT= 8 loops * 1.75 Min/loop = 14 minutes

Two workers

What would the cycle time and output rate be using two workers? Clearly, this depends upon the machines assigned to each worker. Two *Loop Decouplers* are used to synchronize part flow between the two workers. This is also called *Baton passing*. In order to synchronize the two workers and avoid blocking due to starvation, it may be necessary to hold one extra unit of stock in each loop decoupler. If each loop decoupler holds 2 parts and each machine decoupler holds only one part, there will be 9 parts in this cell at any one point in time.

The cell loop cycle times for the two workers are as follows.

$W1_{loop\ time}$ = 41 min (task times) + 12 min (walk times) = 53 sec (Locations 1-2-7-8)
$W2_{loop\ time}$ = 40 min (task times) + 12 min (walk times) = 52 sec (Locations 3-4-5-6)

The time to produce one part and the output rate is now:

CT_{Loop} = Max (53, 52) = 53 Sec
$Output\ Rate_{Cell}$ = 0.88 parts/min = 68 parts/hr

Note that by adding another worker to the cell, that the output rate has doubled, and the required output of 50 parts per hour is easily achieved. In fact, it should only take (68)*(8)=2650 sec to make 50 parts/hr. The remaining time per hour could be spent on maintenance or quality assurance.

4.2.3 Three Workers

A possible three worker assignment is shown in Figure 4.6. It will be left as an exercise to show what the system performance measures would be using 3 workers as shown. We will discuss other ways to create additional capacity in Section 4.6.

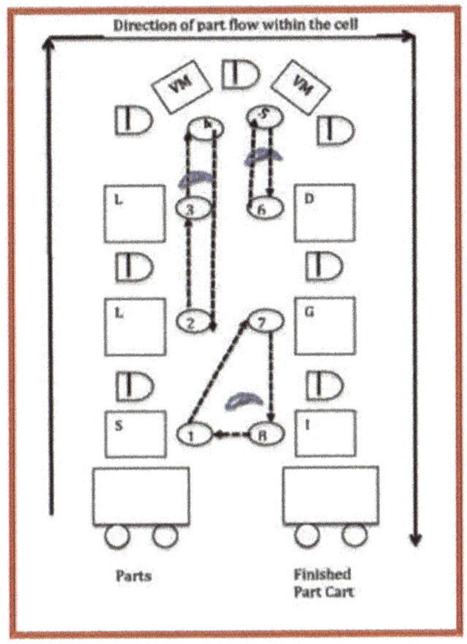

Figure 4.6
A Lean Manufacturing Cell Operated by 3 Workers

It should be clear that the cell can be operated by as few as one worker or by no more than eight workers. The output rate of the cell (1/Takt) is determined by the number of workers and their machine assignments (loop of work), and not by the machines.

4.2.4 Maximum Cell Output

Workers can continue to be added to the cell and production rate will increase until every machine has one worker. When this occurs, the maximum cell output will be dictated by the *Bottleneck Operation,* which will be at the **M**illing machine. The maximum output of the cell would be: 100 minutes per part or 36 parts per hour per machine or 72 parts per hour.

4.3 Continuous Improvement

After machines are first placed into the cell and the cell begins production, they are appropriately (time and cost constraints) modified to support Lean cell production. For example, the classic drilling machine shown in this example might be modified as follows.

- Convert any manual feeds and hand-held tools or work pieces to automated power feeds controlled by specifically designed jigs and/or fixtures.
- Add an automated stop mechanism for the machining process,
- Add automation to make the machine return to a ready-to-be-loaded state after the finished part is removed and a restart button is pushed by the operator.
- Install a walk away switch to activate the next machining cycle once the next part is loaded
- Install an automated unloading device (*Chuku-chuka*).

- Install an automated loading device for the next part: This is a good robotic application for small robotic arms and transfer lines.
- Possibly develop and install automated Quality control checks and error diagnostics (Poka-yoke). These can be fully implemented later.
- Design a decoupler between each machine to hold and position a part for the next machine. At least design a Quality control check; this can be fully implemented later.

While traditional Mass manufacturing companies historically have designed automated machine loading and unloading mechanisms, few Mass production lines use the concepts of decouplers and single cycle machines. Material handling is usually done by fork lift trucks, pusher dogs, or expensive conveyor material handling systems. It is not unusual for semi-finished parts to be stored in ASRS storage systems and retrieved later when space becomes available at point of use.

In Lean U-shaped cells there may be no need for a costly ASRS except after final assembly, and this is not to store in-process-inventory but to buffer final delivery requirements. Decouplers rarely contain more than one unit. In many cells, machines have been modified to both load and unload from decouplers, and almost all second generation Lean cells have some automated machine loading/unloading devices. If automation is used, it is supervised and often activated by workers. Toyota called this *Autonomation*… automation with a human touch

Automation of the machine start and stop cycle supports single cycle automation and unit processing and is essential to the elimination of the *one operator-one machine* syndrome which characterize many Mass production job shops. Partial or full machine automation is accomplished with standard limit switches, sensors and programmable controllers (PLC's). When first initiated by Ohno in Toyota's machine shops in the 1950s, it allowed hands-free operation by replacing manual controls with pneumatic/hydraulic chucks, limit switches, and end-point detection. Modification of a simple drill press is shown in Figure 4.7.

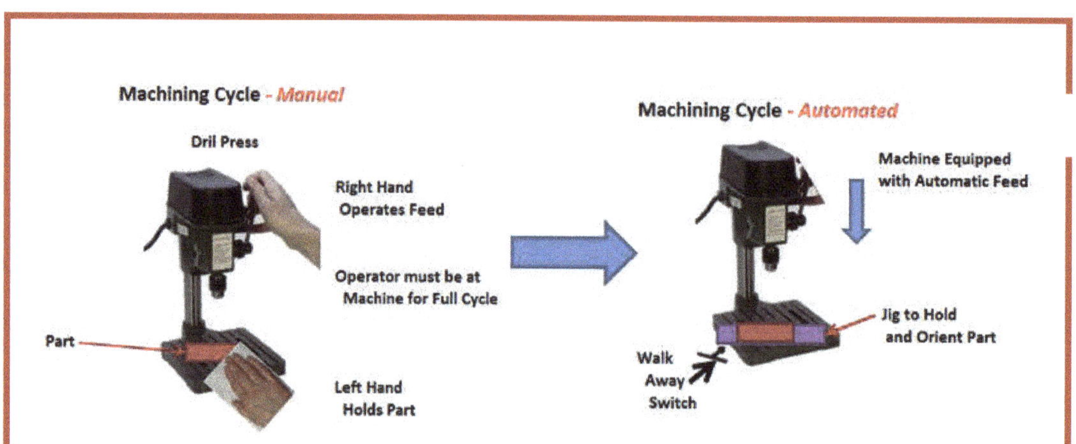

Figure 4.7
A Two- Step Modification of a Drill Press to Automatically Feed and Eject a Part Using A Jig and A Walk-away Switch

Section 4.4 SMED

Ohno spent a great deal of time developing specialized jigs and fixtures (SMED) to support not only rapid setup and changeovers for different part types, but also to facilitate machine automation. Finally, Total Preventative Maintenance (TPM) practices were implemented to render each machine highly reliable, and Total Quality Control (TQC) practices were implemented to enable *zero defect* production.

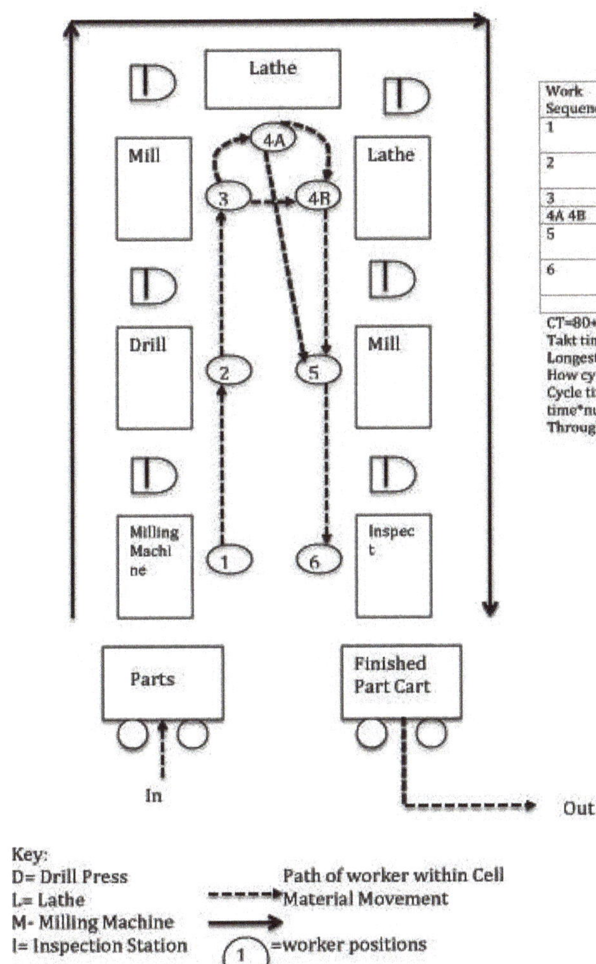

Work Sequence	Operation Name	Time Manual	Time Walking	MT Machine
1	Mill ends on work on HM 1	12	5	30
2	Drill hole on DP	15	5	20
3	Mill Flat	13	5	30
4A 4B	Turn OD	12	5 5	120
5	Mill Steps on VM	13	5 5	30
6	Final Inspection	15	5	
		80	20	230

CT=80+30
Takt time = 2 Min
Longest Machining time=120
How cycle time is calculated for the cell
Cycle time (CT)= (Operator Manual Time *Number of operations)+(walk time*number of walk cycles between machines)
Throughput time (TPT) = (cycle time *Number of cycles the part was in the cell)

Figure 4.8
Lean Cell with Insufficient Capacity

4.5 U-Shaped Lean Cells with Machining Time(s) Greater than Cell Cycle Time

Consider the manufacturing cell shown in Figure 4.8. The Takt time is 2 minutes or 120 seconds. The cell consists of five machining steps and one final inspection station. The Table in Figure 14 shows processing times in seconds, walking times between machines and the manual task times at each station. Using a single worker, the cell cycle time would be 110 seconds. The cell output rate is therefore 0.54 unit/minute or 32.7 units/hour. However, machining time for each part on the lathe is 120 seconds. If only one lathe is used in the cell, Takt time could not be met. This operation will delay the worker for 10 seconds each cycle. This violates one of the Axioms for Lean cell design which is $MT_i < \text{CT}$.

Where:
MT_i = Machining cycle for machine i = 1, 2…M machines CT = Cell cycle Time

Note that the Lathe operation is not only the longest machining time, but it is also much larger than any other machining time. The Lean Engineer decided to install a duplicate lathe as shown in Figure 4.8. This will result in the average time to lathe a part to be about 60 seconds (It will actually be slightly more). Hence, we now have a feasible solution. The machining time of 120 seconds is still at least three times larger than every other machining time, so there is considerable idle time at each machine but the lathe. However, this causes no real difficulty. It simply drives the *bean counters* crazy. Traditional thinking is that machine utilization should be as close to 100% as possible (Why?) A worker will alternate between the two lathes as he/she works the cell loop. This will require a decoupler between the two lathes (Why?).

4.6 Creating Capacity

In reality, if more than one part is scheduled to be made in this cell, the actual time spent by each part on every machine would likely change as the part type changes. In this case, every machine in the cell can meet Takt time, but the Lean Engineer must check to see if the *weighted machining time* exceeds the required cell cycle time. This is always true in a Mixed Model production cell where demand is leveled. In addition, once a leveled production sequence is chosen the sequence of parts which flow through the cell will require multiple setups. The total number of changeover setup times must all be subtracted from the available production time and the actual cell cycle time recalculated. Recall that the cell Takt time must still be met. There are two methods which might be used to determine how long each part should spend at each machine in *designing* U-shaped Lean cells to guarantee feasibility. The first is to use *engineered standards* to determine the time that each part will spend on each machine. The second is to set *stop watch standards.* Good Industrial Engineering methods analysis should always be executed in parallel with machine design to achieve consistency and maximum worker efficiency. One of the most effective ways to restore cell production capacity is to optimize machining times using *mathematical analysis*. For example, the following equation will predict the drilling time at the drill press.

$$MT_i = (d + A) / (F{*}R)$$

Where: d = depth of hole (Inches)
A = Allowance
F = feed rate (Inches/revolution)
R = Revolutions per minute (RPM's) of the drill

The lean engineer must select the two critical machine inputs of cutting speed and feed. The RPM of the drill is related to the cutting speed by the following equation.

$$R = \frac{12V}{\pi D}$$ Where: D = Diameter of the Drill bit

For the drilling operation in this example, suppose that a 1.5 inch diameter hole is to be drilled 1 inch deep with a 0.25 allowance. The material to be drilled is cast iron and the drill is made of carbide. The Lean Engineer selects V= 250 feet per minute and F= 0.015 inches per revolution. The drill RPM is therefore: R = (V*12) / (π*D) = (250*12)/(2*π) = 363.94 Rev/min. The feed rate is (363.94*0.015)=5.46 inches per minute ; Hence:

$$MT_{Station\ 2} = (1+0.25) / (5.46) = 13.739 \text{ Seconds}$$

This drilling operation will take 13.739 seconds to drill a 1.5 inch diameter hole 1 inch deep with a carbide bit. There may be some variance in this prediction, but not much.

As long as the worker loop time is larger than the longest weighted average machining time in the cell, any part in a part family can be made in the cell. The number of workers assigned to the cell, and the number of machines assigned to each worker, is always determined by the required downstream Takt time requirement(s). The sequence used for *multiple parts* must be carefully determined. The *sequence* in which parts are presented to each *machine* will influence cell throughput time, cell output rate and the degree to which downstream production can be smoothed. The cell can continue to function without any physical alterations as long as there is capacity to produce. There is no scheduling required, and the cell is balanced by design. Such a cell is both flexible and *uncoupled* from all other cells except for the required Takt times and the effect of sequencing. In some cases, cell worker balance might require sharing one or more workers between adjacent cells.

Cell capacity is not always increased by machine parallel redundancy, although this is often the only solution considered. if manufacturing is only on a two shift operation, overtime on the 3rd shift can be used. This is contrary to Lean thinking, but it is usually acceptable as a short term solution. Subcontracting can be another short term solution but should rarely be considered as a cost effective long term solution. A permanent solution that is often ignored is to use a *Kaizen Blitz* to improve procedures, reduce set-up times and find other long-term solutions. This is correct Lean Thinking.

4.7 Takt Time and Cell Cycle Time: *Revisited*

As previously stated, the Takt time requirement(s) for a cell dictate how the cell will be operated. Like all other cells (subassembly and component cells), each cell is designed to produce parts only when required (authorized), and only in the quantity needed (Kanban control) for immediate downstream production. The entire system is paced by a Mixed Model final assembly. In the Lean, Linked-cell factory.... the cell is operated to produce parts at exactly the required rate and no faster. Cycle times are usually buffered with a 10%-20% allowance factor to account for any delays or uncertainties, and although it might violate strict Lean Thinking, it is sometimes a good business strategy to maintain a small amount of finished goods inventory to protect customer delivery.

4.8 Flexibility and Cell Design

Product variety demands cell flexibility. It should be obvious that in either single or Mixed Model production systems, any Lean cell must be flexible enough to react to changes in both product mix and volume or demand....while meeting required Takt time. Lean manufacturing systems should not be confused with Flexible Manufacturing Systems (FMS). U-shaped Lean cells are operationally quite different. In reality, Lean cells are designed for flexibility at lower cost, but FMS manufacturing systems actually have limited flexibility at higher cost. Manufacturing flexibility is characterized by four main capabilities.

(1) The ability to rapidly respond to product design changes
(2) The ability to rapidly respond to changes in demand and/or product mix
(3) The ability to rapidly reconfigure a production line (New products).
(4) The ability to accept technological innovation

First, any manufacturing company must constantly design and redesign the production system to respond to customer demands or gain a competitive advantage. Envision the style, appearance and functional changes in a basic pickup truck over the past 20 years. Engineering design changes are simply a fact of life. There are three continuous design activities: (1) *New machines* for part production (2) Part *design* and *redesign* and (3) *Design for Manufacturing* (DFM). Existing end products changed in appearance and to improve performance remain basically the same but intermediate production activities are simplified or even eliminated by such methodologies as *3-D* manufacturing.

Second, to be flexible the process/machine must be able to rapidly change and respond to variation in either product demand (SMED) or product mix (Mixed Model Production) or both. There are two different kinds of flexibility, and in traditional mass manufacturing both can create significant problems. The first is flexibility to support changes in *product performance characteristics* such as improved components, and the second is to respond to different *part configurations*. Both can require modified machining operations or adding or subtracting machines to create technologies to respond. In traditional Mass production lines, production lines sometimes need to be completely shut down and reconfigured. In Lean manufacturing, response to either form of flexibility requirements can be both designed and built into U-shaped cells.

Third, although changes in both demand and product mix in Lean cells can usually be accommodated without changing anything but possibly some worker machine assignments, the number of workers in a cell, or parallel additions to bottleneck machines. In Lean manufacturing cells, these changes are certainly not minor, but because each cell and every machine in a cell is designed to operate independently using simpler, single cycle machines; modification is significantly easier, cheaper and faster than traditional mass manufacturing lines.

Fourth, flexibility can also be defined as the ability to easily accept machine upgrades or new technological innovations. A properly designed U-shaped Lean cell or Mixed Model final assembly line has been specifically designed to easily accept new methods and improved processing workstations. In the case of bottleneck operations, a solution which might be considered other than parallel redundancy is the acquisition of a newer, faster machine.

4.9 Clusters of Lean Manufacturing Cells

Most of the examples so far have focused on individual cell design and operation. However, it is common practice to design a Lean manufacturing system that consists of several linked cells and two or more cells sharing workers. Figure 4.9 shows four different cells; all of which are building up components for other downstream cells. Cell A is linked to Cell B. Cell B is feeding Cell C and Cell C is feeding Cell D. Cell D is feeding the final assembly line. Each cell is being paced by the Final Assembly Schedule and production is scheduled by Kanban cards.

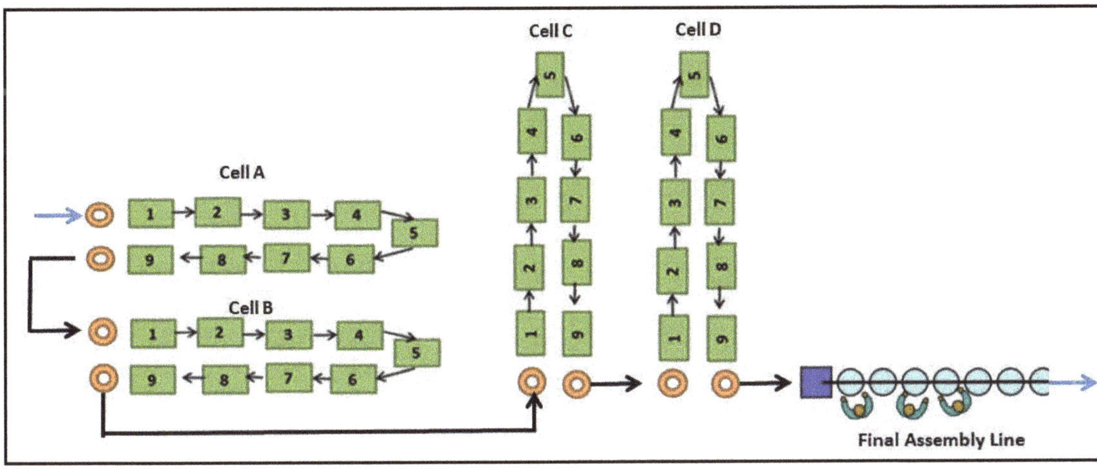

Figure 4.9
Four Subassembly Cells with Cell D Feeding Final Assembly

4.10 How Cells are Linked and Regulated

All U-shaped Lean cells are scheduled and regulated by *Kanban linkage*. Recall that the word Kanban means a *signal* or a *control*. Once parts enter the cell, they flow through the cell one at a time. This one piece movement with (ideally) unit decouplers eliminates queues, maintains the order sequence, saves floor space and promotes zero defects. Scheduling is completely eliminated within the cell. We again cannot overstate the importance of Rapid Exchange of Tooling and Dies (RETAD) which reduces the time for setup and changeovers (SMED). RETAD strategies are presented and discussed in subsequent chapters.

Each cell has an *input station* and an *output station*. These locations maintain temporary storage between cells, or buffer inventory between component and subassembly cells, or buffer storage between subassembly cells and final assembly. Parts move between these storage points and point of usage on the command of Kanban linkages. There are four factors to consider: (1) The number of carriers in the system, (2) the number of parts per carrier, (3) the number of carriers/parts that can be in any one input/output station at any one time and (4) the signal used to produce and/or move carriers/parts. Carriers between system production components should be standardized to carry the same number of parts. The size of each input/output station is determined by the operational requirements of the downstream cell and how frequently parts arrive to the output station; which incidentally is the same rate that parts enter the system since at steady state production cell input must equal cell output.

The number of parts in each Kanban link is the number of carriers times the number of parts per carrier. One of the significant advantages of a Lean, linked cell system is the WIP Cap which is automatically enforced both between cells and in the system. The inventory in each buffer has an absolute maximum allowable value. Note again that the cell itself does not schedule production or control the level of output WIP. Cell production is authorized when a downstream cell needs to receive parts. Cell output only occurs when full carts are requested by downstream demand. A simple way to do this is for downstream production to send an empty cart upstream using a withdrawal Kanban card. This request authorizes the upstream cell to deliver a full cart downstream. When the full cart is removed from the upstream cell, a production ordering card is removed from the cart and placed in the production ordering box. This card authorizes production. An upstream operation will not *produce* unless authorized, and product will not be *shipped* downstream unless authorized with a move card. The normally complex, production schedules produced by MRP systems in traditional manufacturing systems are either dictated by a routing sheet or by priority scheduling. Orders are *pushed* into the system and flow from one area to another as they complete process steps; usually in large lots. The critical functions of lot sizing, sequencing and scheduling are integrated and infused into the Lean system and executed based upon current requirements and not a routing sheet constructed by some computer. Lean parts are *pulled* through the system in small lot sizes. This is such an important concept that Lean Manufacturing has been called Pull Production, but this is a limiting title.

The entire system is under Kanban control. Figure 4.10 shows the area being run by 8 workers; some sharing machining centers.

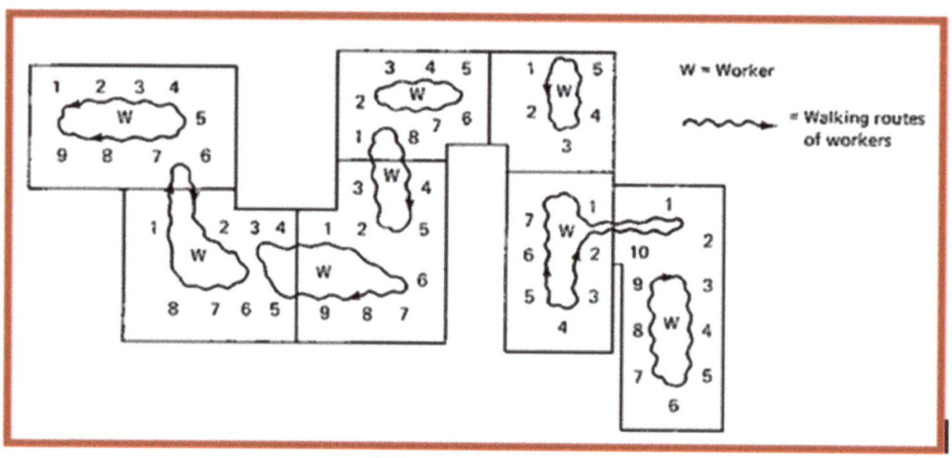

Figure 4.10
Allocation of 8 Workers to the Cells: Some Sharing Machining Centers

We once again point out that some amount of WIP must be designed into the system to promote balancing, leveling and to avoid bottleneck starvation. In some rare cases, a Kanban link might be completely eliminated if production in both cells is producing over the same period of time at the same rate. The cells are tightly coupled, requiring only a conveyor or gravity chute linkage.

It is worth noting once again that changes in either product mix or demand will likely alter the cell cycle time, number of workers required, capacity requirements, machines required.. or all of these things. The ability for any one cell to respond to these changes is what we have called *cell flexibility.* In Section 4.2 we have seen how the allocation of operators can be changed to accommodate a change in cycle time in a Lean cell. In Lean Manufacturing Cells, changing a machining time will *usually* have no effect on the cells production rate, because *the cell cycle time is determined by the time it takes a worker to perform all manual tasks at each machine and walk around his/her loop.* This is perhaps the most important key to understanding cell design and operation. The basic cell design decouples the production rate of the cell from the processing time required by each machine tool (provided capacity exists).

If the same family of parts was made in a traditional Computer-numerical-controlled (CNC) machining center, the cycle times to produce different parts would directly affect product cycle times. It is also likely that the cycle time in a CNC machining center would be longer because machining operations are performed serially by one spindle with possibly multiple tools, other than serial cell production on single cycle spindles.

4.11 Plant Design and Lean Cell Design

Operationally, we have discussed how Lean cells can usually respond to change by using several methods. As cells mature an opportunity to design Lean into each cell increases. Traditionally, the Industrial or Manufacturing engineer designed a new *Greenfield* facility using Facilities Layout or Plant Layout tools such as Corelap, Aldep, Craft and other techniques. These historical methods will not produce a new Lean Plant design. Today the modern term is *Facilities Design*, but the general landscape of computerized plant design has not changed very much. Once a plant was laid out, the job was considered finished. Management and production control who doggedly held to this philosophy created *ivory towers* and *legacy systems* which were very inflexible. There is a great need for modern Industrial and System Engineering departments in Academia to execute funded research in this general area. A new computerized software package needs to be developed to provide an alternative to older plant layout packages. This new software should not only embrace Greenfield Factory Design but Factory Redesign using Lean Principles.

The Lean Engineer is fully aware that any manufacturing system must be continuously modified and changed. This is what continuous improvement is all about. The Lean Engineer will design a Manufacturing system and Lean Cells that seldom correspond to historical plant layout algorithms. In fact, it is time that all of the Industrial Engineering plant layout programs are put into a museum. This is fertile ground for Industrial and Systems Engineers to learn Lean Philosophies and build new plant layout tools . Processes and manufacturing production designed from Lean Engineering principles promote flexibility and will reduce life cycle costs, improve quality, accommodate product variability and reduce system WIP. The Lean engineer and appropriately educated product designers should know and understand the processing/machining requirements of all parts and the way the manufacturing system operates (Cellular design). This is truly *Design for Manufacturing* and is a new frontier. This will not only design a Future Factory but a Factory with a Future.

4.12 Variance vs. Variation

Each Lean cell operates on the principles previously defined. In all cells, quality will be the responsibility of cell workers aided by automated QC inspection. In mature, Lean cells automated inspection devices are often installed in a decoupler. Sensors and monitoring devices are installed on machines to monitor the machining cycle. Most quality problems are not created by faulty machines, but by our old enemy… *Variance*. Ever since Motorola Corporation first introduced the concepts of 6-sigma engineering, Manufacturing engineers have become acutely aware of the *corrupting influence of variance*. There are two concepts that are almost universally misunderstood: One is *variation* and the other is *variance*. Variation is necessary to sell finished products. Variation corresponds to the number of different ways that a product can be configured

(Color, add-ons, appearance, etc.). Variance is created both within production of a particular product, and between production of different product types. For example, Henry Ford was adamant about reducing variance by eliminating variation. That is why he said: *You can have any colored Model-T you want, as long as it is black.* However, he confused variation with variance, because there were always differences in appearance and performance among his black Model-T's and Model A's. Variance caused by Variation is even more important today due to customer demands for different colors, options...Etc. Even within part types, two automobiles that look alike and were built the same way can perform vastly different. Some may even fail to perform as designed. In the extreme, we call these automobiles *lemons*. Variance is the culprit, not variation… although variation in itself can contribute variance components.

4.13 Capacity Analysis and Lean Engineering: *Revisited*

It should be obvious that when a Takt time requirement is given to any individual cell, the cell may not have enough capacity to attain the required output rate. We have already discussed different ways to increase capacity, but this is such an important concept in Lean Manufacturing that we will revisit this concept in greater detail. This will occur in a Lean manufacturing cell when one or more machines *bottleneck* the cell (Machining time and/or worker Task time is greater than the required cycle time). In this case, we offer several suggestions to increase capacity. Some will be more cost effective than others: Some may simply be a temporary fix. The situation and cost will dictate the remedy.

(1) The process/machine is duplicated to effectively double the output rate. Note that as we discussed earlier, this will introduce variance into the system.

(2) The process is engineered to reduce machining time(s). We have previously pointed out that depth of cut, speeds, feeds and type of tool material are all important factors to consider.

(3) Schedule overtime production for the cell. This is only a temporary solution, but one that is commonly used. Working overtime will necessarily increase inter-cell WIP, throughput time and cost per unit.

(4) Sometimes it is possible to separate the bottleneck processing time into two or more stages on sequential machines. Sometimes it may be possible to shift operations to another machine with enough idle time per cycle, but this is not always technologically or operationally feasible.

(5) The product is redesigned by using *design for manufacturing* methodologies.

(6) Redesign the entire cell (Call for a Kaizen event).

Finally, if demand or product mix cannot be accommodated in any existing cell, build a new cell to produce the parts and consider replacing machine(s) which cause a bottleneck. This is a long term solution that usually requires designing and building new machines, so this approach can be costly; but it may be the only feasible solution. If an entire cell is simply cloned, then the capacity immediately almost doubles. This may be particularly true in Mixed Model production. To illustrate this observation, suppose that a U-shaped Lean cell is making two parts, A and B. Further assume that business is good, and that forecasted demand is twice what the cell is currently producing. Further assume that the cell is producing at near its output capacity. The first inclination the engineer might have is to clone the existing cell and continue Mixed Model production. . Enter the Lean Engineer, who proposes that one cell be dedicated to producing Part A and one to produce Part B. The traditional Manufacturing manager may not realize the difference. However, by dedicating each cell to a single product all setups and changeovers are eliminated; labor is specialized to one part; machines can possibly use dedicated jigs and fixtures; and result in a number of different time/cost saving advantages. Perhaps more important, dedicated cells will reduce variance in the system.

4.14 Standard Operations Routine Sheet (SORS)

Standard operations are critical to maintain consistency of production and to make sure that every operation follows a standard procedure. Standard procedures help to reduce variance in any required manual operation. There are three key elements in a standard operations routine sheet.

1. Cycle time and production rates are all determined by Takt time set at final assembly...Possibly modified by mismatched assembly (one table top requires 4 legs)
2. There is a standard sequence of operations needed to complete work in a cell and at final assembly.
3. There is a fixed quantity of stock maintained in the cell (WIP plus Supermarkets plus material in decouplers

In addition,

- Each manual task is standardized by using approved methods, standard work practice, and standard operating procedures
- All material handling carriers are standardized within the scope of their use (cell to cell in fixed quantities)

Route Sheet : XYZ Machine Shop Inc.						
Part No. 081099	Part Name Shaft, generator		Planner MPGrrover	Checked by : N. Needed	Date 08/12/XX	Page 1/1
Material 1050 H18 A1	Stock Size 60 mm dia., 206 m length		Comments :			
No.	Operation Description	Dept.	Machine	Tooling	Dept.	Std.
10	Free and (approx. 3 mm). Rough turn to 52.00 mm diam. Finish turn to 50.00 mm diam. Face and turn shoulder to 42.00 mm diam. And 15.00 mm length.	Lathe	L45	G0810	1.0 hr	5.2 min.
20	Reverse end. Face end to 2000.00 mm length. Rough turn to 52.00 mm diam. Finish turn to 50.00 mm diam.	Lathe	L45	G0810	0.7 hr	3.0 min.
30	Drill 4 radial holes 7.50 mm diam.	Drill	D09	J555	0.5 hr	3.2 min.

Courtesy of Expert Minds.com

Figure 4.11
A Partial Routing Sheet

A partial example of a blank standard operation routine sheet (SORS) is shown in Figure 4.11. The SORS provides a reference point for manufacturing instructions for each part produced. It can be expanded to include machining instructions, settings of machining parameters, Quality Control and Preventative Maintenance tasks. All machining times, walking times and manual tasks performed by the worker are usually set using Methods and Standards by the Industrial Engineering group. These standards form the basis for machining and assembly times in each

Lean cell. These standards are used not only to standardize work in each cell, but they are also used for product costing.

4.15 Lean Manufacturing Cells with Manual or Automated Machining and Manual Assembly Operations: *Revisited*

A U-shaped Lean cell can be entirely composed of automated machining operations, or contain a mixture of automated machining tasks and *manual* tasks such as assembly or manually operated machining operations. The latter is often found in cells which have a small demand. This will not change the basic operation of a cell, but might drastically affect cell cycle time and worker activities. Figure 4.12 shows a Lean cell with a combination of manual and automated operations. Let MT_j designate an automated machining cycle at machine j, and HT_j designate a manual operation at machine j. Note that any one process step might have both automated and manual operations. The cell has 8 different workstations; 5 are automated machines and 3 are manual. The calculation of cell cycle time proceeds exactly as before, but all manual times in a worker's loop are added to the worker task times and walking times. Using only one the cell cycle time is calculated as follows.

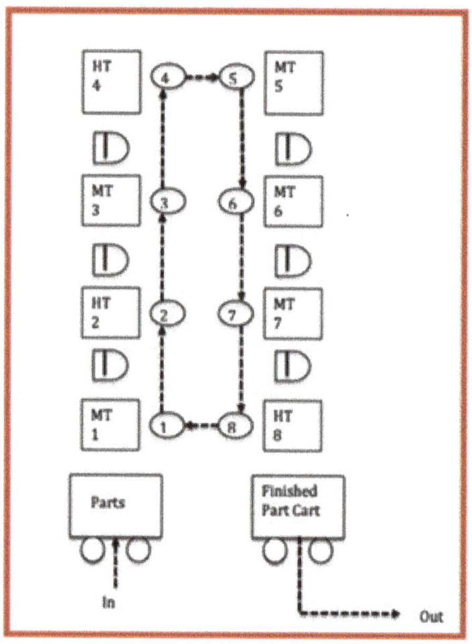

Operation Number	Load/Unload (Sec)	Total Manual Time (Sec)	Machine Time (Sec)
1	3	•	20
2	•	17	35
3	4	•	30
4	•	4	80
5	4	•	80
6	6	•	40
7	9	•	30
8	•	5	0
Total:	26	26	

Figure 4.12
A Partially Automated Lean Cell

The time required to walk from one machine to another is 3 seconds.

$Cell_{Cycle\ Time}$ = (24 sec. walk time) + (26 sec. manual operations) + (26 sec. task times)
$Cell_{Cycle\ Time}$ = 76 Seconds/Part
The cell cycle time is greater than any processing time (MT).
The cell throughput time per part is given by:

$Cell_{Throug\ hput\ Time}$ = (76 sec)*(8) = 608 Seconds Cell Production Rate = 6 Parts/Hr

4.16 Design for Customer Satisfaction

Product and system designs are done using two fundamental goals: (1) Design for manufacturing and (2) Design for customer satisfaction. Note that both design goals are focused to different customers. Design for manufacturing is aimed at the *internal customer* and design for customer satisfaction is aimed at the *external customer*. The first design goal is to *minimize non-value added activities* and the second to *maximize value- added activities*. The first design activity is directed to lower production costs, while the second activity is directed to maximizing profits (sales). In a real sense these two design activities are tightly coupled, since the external customer demands the highest quality product at the lowest cost. A major problem that has developed over the last 20 years is the separation of the design engineer from the manufacturing engineer. The use of CAD/CAM, virtual reality design tools and other computerized techniques has created what we will call *Functional Foxholes* and *Islands of Automation.* This has created tremendous problems for floor manufacturing. It is an all too common practice for the design engineer to *lob* a new product design over the fence and simply say *make it*. This is a sure recipe for manufacturing cost escalation and customer dissatisfaction. A good example is as simple as replacing an automobile starter. This should be simple, but American cars or trucks sometimes require the removal of the fan belt, the cooling fan and a morass of other things just to get to the starter, so the task might take hours. At least one clever European car manufacturer has moved the starter to the top of the engine block. Replacement can be done in 15 minutes or less.

One activity that must be done is to execute *benchmarking* of the market competition. Although not covered in this text, the use of market surveys and the *House of Quality* is emphatically recommended. It would be a good exercise for the student to go to the internet and write a short essay on each topic. It is a serious mistake to ignore the customer and the world around us in the design function. The world that surrounds us is more and more beginning to influence and even dictate the way manufacturing will be done in the future. Product design is now being influenced by more than just placing machines and moving product through the plant. Environmental engineering, Safety engineering and Green engineering are now playing a major role in the product design function, and any major change in traditional plant layout factors will in turn affect the way parts and products are manufactured. Another major factor which will begin to influence how products are being designed and manufactured is how products will be retired and raw materials are disposed. For example, one day when an automobile is ready for the junk yard, there will be no junk yard. Every piece of the automobile will be recovered, reused or recycled. The next factory revolution will be the mandate to consider all of these areas when building commercial products. The flexibility of Lean systems will be ideally suited to adapt and retrofit to meet new social and legislative constraints. In the Factory of the Future, or more appropriately in a *Factory with a Future* (Black), there will be fewer workers than machines or processes. Hence, the Lean production system will provide each worker with a natural environment for job enrichment. Greater job enrichment will result in a satisfied worker who will produce a higher quality product.

4.17 The End Game: Benefits of Converting to Lean Manufacturing and Lean Cells

The process of applying Lean principles, eliminating non-value added activities, and simplifying production avoids and mitigates many unknown risks and makes automation easier. Conversion to a linked-cell lean manufacturing system can result in huge savings over a 3-5 year period of time. Many companies who have begun and completed this journey report significant reductions in WIP, setup costs, cycle time and throughput time along with higher quality. Lean thinking and a Lean reorganization also has other benefits. It simplifies management functions, eliminates many traditional production planning and control mechanisms, and paves the way for automation. Conversion and progression from job shops, flow shops and mass production systems to manned linked cell, Lean systems must be accomplished in logical, economically justified steps.... each stage of conversion building upon the previous stage. Even in early conversion activities, *interim cells* offer many advantages, including:

- Significant WIP reduction
- Greater worker productivity and improved moral
- Balanced production rates
- A reduction in material handling costs and movements
- No expediting or changing priorities
- Setup and changeovers simplified and less time consuming
- Better quality and less rework/scrap
- Smoother, faster flow of products (shorter TPT)
- Less bottleneck problems
- Reduction of variability
- Greater production flexibility
- Cheaper automation costs

Finally, again we issue a stern warning. The transition from traditional manufacturing to Lean Manufacturing is a long and arduous journey. Transition will not occur overnight, and piecemeal implementation is not desired but sometimes necessary. There will be tremendous opposition to change and be aware that *things will probably get worse before they get better*. Management will need to adopt a new world view, and many traditional corporate functions will turn from control to support. Maintenance, quality assurance and order release will need to be redefined. The ideal transition to Lean is through a Greenfield Factory where predisposed opposition can be overcome more easily.

4.18 The Lean Supply Chain

We have not discussed the advantages (and sometimes necessity) for a Lean supply chain, but a major part of migrating from traditional manufacturing to Lean manufacturing is the application of Lean principles to all subcontractors and activities in the supply chain. Lean production facilities usually change to a supply chain with sole source suppliers. These vendors are also encouraged and sometimes forced to not only supply key parts, subassemblies and resources but to hold those resources on-site and actually function within the Plant itself. Production support through Supermarkets is a good example. Resources at the right place in the right quantity become the responsibility of partners. Here is where mature vendors become unique, developing specialized equipment and tools to support *Just-in-Time* delivery at *Point-of-Usage* at minimum cost. Because the main production lines cannot tolerate defects or rejects, sole source vendors are forced to supply defect free products to the end user. A main requirement for migrating from interim Lean cells to mature Lean cells is to fully integrate supply chain vendors into the Lean factory, quickly followed by Lean marketing, warehousing and distribution. At this point, the *Lean Enterprise* begins to emerge. There are many other steps needed to complete a full transformation from traditional manufacturing to Lean Manufacturing. Of course, the development and full partnership with an outside supplier should be an economic decision.

4.19 *Quality Revisited*: Poka yoke Devices

Poka-yoke (ポカヨケ) is a Japanese term that means *mistake-proofing*. A Poka-yoke is any mechanism in a lean manufacturing process that helps and sometimes prevents an equipment operator from making a mistake. Its purpose is to eliminate product defects by preventing, correcting, or drawing attention to human errors either as they occur or before they occur. Artificial Intelligence can play a major role in Poka Yoke devices. The concept was pioneered and the term first used by Shigeo Shingo as part of the Toyota Production System. It was originally described as *Baka-yoke*, but as this means i*diot-proofing* the name was later changed to *Poka-yoke* which means *mistake proofing.*

Shigeo Shingo recognized three types of Poka yokes for detecting and preventing errors in a mass production system.

1. The *contact* method identifies product defects by testing the product's shape, size, color, or other physical attributes.
2. The *fixed-value* (or *constant number*) method alerts the operator if a machining step has been terminated due to some anomality. The *motion-step* (or *sequence*) method determines whether the prescribed steps of the process have been correctly followed.
3. An operator is alerted when a mistake is about to be made, or the Poka-yoke device has actually prevented a mistake from being made. In Shingo's lexicon, the former implementation would be called a *warning Poka-yoke*, while the latter would be referred to as a *control Poka-yoke.* Shingo argued that errors are inevitable in any manufacturing process, but that if appropriate Poka-yokes are designed and implemented then mistakes can either be completely eliminated or caught quickly; hence preventing defects. Poka-yokes proactively eliminate defects at the source, as opposed to using reactive Control Chart or inspection procedures. Figure 4.13 illustrates a Poka-yoke device for a drill press to prevent drilling a hole too deep (Black).

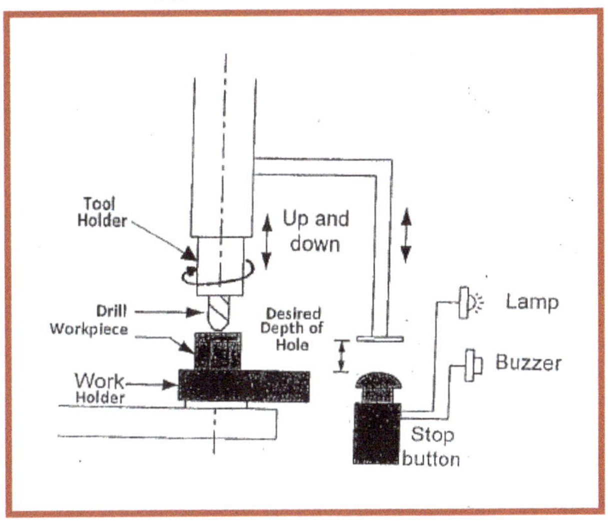

Figure 4.13
An Example of a Poka yoke Device to Control Hole Depth in Drilling

4.20 Ergonomic and Human Factors Considerations in Lean Cell Design

The following Ergonomic and Human Factors features are representative of what might take place in a superior Lean cell design.

- Regardless of whether workers are seated or standing, design a workplace to avoid stress, fatigue and unusual tasks that might be physically harmful.
- Eliminate any loop/worker/machine assignments that cross other workers cells or workstations. Mark walking paths.
- Provide good lighting at all machines to support visual inspection
- Follow good ergonomic design rules that dictate how much can be safely lifted, angles of extension and span or reach
- Safety in the workplace is the most important design factor to consider
- Design walk-away switches to be easy to use and easy to access
- Encourage worker interactions. Periodically hold brainstorming sessions to discuss problems and problem solutions.
- Provide comfortable, ergonomically designed shoes for all workers; cell workers walk a lot on concrete floors
- Task a trained Industrial Engineer to establish standard times and standard operating procedures for all operations.
- Task a trained Manufacturing engineer with developing customized jigs, fixtures and material orientation devices to speed production and save time

4.21 Summary and Conclusions

This Chapter has addressed a wide range of general topics associated with U-shaped, Lean cell design; and provided an overview of cell design considerations and how Lean cells play a major role in Lean manufacturing systems. We have attempted to unify Lean thinking and show why Lean manufacturing is superior to traditional Mass manufacturing systems. The particular focus of this Chapter has been to discuss the design and operation of U-shaped Lean manufacturing cells, but we have also touched upon both the supply chain and subcontractor roles. Lean cells can be configured for both manual assembly operations or automated machining operations or both. Creating worker loops with specific workstation requirements provides a way to provide cell flexibility, and to respond to changes in both product demand and types of products. The maturation of Lean cell design over the past 30 years has destroyed the myth that Lean manufacturing only works in production systems with steady demand and single part production. Another common myth is that Lean systems cannot be designed and built for job shop operations… nothing could be farther from the truth. The same design principles can be used in both single and Mixed Model production systems. Leveling and balancing production are two key issues which largely influence how well a Lean system performs. Both work-in-process and production rates are closely tied to simple Kanban control mechanisms.

To summarize, the following fundamental goals must be met to realize the full benefits of Lean cell-based manufacturing systems.

- All setup and changeover times must be reduced to a minimum
- Zero defects are mandatory along with zero rework

- Takt time and production rates for all operational components must dictate production and be determined by final assembly and intermediate subassembly requirements
- All final assembly lines must be converted to Mixed model production lines
- All work must be balanced and leveled to smooth production
- All component and subassemblies should be converted to U-shaped Lean cells
- Everyone in the company must be trained in Lean thinking, and realize their role in the transformation
- Every cell worker must be cross trained to work any machine in the cell
- Every cell worker must be trained to be a quality inspector and perform simple preventative maintenance tasks
- All subcontractors in the supply chain must be converted to Lean subcontractors. Each must strictly adhere to just-in-time supply strategies
- Material handling carriers must be standardized by area of usage and equipped to carry Kanban control cards

The motivation and reasoning behind the conversion of all component and subassembly Mass production lines into Lean cells are summarized as follows.

- U-shaped Lean cells improve productivity and flexibility
- Workers are more motivated and focused in Lean cells
- Lean cells provide the mechanism to rapidly adjust to changes in product demand and model types
- Cells encourage teamwork and allow operators to assist one another if needed
- Cells promote higher quality products
- Cells are by design self-balancing
- Cells are designed with a WIP cap
- Cells cannot overproduce: Production and output WIP is tightly controlled by Kanban signals
- Lean cells promote one-piece-flow and eliminate large lot and batch production
- Production problems and machine malfunctions are more easily detected
- The cost of tooling can be significantly reduced by lowering key machining parameters such as speed and feed. Lower cost tools can sometimes be used under these conditions.
- Cell cycle times are independent of component machining times
- Lean cells significantly reduce the floor space (footprint) required
- Historical reports from companies who have successfully converted to Lean manufacturing reveal huge benefits.
- 30% to 40% fewer operators do the same amount of work
- There is 80% to 90% less work-in-progress
- There is a 50% to 90% reduction in rejects and rework. This significantly improves as interim cells become mature Lean cells.
- A 20% to 30% reduction in floor space is required
- The cost per unit of production is reduced from 20% tom 30 %. This enables a company to maintain the same profit margin while investing more money in R&D and technological innovation.

It is the premise of this book that if any company decides to begin the long journey to Lean manufacturing, that a new breed of Manufacturing, Mechanical and Industrial Engineer is needed to lead the transition. We call this new engineer a *Lean Engineer.*

Finally we exit this Chapter with a word of caution. The transition from traditional Mass manufacturing to Lean manufacturing is not instantaneous, nor is it simple. A meaningful transition will usually take from 3-5 years, but the journey never ends. Changing market forces and the need to introduce new or better products demand that Lean transition and Lean thinking is not only in the design phase, but is evolutionary. Also be aware that once the transition is started, things will usually get worse before they get better. There will be resistance to change at all levels, but the leadership must stay the course. As cellular production kicks in, Kanban control replaces traditional scheduling and order release; and the work force is retrained... there will eventually be significant gains in cost reduction, responsiveness and versatility.

Review Questions

1. In designing single-piece flow cells, what will be the consequence of all the hidden variations that existed in the job shop design?
2. How will you prepare the workforce to deal with the conversion to the lean shop?
3. What can be done before cell formation that will minimize the variation in cycle time?
4. Who has the best ideas about how to improve the operations in the cells?
5. Who should be tasked with sustaining the maintenance of the standard work instructions?
6. What is standard work?
7. Why is standard work necessary?
8. What is Takt time versus CT for the cells?
9. If a machine's processing time exceeds cycle time in a manufacturing cell, what alternatives might the manufacturing engineer consider to ensure that a cell meets the necessary cycle time?
10. What are the advantages of single piece flow in a manufacturing cell?
11. What are the machine design changes for interim cells (for example a walkaway switch)?
12. What is the key role of the worker in the cell?
13. Assume that the Takt time is 60 seconds for a cell, but a finishing process has a cycle time of 65 seconds, and there is no way known to reduce that time.

    a) What would you do, ideally and why?
    b) If a new finishing process machines costs $5,000, what would you do?
    c) If a new machine costs $50,000, what would you do?
    d) How are the parts pulled through a cell?
    e) What is an example of mistake-proofing devices in common practice?
    f) Why is mistake proofing so important in a cell's operation?
    g) Assume that you are designing a machine cell for a product that has an expected life of one year. What would you automate? What would you not automate?
    h) Why do cells use single-cycle automatic machines?

Problems

1. A Lean Cell was been designed with five Workstations and a final inspection station. The original lathe operation took 100 minutes. A decision was made to add a second lathe which would operate in parallel with the original lathe. The modified cell diagram and the Workstation machining times are shown below

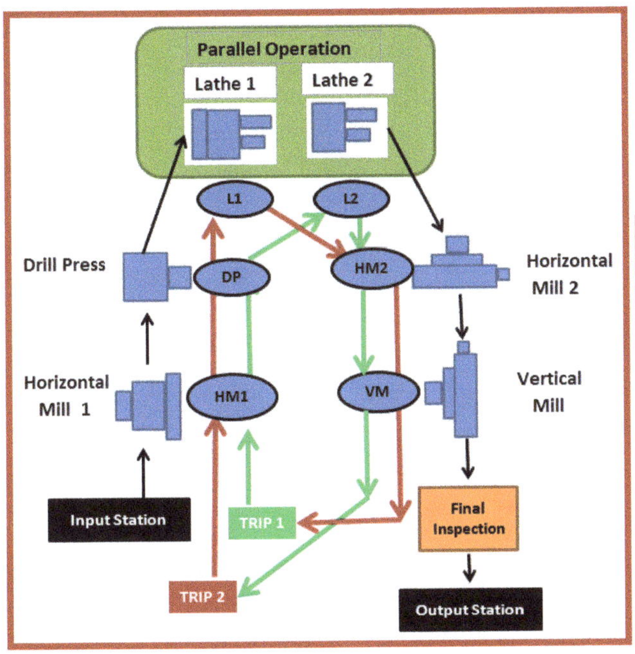

Sequence Number	Operation ID	Worker Time (Sec)	Machine Time (Sec)
1	Horizontal Mill 1	20	30
2	Drill Press	25	20
3-L1	Lathe-1	20	50
3-L2	Lathe-2	20	20
4	Horizontal Mill 2	25	30
5	Vertical Mill	25	25
6	*Final Inspection*	20	40

It takes 12 seconds to walk between any two workstations. Answer the following questions.

PART I: Describe how a single worker will run this cell.

a. What is the cell output per hour?
b. What Takt time can this cell support?
c. What is the throughput time for a single part?

PART II: By inspection, specify three different worker loop assignments using two cell workers.

a. How many 3 worker loops assignments exist?
b. For each of the three loop assignment: answer Part I, Questions b. and c.
c. Is there an assignment which will result in the lowest Takt time? Explain your answer.

2. A company has contracted to make 5000 custom wheels for Generalized Motor Corporation each year. The company currently works 260 days per year, and each workday is 8 hours long with a 30 minute lunch break and a 15 minute rest break. What is the Takt time requirement to fill this order in terms of minutes and seconds?

3. The Phillips-Black *Make-a Brake* company is a first tier subcontractor to Custom Car Corporation for brake sets used on three different car models. The monthly demand rates are:

Type A= 800 per month; Type B= 500; Type C= 700 per month. A Lean, U-shaped cell is going to be designed to manufacture all types of brakes in the same cell. Each type of brake follows the same exact manufacturing sequence, but takes different processing times at each workstation. The processing times for each brake type are given in Table 3-A. All workstations are single-

Table 3-A

Processing Times (Minutes/Part)

Part	Machine 1	2	3	4	5	6	7
A	7.00	6.50	5.20	6.10	11.20	9.20	4.50
B	10.00	5.20	5.80	7.00	6.30	7.80	5.20
C	8.00	5.70	5.00	6.50	7.80	8.10	6.00

cycle automatics with a walk-away switch and automatic discharge. Worker assignments at each machine take 2 minutes. Walking times between machines are all 10 seconds. Changeover times are given in Table 3-B.

From \ To	A	B	C
A	0	1.2	1.5
B	2.4	0	2
C	2	1.5	0

Times are in minutes

Table 3-B

a) Calculate the required Takt time and output rate in minutes.
b) Calculate the composoite workstation processing times
c) Determine the number of workers required and machine loop requirements loop assignments for 1 and 7 workers. Can the cell meet theTakt time requirement? If not why not?
d) Determine the sequence of work for Parts A, B and C that.....
   a. Minimize set-up times.
   b. Result in the maximum set-up time(s) possible
e) As a Lean engineer, suggest a reasonable production sequence that would level downstream production and result in a reasonable amount of set-ups/set-up time(s).
f) Determine the minimum number of workers required to meet Takt time requirements, and loop assignments to balance worker workload(s).

4. The manufacturing cell shown in Figure 4-A has seven machines and can be manned by one or two operators. These operators can work at any machine in the cell. Each of the machines is at least single cycle automatics; that is, they can complete machining cycle automatically once it has been initiated by the operator (using a walk away switch). The machining times for the processes are in seconds shown on the machines beginning with the saw and going CW around the cell to the grinder. The final inspection machine is also automatic and takes about 15 seconds to load in a part, depending on which component from the part family is being inspected. The time for the inspection process is 15 seconds. The machining times given above are for part A from the part family. Machining times (MTs) for some operations are somewhat smaller for parts B, C and D. The operator (or operators) take about 10 seconds at each machine to perform various manual operations like unloading the machined part, checking the part, perhaps deburring the part, and loading the next part to be machined. The operators spend about 3 seconds walking from machine to machine. The aisle between the machines is about 4 feet wide.

Answer the following questions based on the cell design above and the information provided.

a) The grinder is a single-cycle automatic. How long does it take to complete its machining cycle? Why is it important?

b) For the Drill, the Machine Time (MT_{Drill}) can be calculated from the equation on Page 125. Suppose that a 2 inch diameter hole is to be drilled 1 inch deep: V=200 feet/minute, the Feed rate is 0.01 inches per revolution and the hole Allowance is 0.25. What is the Drill Press machining time?

c) For the cell with one operator, what is the cycle time in minutes?

d) For the cell with one operator, what is the production rate?

e) For the cell with one operator, what is the TPT in minutes?

f) For the cell with two operators, what is the CT?

g) For the cell with two operators, what is the TPT?

h) Suppose the time on the grinder needs to be increased to improve surface finish. The new MT is 1 minute. What is the impact of this change on the throughput time for the cell?

Machine Tool	Operation or Process	Machining Time (min)	Operator Manual time (Min)	Walk Time (Min)
Saw	Cut bar to length	25	6	3
Drill	Countersunk hole	28	20	3
Lathe 2	Finish turn	25	10	3
Horizontal Mill	Mill Step	30	10	3
Vertical Mill 1	Mill slot	30	15	3
Vertical Mill 2	Drill and tap holes	20	10	3
Surface Grinder	Grind slot	45	10	3
Final inspect	Final Inspection	15	15	3

Table for Figure 4-A

Chapter 5
Subassembly Cells

I ain't supposed to do nothing unless somebody tells me to
General Motors line worker

5.1 Introduction

The Toyota Motor Company has risen to a position of world prominence in the automobile industry by redesigning the traditional Mass manufacturing system into a Linked-Cell Manufacturing System (L-CMS), or what is now known worldwide as *Lean Manufacturing*. In redesigning the mass production system, final assembly is converted into a mixed model production line, and many subassembly operations into U-shaped lean production cells, unless they are in sequence at final assembly. Final assembly is set up to produce to Takt time, and requires line balancing for any mix or volume of parts. Subassembly cells are required to have a cycle time slightly less than the Takt required for downstream operations. All machines in a lean cell are redesigned to operate on a basic **L**oad-one, **C**heck-one and **M**ove-**O**ne-**O**n (Mo-Co-Moo) basis. One or more workers are assigned to each cell, and one or more machining operations are assigned to each worker in such a way to satisfy Takt time requirements. This is called a *loop*. Decouplers are placed between each workstation and at loop completion points (*baton passing*) to make each machining operation and each worker loop independent of any other. The term *decoupler* is derived from the axiomatic-design rules developed by Nam Suh.

Concurrent engineering (CE) and Design For Manufacturing (DFM) methods have traditionally focused upon the relationship between product design and process design, but have often ignored system design (SD). Hence, a true integrated manufacturing system design has been an important but neglected aspect of CE and DFM. If a company would focus on systems design using Lean Manufacturing Principles, the gap between CE, DFM and SD would be narrowed if not closed. Both need to fully integrate the fundamental role of humans using sound human factors and ergonomic designs. The result would be to position any manufacturing company to more successfully compete in a world market by adding flexibility, saving production costs and reducing time to market. The results would be reduced production costs, improved flexibility, versatility and greater market penetration.

A *Manufacturing Enterprise* (ME) is a complex organization of supply chains, manufacturing lines, final assembly and product delivery. However, the *manufacturing system* is the beating heart of any manufacturing enterprise. The manufacturing system has been defined as *a complex arrangement of physical elements characterized by measurable parameters* (Black, 1988). The physical elements are machines, people, material handling systems and information systems. Ohno asserted that the most important element in the system is skilled laborers (people). The different processing centers required to make a final product are linked to both upstream (input) and downstream customers (output) by simple Kanban, and visual control mechanisms such as empty or full carts of work-in-progress. Each processing center is therefore a customer (*internal customer)* and a supplier (*internal supplier)*. The final assembly line is producing products to satisfy the demands of *external customers*. In terms of a harmonious and successful system design, this is a key concept. *A manufacturing system must be designed to satisfy the needs of both internal and external customers.* The Lean system is designed for two purposes: (1) Eliminate all forms of waste, and (2) transform an extremely complex system into one that is simpler to operate and understand, while simultaneously satisfying the needs of these two customers.

5.2 Subassembly Cell Design and Operation

Standard work is the basis for maintaining productivity, quality, and safety, and provides a consistent structure for performing the tasks in such a manner as to meet required Takt time, while simultaneously uncovering opportunities for making improvements in work procedures. There are four aspects in structuring standard work in a cell: (1) Takt time, (2) The sequence of required cell operations (3) Every worker cross trained on each machine in the cell and executing best-practice methods and (4) minimizing cell WIP.

5.2.1 Takt Time and Cell Operation

Takt time reflects *the pace* of sales in the marketplace. It is based upon a combination of forecasted plus actual demand. The Takt time required for final assembly dictates the cycle time for each cell. The working *sequence* is the sequence of operations that is the best way to perform the subassembly job. The *stock-on-hand* is the minimum number of parts needed in a cell to maintain a smooth flow of work.

Within a cell, a set of process steps is performed by each worker. This set is selected to make each workers loop or sub cycle time equal to the required output Takt time as close as possible. If the cycle time of each worker is equal to the required Takt time, this creates a perfectly balanced cell. Takt is the German word for a conductor's baton which is used to keep every musician in time. Takt time is calculated by dividing all available production time during any one day by the daily demand (DD).

$$\text{Takt Time} = \frac{\text{Production Time Available/Day}}{\text{Daily Demand}}$$

Takt Time (TT) Example

Suppose that over the next 6 months a certain item has confirmed and forecasted demand of 60,000 units. This is 10,000 units per month on the average. The normal work month is 20 days, two shifts per day with 30 minutes off each 8 hour shift for a meal. This results in a required production rate of 500 units per day, or 33.3 units per hour. The required Takt time 1.8 minutes per part.

Takt time is the *drumbeat* for a Lean cell. Every Lean Cell tries to produce to Takt time to meet the functional requirements of producing the right quantity at the right time. The L-CMS groups the operations or processes into U-shaped lines called *Lean cells*. The processes can be changed over rapidly so products can be turned out in greater variety in an almost customized fashion with no cost penalty for small production runs (small lots). The entire manufacturing system is designed such that the throughput time (TPT) is as short as possible. This requires a continuous redesigning of the manufacturing system (which is just another way to say *continuous improvement*) to shorten the TPT, the total time it takes each part to move through the plant.

5.2.2 Sequencing Cell Operations

Just as in Mass manufacturing, the sequence of operations is determined by the process routing sheet. The first operation to be performed in a cell proceeds the second, the second the third and so on until the entire sequence of operations is completed to build a component or subassembly in that cell. However, the order in which each machine is placed in a cell is different from the order that each cell worker follows as he/she proceeds through the loop of one or more machines in the loop. Each worker has an order in which each machine is visited, but that order does not need to coincide with the build sequence.

5.2.1.3 Minimizing Cell WIP

One of the fundamental differences between a U-shaped Lean cell (LC) and a traditional Mass production system (MPS) is the way that cell WIP (Work-in-Progress) is controlled and managed. In a Lean Cell, WIP can be found in machine decouplers, or in a loop decoupler which is positioned between two or more workers in a single cell. A loop decoupler is placed at the point where two workers interface. In either a machine or loop decoupler, there is usually no more than one unit. Hence, the amount of stock-on-hand in each U-shaped cell is tightly controlled and minimized. In a *rabbit chase*, it can be as low as only the number of workers in a cell (See Section 5.7.2).

In a L-CMS, the manufacturing cells and subassembly cells are designed with cycle times slightly less than the TT of final assembly. This requires *balancing* the entire manufacturing system. Some subassembly lines and cells are *synchronized* with final assembly, so if final assembly stops (for a problem), they stop and then restart with final assembly. The balancing and synchronization is accomplished by a *linked-cell Kanban control system* which will be described later. For now, we will simply note that a linked-cell manufacturing system (L-CMS) synchronizes all manufacturing and subassembly cells to final assembly by means of Kanban inventory links. The Kanban links serve to carry production control information upstream to supply cells while delivering the inventory downstream to subassemblies and then to final assembly. The Kanban links represent the supply chain for sequential Lean Cells and the control of inventory is automatic.

In subassembly cells, when an operator arrives at a machine or station, the operator usually remains at the station or machine and in a *Manual* operation completes assigned processing duties before moving on to the next machine. Some Machines are *automated* such that the cycle is controlled by computers. Cells are typically manned by multifunctional operators trained to be able to perform many tasks and operate every process at every station in the cell. The machines in a cell are placed next to each other in a U-shaped or parallel row design to reduce walking times between machines and achieve one-piece flow within the cell. So: *How are cells different from flow lines?*

Redesigning a Mass production flow line into a U-shaped Lean cell using multiple workers automatically achieves a balance in task times by letting the workers sequentially visit a group of machines assigned to him/her in a walking cycle called a *loop*. This type of cell design is more flexible than the traditional Mass production line, and it is easier to change the output as needed by changing the number of workers in the cell than the number of machines or processes in the manufacturing sequence. We will see *it is not necessary to balance the cycle times at individual workstations* (no line-balancing) and the design leads to self-balancing by the operator.

From the perspective of facilities design, a linked-cell manufacturing system is fundamentally composed of multiple U-shaped cells. A cell is usually a connected set of machines, processes, or stations in which a single worker or multiple workers perform all the tasks needed to make a component or subassembly. The shape and size of designs vary greatly. Cell designs can be laid out in U-shapes, L-shaped lines, parallel lines or even more twisted lines. However, the U-shaped design is by far the most frequently adopted design for the L-CMS.

5.6 Cell Performance Measures

Different system designs will result in different levels of measurable parameters. *Cell cycle time* (CT) is the most critical measurable parameter for evaluating cell performance. Closely coupled to cell cycle time is *throughput time (TPT)* and the amount of inventory in the cell. By systematically reducing TPT, a company can achieve world class status. In the lean manufacturing system, the manufacturing TPT is reduced by the systematic and gradual removal of inventory (cell WIP) as proved by the well-known Little's Law, which mathematically relates TPT and CT to product demand. Little's Law is stated as follows:

$$\text{WIP} = \text{TPT} * (1/\text{CT}) = \text{TPT} * \text{Production rate}$$

Since the cell production rate is fixed by customer demand, we see that TPT and WIP are directly related to each other. Inventory in the system is held in links (between the cells) in the system or in inventory storage locations called Supermarkets or Heijunka Boxes. Finished goods inventory temporarily holds finished products prior to shipment. All work-in-process inventories are moved by the demand of internal customers (other cells or final assembly). Movement of WIP is triggered and controlled using the Kanban system of production control. Kanban is discussed in detail in Chapters 4 and 5. The linked-cell manufacturing system (L-CMS) can have many cells linked to one another by Kanban inventory control.

The material in a Lean cell is usually referred to in the Lean world as *stock-on-hand* (SOH), and is not considered to be inventory or traditional WIP. However, the Operations Research area called *queuing theory* will neither recognize nor use this term. Little's Law uses the term WIP for any inventory in the system, whether it is in a cell, in storage, or in a process step, and it can be used to predict throughput time for any cell or the entire system given the demand rate. SOH material is either in the stations or in the decoupler elements. WIP in a cell (SOH) is strictly controlled and is a major factor in how a cell is operating. The SOH held between machines result in a *decoupled design*. The design of the factory and its physical elements must precede the design and manufacture of the product. Because the design includes people, ergonomic and human factors issues must also be considered. In the links between cells, the inventory is standardized, controlled, and minimized by never producing, storing or moving inventory unless required by internal customers. Final assembly triggers the entire system by either producing to final goods inventory or directly to customer demands. Kanban controlled links also provide production control information to the upstream processes and suppliers, telling them *what* to make, *when* to make it, and *how many* to make on a daily basis.

5.7 Operating Lean Cells

There are three basic methods for operating Lean cells. They are: (1) *Single or multiple workers* (2) *Rabbit chase* and (3) *Bucket Brigade*. None of these methods require that the individual workstation cycle times be balanced. We will describe these four methods within the context of a manual subassembly cell. In this case, the cycle time for each worker is composed of assigned tasks at each machine or workstation, walking times between machines in their loop, and the times to perform other tasks such as Quality Control or Preventive Maintenance.

5.7.1 Single or Multiple Workers

A Lean Cell operated by single or multiple workers responsible for one or more machines in a loop arrangement has already been described. A lean cell managed and run by a single worker is simply assigning one worker to completely perform every assigned duty in a U-shaped cell over a series of sequential operations. Machines and workstations stations can be automated or manual. Rather than use only one worker, the cell operations can be divided into sub cells, each managed and executed by a single worker. We call these sub cells *loops*. The time it takes a worker to complete all operations in a loop is the walking time between stations plus the time it takes to complete all assigned duties in the loop, including loading and unloading machines and support activities (QC or Maintenance). An assigned worker will also be responsible for loading parts into the cell and removing parts from the cell. The maximum possible number of loops that can be created is one worker at each station. This also yields the fastest possible output rate for the cell. In this case, the cell output rate will be that of the slowest machine cycle time or the *bottleneck machine*. The slowest output rate for the cell is when only one worker is used in the cell. Since the time required for each worker in a multiple worker cell can be slightly unbalanced, blocking can occur at the point where each loop worker hands off a part to another loop worker. To eliminate this form of blocking, a separate decoupler is usually placed between worker loops to prevent delay. With only one unit of WIP in each machine decoupler and in each loop decoupler, there is usually almost no blocking with three or fewer workers in the cell. If there is significant variation about the average processing time, and as the number of cell workers increase, blocking is more likely to occur without loop decoupling. If blocking becomes a persistent problem, it can be temporarily eliminated by increasing the amount of WIP in the cell decoupler..but this is just "patching a problem". The permanent solution is of course to balance all loop sub cell cycle times with *Kaizen* event(s). A Kaizen Event is a special investigation of a known problem by assembling a team and resolving a known problem or one which has surfaced in system design.

5.7.2 Rabbit Chase

In Figure 5.1 two worker are executing all the cell operations using a *rabbit chase*. The Rabbit chase was one of the 1st attempts to use multiple workers in a single cell (multiple workers follow each other around a single loop). The beauty of a rabbit chase is that it completely eliminates the need to balance worker cycle times....it is self balancing. In a rabbit chase, the 1st worker will begin at Workstation 1 (Cell input station). He will *take the same part* to Workstation 2 and immediately begin work. Meanwhile, Worker 2 has been waiting on Worker 1 to leave the input station. As soon as Worker 1 departs, the second worker will mount and start his part. The 2nd worker will " chase" the 1st around the cell. If there is a 3rd worker, he will follow the 2nd in like fashion. Blocking can and will occur if the 2nd worker is faster than the 1st. If this occurs regularly, the solution is to train the slower worker and increase his speed until it matches the faster worker. The main advantage of a Rabbit Chase are that it is easy to add or subtract people from the line to increase output, and that you do not need to balance the line because people self-balance as they develop a pace. The 2nd advantage is that since a single worker is always working on the same part, there is a sense of ownership from the first station to the last. In addition, each person could gain a sense of satisfaction from learning how to operate each process.

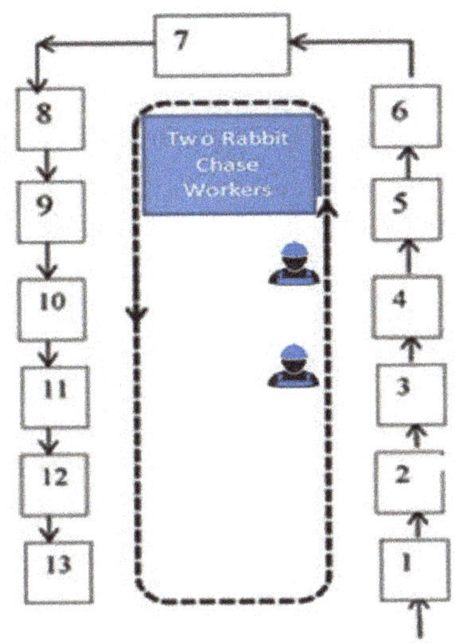

Figure 5.1
Design of a U-Shaped Assembly Cell with 13 Assembly Stations Operated by Two Workers in a Rabbit Chase

Another significant advantage is that there is no need to maintain WIP in any decoupler between machines. In this example, two workers follow each other around a cell, doing all the processes in sequence. This method is usually not used in automated machining cells. The need for precise line balancing for the entire cell is eliminated, and the cell uses no decouplers because the work is always with the worker. After completing a part at the last station, the operator returns immediately to Workstation 1 and begins another piece of work. The rabbit chase method requires every worker to operate all the machines or processes, which may increase the cycle time, the possibility of making errors, and the training time for new workers. If there is a systemic quality control problem that is inherent in any one worker, it is easy to identify.

The main disadvantage of this method is that the slowest worker dictates the cell's output. Blocking by the slowest operator usually occurs because all workers operate (assemble) at different rates. There will also be inherent variation in their average assembly times for every task. Also, workers may be blocked by the workstation which has the longest average processing time. This is called a *bottleneck* workstation. A bottleneck workstation will increase idle time and decrease the output rate of the cell. However, the cause of a bottleneck should almost immediately be discovered by anyone waiting on that operation. To Summarize:

Advantages

- Easy to set up and manage
- Low inventory
- Fast lead time
- Psychological ownership of apart in that cell

Disadvantages

- The slowest worker will dictate cell production
- A traffic jam will develop behind the slowest worker.
- Significant walking distances. Each part requires a worker walking the entire line Every worker needs to know all processes from the beginning, and training of new workers may be more difficult.
- Not as scalable as adding more workers to unique loops within the cell. Adding more workers will increase output, but not linear.

All About Lean.com by Christoph Roser

5.7.3 *Bucket Brigade* (Bartholomew and Einstein)

The bucket brigade evolved from the basic philosophy of a rabbit chase. It is operationally very similar to rabbit Chase. The name *Bucket Brigade* was taken from the ancient practice of citizens attempting to put out a fire. They would assemble in a line from a water source to a fire, and then pass buckets forward to the fire and return empty buckets to the source. The motivation for developing a Bucket Brigade was to improving the Rabbit Chase: It was to minimize waiting times when workers are assigned to a cell with different skills and capabilities. A partial Lean cell is shown in Figure 5.2. The operation of a manufacturing bucket brigade is as follows.

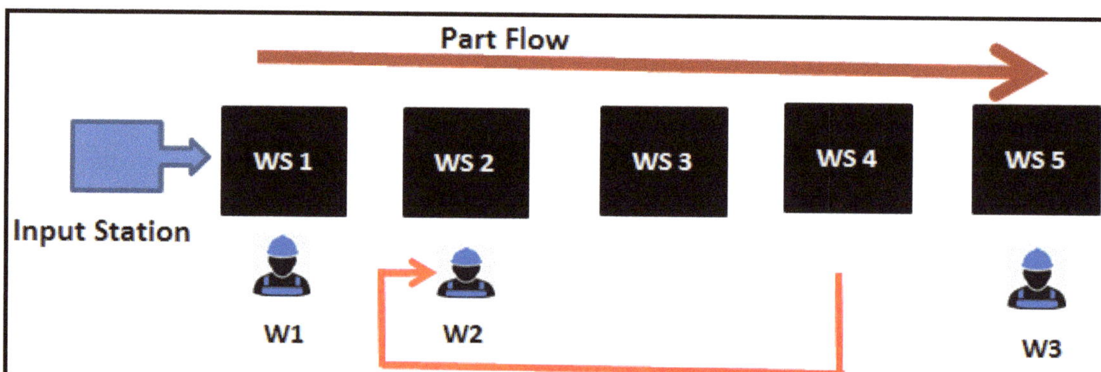

Figure 5.2
Bucket Brigade-I

A worker can seize a part in one of two ways : (1) At the cell input station (2) From another worker. In Figure 5.4 the 2nd worker has just completed a part at WS 4 and cannot move on to WS 5 because the 3rd worker is already there. He then walks back until he encounters the 1st worker at WS 1. The 2nd worker will take the part from W1 and start to move it up the line. W1 will pick up a new part from the input station and begin again. W2 will take his new part on down the line. One of two things will happen...He will either encounter W3 again...give his part to W3..and walk back to W1 again; or he will finish his part at the last workstation, walk to the input station and begin again. It is obvious that each worker will usually work in a loop of varying size.

There is possible bottlenecking in a Bucket Brigade. If the last workstation is a long cycle station, then all workers will stack up waiting for that process to finish. There are two solutions to this problem: (1) Automate the last workstation or (2) break the last workstation cycle time down by dividing into one or more separate workstations. The Bucket Brigade is most effective with short machine times and a balanced line.

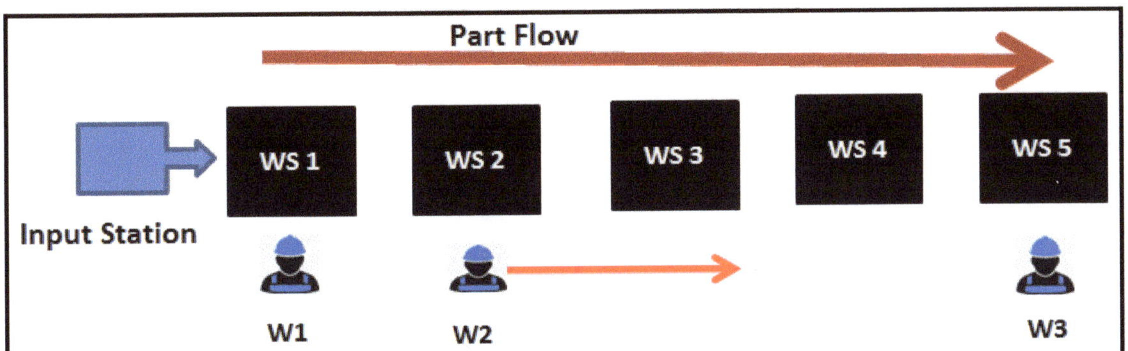

Figure 5.3
Bucket Brigade-II

Notice that workers must maintain their sequence: No machine can be skipped and so it can happen that a worker is blocked by his predecessor, in which case we require that he simply wait until he can resume work, after his predecessor has moved out of the way. (This waiting is not necessarily bad because it is the means by which the workers migrate to their optimum locations). There are several potential benefits in using the bucket brigade cell staffing system.

- There is a reduced need for planning and management because bucket brigades make the flow line self-balancing.
- Production becomes more flexible and agile because bucket brigades 'tune' themselves, without time-motion studies or the other cumbrous endeavors of assembly-line balancing.
- Throughput is increased because bucket brigades spontaneously generate the optimal division of work.
- Secondary labor is reduced and quality improved because bucket brigades operate with the absolute minimal work-in-process.
- Training and coordination are simplified because it is easy for workers to know what to do next.

5.8 Linking Cells

Cells may be linked to each other in serial or parallel flows, and flow may be split into different branches or joined as shown in Figure 5.4. The linking of the cells may result in a large variety of cell designs which are given names like *spider line, spiral line*, *escargot line* or *heart line*.

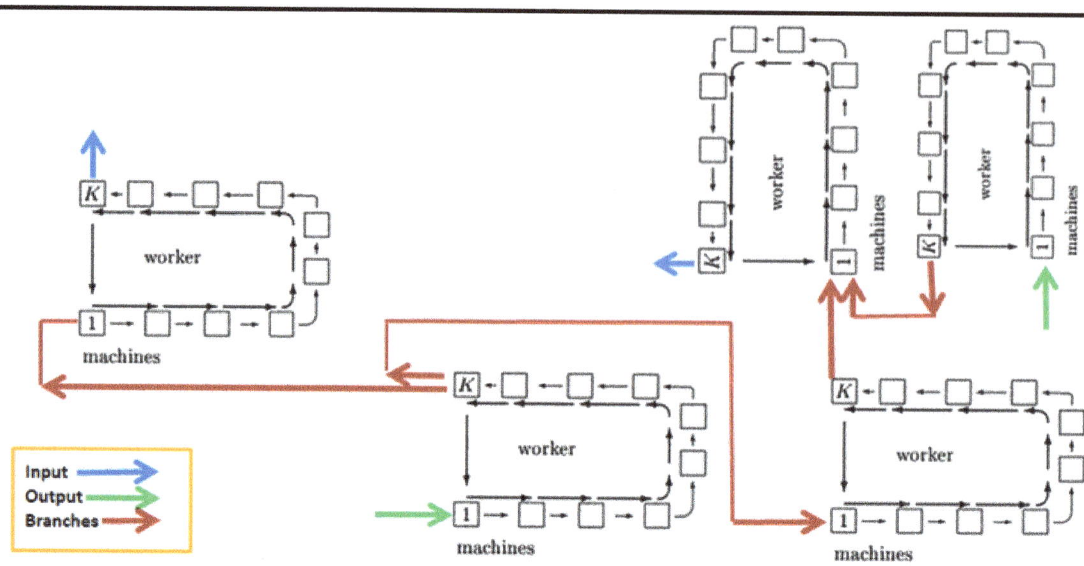

Figure 5.4
A Linked Cell Subsystem

A principle that guides the design of the L-CMS is to make the cell as compact as possible by placing all machines in a U-shape, and narrowing the distance between different stations to minimize walking. Other motivations include the minimization of conveyors , maximizing the use of limited floor space (footprints), and minimizing total WIP. In Lean production, conveyor lines are usually eliminated. Parts are transferred from one workstation to another by passing through small in-process inventory locations called *decouplers*. Large products can be moved to decouplers by overhead cranes or robot arms. Small parts may have either gravity chutes, manual or robot transfer. A Kanban system is used to control the withdrawal of finished parts from the cells by another downstream cell or a final assembly line.

The linked-cell manufacturing system (L-CMS) is firmly grounded on fundamental Lean principles based upon the concepts of producing only to customer demand (actual or forecasted), eliminating all forms of waste (*Muda*), reducing Work-in Process (WIP) to a minimum, and scheduling production with a *pull* system. The commitment to continuous process improvement by systematic identification and elimination of waste through *Kaizen* events (Section 2.17) leads to superior quality levels. Many manufacturers in Japan's electronics industries rely on guidance or support provided by consultants who are expert in TPS to introduce Lean Cells into their plants. These individuals are the forerunners to what should eventually be called *Lean Engineers*. The physical configuration of U-shaped cells, which are the elementary production units of the

5.8 Traditional Conveyorized Production

Automated conveyor lines have historically been considered by most modern Manufacturing engineers to be the key element in implementing a mass production system. Many leading manufacturers are now revisiting this traditional approach while implementing a Lean manufacturing strategy. Figure 5.5 shows an example of a typical final assembly with 9 workstations, each having one operator. This line will require balancing. This topic will be discussed in Chapter 8 in some detail.

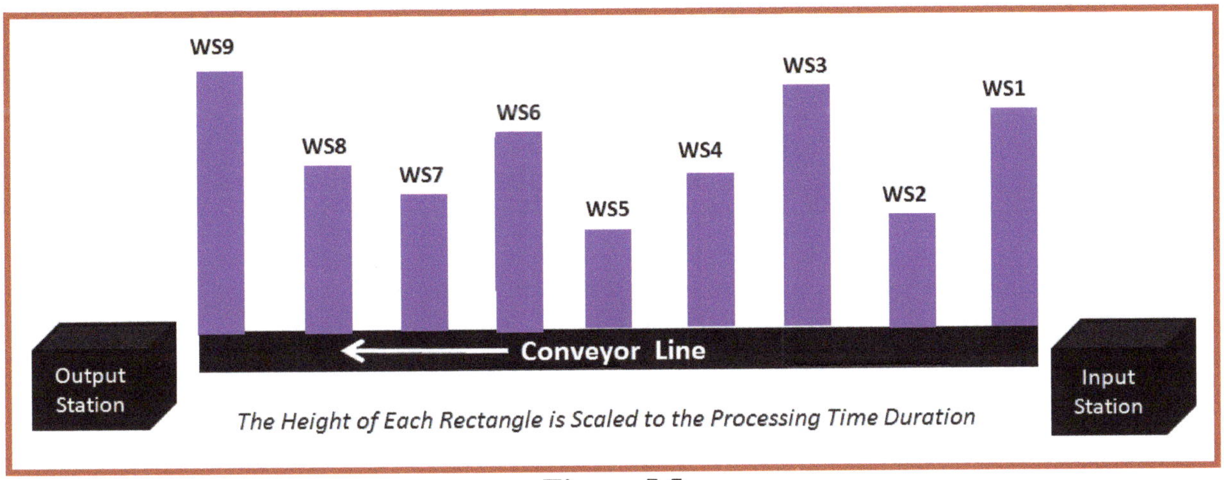

Figure 5.5
Mass Production Line with Conveyor Paced Material Handling

This conveyor line presents a set of problems which are typically plagued by the following categories of waste (*Muda)*.

- Underutilization of workforce due to the fact that line cycle time is bottlenecked by the workstation with the longest processing time (WS9).
- Waste of time in reaching for the work piece on the conveyor and returning it to conveyor after task completion.
- Waste of inventory due to the holding of material between successive stations in buffers. In manned conveyor lines, keeping a buffer of parts at workstations is indispensable because although all the line workers try to follow the pace imposed by the automated conveyor, it is not possible to keep perfect balance in all workstations on the line.
- Waste of resource capacity during product model changeover. The longer a task time becomes when dividing the entire line into multiple workstations, the more time-consuming it is to deal with product model changeover readjusting and realigning tasks at each workstation to balance the entire line might require many technological and operational changes, and can take a long time. The Mass production, conveyor paced design generally does not allow for handling multiple stations; so there is also waiting time for workers operating partially-automated short-cycle processes. Manufacturing with conveyor lines have other types of systemic problems such as:

- The line only runs where all stations (and operations) are available and ready to operate
  - The line can be ergonomically unsound with adoption of predominantly one-sided physical motions.
  - Considering the conveyor line system from another point of view, larger lines are very problematic when production volumes decline. Given the huge investments needed for

the installation of a conveyor line, management wants high utilization rates, and quite often these result in large work-in-process inventories and perhaps large final product inventories.

- Reliance on huge and expensive investments in facilities
- Reliance on indirect and support personnel who generate to added value
- Not fully utilizing the workers' intellectual capacity, and
- No flexibility of the production facilities.

This last problem, the lack of *flexibility*, is really the main disadvantage of Conveyorized Mass production lines.

- Product model design changes; introduction of new products; and changeover of work holding devices like jigs and fixtures are costly and time consuming.
- Reconfiguration of the design to respond to changes in demand is extremely difficult, requiring rebalancing of the entire line.
- Technological innovation which will result in shorter cycle times at each workstation will require line rebalance

5.9 Converting Conveyor Lines to Lean Cells

An example of converting a conveyor fed serial production line into a U-shaped, Lean production cell (from Sekine) is shown in Figure 5.6.

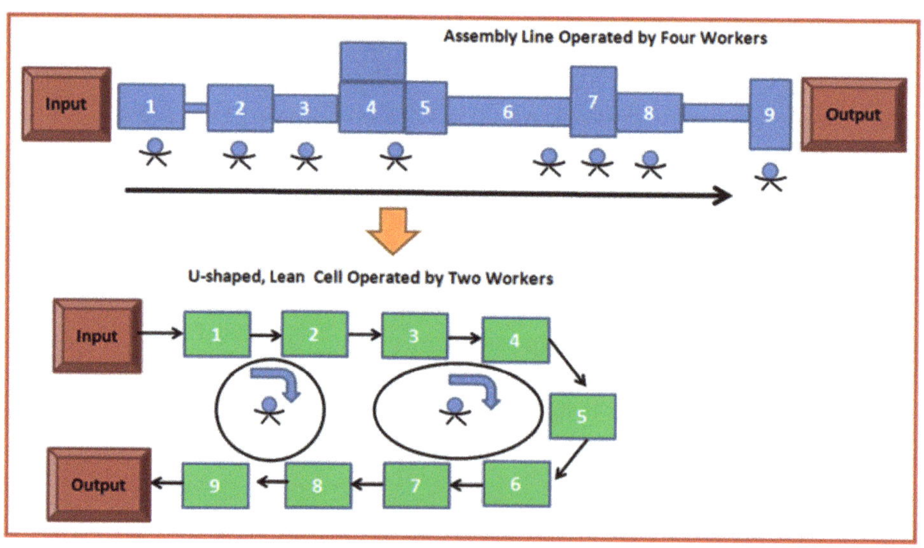

Figure 5.6
Converting A Conveyor Line Into A U-Shaped Cell
Sekine. Modified by Phillips

The original Mass production line used 10 workers. After conversion and redesign, the same product was made using only 4 workers in Cell 1 and two workers in Cell 2. The first cell uses a cell staffing methodology called *rabbit chase*, The second cell design uses only two standing, walking workers who work in two sub cells (loops). The Lean cell with only two workers was made possible by modifying the four presses and the plating process. An *inspection station* was added that is shared between the two workers. Note that the *shaft BB* station was a short cycle operation that was also shared by both workers. The details of this conversion were made proprietary by the company, but the results of Lean conversion are impressive. These are examples of self-balancing subassembly (SBS) cells.

The output can be changed by changing either the number of workers in a cell, or by reassigning machines to each worker. There is no need or requirement to equalize and balance the processing time cycles at individual stations. The operations in a Lean, U-shaped *assembly* cell are usually all manual so the operator must stay at the station until the task is completed. In a *machining* cell, the machining cycle time is separated from the workers loop time. The worker can start a machining operation and it will automatically run to completion. Of course, cells usually have a mixture of manual assembly operations and automated machining operations. Converting from Mass assembly lines to Lean cells generally results in a marked improvement in important measures of productivity such as lower cost, lower cycle times, higher output and less throughput time. Contrary to large lot production and long setup/changeover times reflected in traditional manufacturing, long setup/changeover times must be vigorously attacked and reduced so that the processes can be changed over quickly from making one product to making another. This makes the manufacturing system *flexible,* and enables manufacturing cells to make piece parts and subassemblies for Mixed Model final assembly. Again, Lean cells are designed to manufacture specific groups or families of parts.

Chapter 6
Push and Pull Systems

__Push, Push, Push__ to the limit, never quit for even a minute, just do it, do it!
__Push, Push, Push !!__ Unknown Poet

6.1 Push vs. Pull: What are the issues?

The purpose of this textbook is to create a framework for designing, implementing and managing a Lean Manufacturing System.....A Factory with a Future. A cornerstone of Lean Thinking is the deliberate and aggressive migration from traditional Push Manufacturing philosophies to Pull Manufacturing philosophies. But: *What exactly are the basic concepts of push and pull?* In this Chapter we will try to de-mystify these two terms and show why pull systems are superior to push systems by almost any measureable yardstick.

If you ask any student who is studying manufacturing systems or the average manufacturing shop floor worker, they will acknowledge that they have heard of *push vs. pull*. However, if the issue is pressed to explain why one is superior to the other and fully discuss how... you might get 10 different answers from 10 different people. In recent years, the authors of this textbook have observed a growing body of confusion and diversity as to how a pull system can be defined and what actually makes it work. In a recent technical presentation, a practicing *Lean expert* related a story of how he posed this question to a line foreman in a company which had only recently began a lean movement and started to implement Pull systems. The foreman's response was short and to the point.

"I don't do nothin at any time unless someone has authorized me to do it with a card"

Amusing as this may seem, it is indicative of how some professionals view pull systems. This prompts us to address the following set of questions.

- What is a formal definition of Push and Pull systems?
- How does Pull actually improve system performance measures?
- Why is Pull superior to pure Push methods?

To fully understand and answer these questions it is appropriate to quickly address the historical development of Push vs. Pull, and try to unify concepts.

6.2 Historical Perspectives

In their landmark textbook, *The Machine that Changed the World*, by Womack, Jones & Roo, the authors put forth the concept that the modern automobile was the catalyst that changed the entire landscape of manufacturing. We all agree that the father of mass production and assembly line manufacturing was Henry Ford. His dream was to build an automobile which any American citizen could afford to own. Currently, the average American household has 1.7 cars, so someone must have been successful. In order to accomplish this goal, he developed the *modern mass production facility* built around an *assembly line*. Ford constantly sought ways to produce a high quality product in a minimum amount of time at the lowest cost possible. In this respect, he was the first *Lean Engineer*. If Henry Ford was alive today, he would undoubtedly be at the forefront of Lean Thinking, and a champion of Lean Engineering principles. Womack, Jones and Roo were correct, the automobile and Henry Ford did change the manufacturing world, but in his quest to achieve manufacturing efficiency Ford failed to recognize that the customer would eventually expect a wide range of choices and options to suit individual wants and needs. Ford's philosophy was: *I will sell you any color of automobile that you might want as long as it is black*. This short sided philosophy was what caused William Durant in 1920 to purchase several automobile manufacturing firms and incorporate General Motors. Durant added diversity and choice to the automobile, and the race was on. Today, there are literally thousands of shapes, sizes and colors for the consumer to choose from. The system of *Mass Customization* that exists today was built upon the shoulders of Ford & Durant, but as complexity and variety increased the role of Production Planning and Inventory Control increased proportionately. It is our belief that the *Machine that Transformed the Manufacturing World* was undoubtedly the advent of the mainframe computer and desktop computing. Without the local and distributed powers of desktop computing, coupled with globalization, managing a modern Lean Enterprise would be a daunting if not impossible task.

To fully understand the pivotal role of Pull Systems, it is necessary to understand how distributed computing was an enabling factor. Prior to computerized Production and Inventory Control, manufacturing systems were largely controlled by individuals in remote offices (sound familiar?) using inventory control techniques such as Reorder Point (ROP) and Reorder Quantity (ROQ) methodologies (sound familiar?). As computational power increased and things got increasingly complex, enter a new corporate strategy called *Material Requirements Planning* or MRP. MRP Fueled by IBM and the APICS Society, the MRP movement swept through manufacturing like a hurricane. By the mid-1980s, yearly sales of MRP software and technical support exceeded $1 Billion annually. MRP was a good idea. It attempted to integrate three Production Planning Functions into one package: (1) The quantity, type and mix of raw materials needed to support production, (2) the schedule of when to release production requirements and (3) the timing of release to operational components. MRP established time-sensitive production completion dates for each manufacturing component based upon both forecasted and actual demand. A bill-of-materials usually accompanied these releases specifying what resources were required to support production, when they were needed, the quantities required and a production schedule. As good as these sounds, initial MRP systems either failed to recognize or ignored the shop floor status of machines, facilities and personnel; and virtually ignored real-time production capacity. Worse, they depended heavily upon achieving an accurate forecast of product demand. In a nutshell, initial MRP systems initially failed to address several critical Manufacturing and Inventory Control functions.....and it was terribly labor intensive and expensive.

To plug these gaps, Oliver Wight in the late 1980s proposed an expansion of MRP to accommodate all of the functions of the manufacturing enterprise; including product costing, capacity planning, inventory Control, order release and scheduling. While this new system might sound a lot like MRP, it went far beyond initial vision and implementation. The new system was envisioned as a corporate planning tool to execute ALL of the production planning and control functions of the manufacturing enterprise, a goal which was fueled by distributed processing and LAN lines. Wight proposed that the new system be called MRP-II or *Manufacturing Resource Planning* ...and the new name stuck. MRP-II was soon steam rolling American manufacturing, but it became evident that to support and use such a monster required serious capital investment and integrated, real-time data. Today, MRP-II has evolved into what is called Enterprise Resource Planning (ERP).

Both MRP II and ERP require millions of lines of computer code, large corporate staffs and a virtually unlimited array of *plug-ins* to support applications. Like its forerunner MRP, MRP-II and ERP systems attempted to unify three important components of manufacturing management and control (Sipper and Bulfin). These are broadly speaking (1) Management planning and coordination (2) Operations planning and (3) Operations execution strategies. The *management planning* functions are strategic in nature but include product development, economic analysis, workforce planning and capital acquisitions. *Operations planning* involve sales, forecasting, production planning, production control and order release. *Operations execution* involves shop floor control, sequencing, scheduling and WIP management. Business objectives are translated into corporate goals, corporate goals into production planning and production planning into execution. The lines between strategic and tactical planning were ideally meant to be transparent. All Manufacturing Planning systems attempt to establish a forecast for product consumption and set due dates for product delivery; using a break-back, hierarchical planning module to determine when each step of the manufacturing process needs to be completed, what raw materials are required and when to release work to the shop floor with appropriate shop floor instructions. The product start date is determined by estimating the product due date, and then subtracting the manufacturing, waiting and move times plus any *lead time* factors. MRP and MRP II both had an important *Achilles Heel*....Forecasted demand needed to be accurate and current shop floor status had to be known in real time. When product is sold to a customer it triggered a critical sequence of strategic planning activities. The customer delivery date was called the *Product Due Date*. Three important activities would then take place: (1) Raw materials would need to be procured and delivered to point-of-usage where they would be transformed into parts, subassemblies and final assembly (2) Production resources (machines and people) would need to be scheduled and (3) A product routing sheet with assembly and manufacturing instructions developed. A critical question to be answered is :*When should the order released into production?* Of course, the answer is easy to define but difficult to determine. The cycle time from start to finish is as follows (Simplification).

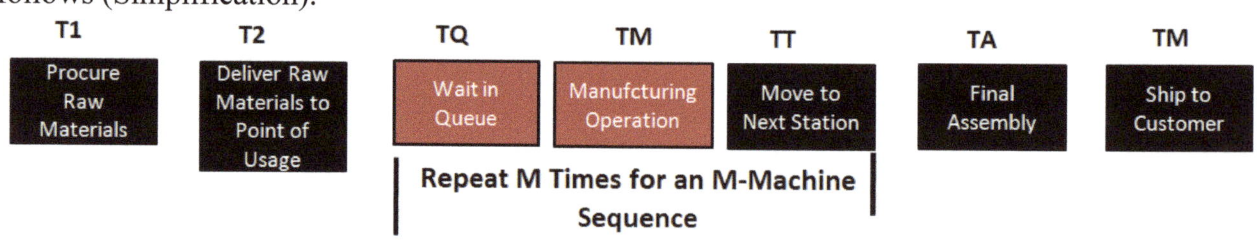

Figure 6.1
Manufacturing Cycle time

This seems straightforward enough, but it requires a lot of data. In addition: What if a material Handler is busy? What if a Machine is down? What if a worker is not available? It may be necessary to accumulate raw material to start production on time and Finished Product inventory to facilitate On-Time-Delivery? The Cycle time for an order is critically dependent upon Queue Time.

In addition to requiring accurate, up to date, shop floor data; this planning cycle time is subject to our old enemy ***variance*** and Murphy's Law...If something bad can happen, it will....at the worst time. The product flow time or throughput time is how long product actually takes from order release to final delivery, and this can be significantly different than the planned time. The result is slipped schedules, reprioritization, expediting, variable lot sizes, lot splitting and scheduling overtime to make up for missed due dates. Once released, work usually flows through the system as fast as it can from one operation to the next, *pushed* from Workstation j-1 to Workstation j , sometimes regardless of WIP status. Customer orders are typically released and processed in large lots, or in split-lots. Hence, MRP/MRP II systems have become known prototypically as *Push Production Systems*. To visually summarize:

A Shop floor worker is assigned to a machine and Raw Material is delivered to the first machine in sequence. The Operator begins to transform the raw material into finished parts (finished at that machine). Eventally, a bin of parts is filled.
He then places the part in a Material Handling device where it will be moved with other parts to the next operation in sequence.

This will immediately create a large Queue at the next processing station.

The worker and his machine will transform the raw material into an interim part and the process step is started over again. Evevtually the Manufacturing Lot size is completed and parts stored in a transport bin.

Material Handling is then called to move the full bins to the next operation in sequence.
This is repeated at each workstation in a manufacturing Process. This is repeated at each Workstation and then Final Assembly will complete the item purchased. It is rare that the final product will be immediately shipped to the customer, so the order is usually stored in an ASRS at *Final Goods Storage.*

Due to all the uncertainties in filling a customer order on time, an **A**utomatic **S**torage **R**etrieval **S**ystem (ASRS) might be used to store finished goods as a buffer to quickly fill customer orders, and then store the new batch to replace the old. More inventory....More waste.

6.3 The Toyota Production System Arises

As MRP-II and ERP began to define American manufacturing, there were other things developing across the *Big Pond*; particularly in Japan. Supported by a universal drive to higher quality products by Juran, Deming, Feigenbaum and others (America initially rejected this group), the Japanese automobile industry set their site on gaining a share of the American Automobile Industry. As late as 1955, the *big three* American automobile companies accounted for 95% of approximately $7 billion in annual sales. The Japanese had limited resources, and were not as yet on the cutting edge of computer technologies worldwide. The Japanese were not interested in controlling production in *corporate offices* but wanted to develop a shop floor system that would rely heavily upon skilled labor, shop floor status and customer demand.......not forecasted schedules or make-to stock policies. A deliberate decision was made to empower the average worker to manage and execute shop floor control.

Enter two historical figures that more than any other force were responsible for a new industrial revolution called the *Toyota Production System* (TPS). These two individuals were Mr. Taiichi Ohno and Mr. Shigeo Shingo of the Toyota Automobile Corporation. To be sure, they were supported by a talented set of dedicated manufacturing engineers, but it is generally conceded that Ohno should be credited with conceptualizing the TPS, and Shingo was the engineering genius who made it work. It is recorded that Ohno struggled for some time to develop a system which would/could be executed by people on the shop floor, and would quickly respond to system status and resource availability. The story goes that he visited America to get fresh ideas, but found American Manufacturing mired in a *push world* with excessive work-in-progress. He also observed that enormous resources were being consumed to develop and execute *super* manufacturing machining centers which were very expensive and highly automated. One day he visited a modern American supermarket and was impressed by how goods and services were supplied to the customer in real-time, and groceries replenished according to customer induced stock levels and not ROP/ROQ rules. Customers would take only what they needed in the quantities desired from shelf stock; variations in customer demands were buffered by *Supermarket storage areas* at the store level; and the store would order only what was needed when it was needed from a host of outside suppliers to replenish low Supermarket inventories. The customers essentially ***pulled*** items through the system according to whatever the demand pattern might dictate, including seasonal variations and holidays. Hence, Ohno galvanized his thinking along one line of thought...*pull not push.* With a push system, work-in-process is released into and moved through the system according to a predetermined schedule ...*damn the torpedoes (system status), full speed ahead.* Conversely, in a pull system, manufacturing activities are in *theory* initiated at the final stage of the process, hopefully in direct response to a customer request. There is no forecasting, scheduling or *slack time* used. Ideally, parts or groups of parts (small lots) are moved only when needed in response to shop floor conditions. Hence, the term *Just –in-Time* became almost synonymous with pull manufacturing. As Ohno later admitted, the term Just-in-time was fabricated to confuse the real meaning of what we now call *Lean Thinking*, which embraced all of the principles described in this book. JIT can be implemented in many ways without a pull system, but a pull system is strategically dependent upon JIT philosophies. The entire set of production, management and control strategies became known as the Toyota Production System (TPS). Many of the elements of TPS are discussed in this book: TQM, TPM, Poka-yoke, SMED, 5-S, PDCA, Autonomation, Kaizen events and Jidoka to name a few (See Appendix A for concise definitions of these terms)

6.4 Operational Characteristics of A Pull System

In a typical *Push System*, customer orders are released to the Manufacturing System and follow a predetermined *Routing Sheet*, they are *pushed* as fast as possible through the system regardless of System Status. If an order gets behind schedule, it is often expedited or reprioritized....sources of variance and unplanned set-ups. In a *Pull System*, the final assembly is scheduled and a request is made to all previous workstations to deliver the parts or subassemblies needed for final assembly. This request and all subsequent production is driven by some signaling system.

The production control vehicle to achieve these goals is called a ***Kanban***, which is usually associated with a card system, but can in fact use lights, colored balls, or any other communication device to trigger shop floor production, material handling, raw material supply and a host of other functions. Some modern facilities use LAN lines and computer terminals. The Japanese word *Kanban* literally means a *visible record* or a *card*. There are many types of Kanban Cards but only two will be described: one which requests **P**roduction (P-Kanban Cards) and one which requests **M**aterial movement (M-Kanban Cards).

Move Kanban Card (M-Card)

Move From:
Lathe 62
Area 5
Move to:
Drill 23
Area 6

Part Name:
Right Wing 20
Part Number:
X408-99

Carrier Capacity:
10

Production Kanban Card (P-Card)

Machine:
Drill 23
Area 5
Manufacturing Instructions
Standard P-45

Part Name:
Right Wing 20

Quantity:
10

Carrier Capacity:
10

There can be many pieces of information on a Kanban Card, and only a few representative fields are shown above. On a M-Card, The key information is origin and destination. On a P-Card it is how many parts or subassemblies are to be made and the Part Name/Part number. Pull systems using Kanban cards are a way to control and coordinate shop floor activities such that both WIP and Throughput time is reduced. Quoting Hopp and Spearman, there is a fundamental difference in push vs. pull systems.

"A push system controls order release to the shop and material movement based upon schedules, while pull systems control order release and material movement by system status".

An important principle which distinctly separates a Push system from a Pull system is that a Pull system by design puts a ***WIP Cap*** upon all Manufacturing cells and all manufacturing work centers. Within each cell, there is usually no more than (2*M +28 W) units of work in an M machine cell with W worker loops. W<M. This results in a significant reduction in WIP (Work within Manufacturing cells). There is also a corresponding reduction in Part Flow Time (CT) by Little's Law. A WIP Cap and Pull Production is what defines the *Magic* in Lean Production Systems.

Through time, many cards were developed to trigger manufacturing activities, including conveyance Kanbans, signal Kanbans, SP-Kanbans and a host of others. The interested reader is directed to the excellent textbook of Nicholas for a comprehensive discussion. The issue of lot sizing is an operational issue, and will not be covered here. Drawing heavily from the work of Suzaki, Hall and especially Nicholas we offer the following characteristics of a pull production system.

6.5 A One-Card Kanban Scheduling and Control Methodology

This example will describe how to use only one Kanban card for the schedule and control between two Workstations in a Manufacturing system. The basis assumption is that a single type of finished product is being sold. Consider the following diagram of two production workstations in series.

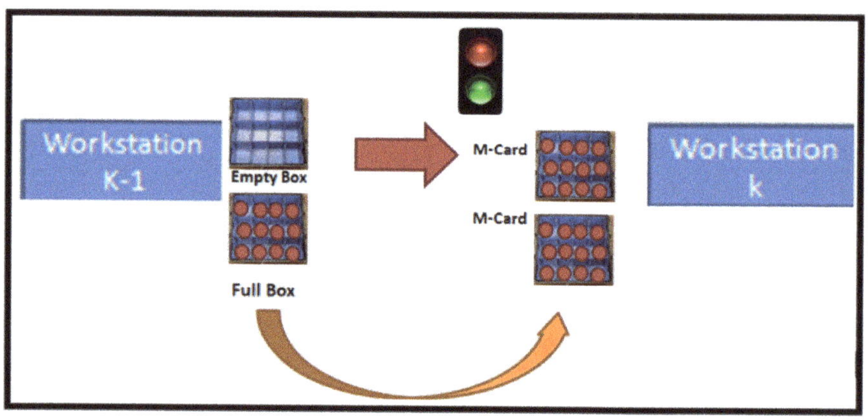

Figure 6.2
Two Workstations in Series

Workstation K-1 produces parts for Workstation K, and Workstation K produces parts for Workstation K+1, etc. Each Workstation could be a multipurpose machine or a Lean Manufacturing Work cell. There is normally a Kanban Station where either M-cards or P-cards or both are posted to authorize movement and/or production of parts, but we will use another type of sign in this example...a stand with a light that can turn either red or green. Since there is only one finished product being manufactured, the manufacturing sequence at Cell K-1 and Cell K is known in advance. Because the rate of production is set by final assembly, both workstations produce to Takt time as dictated by Final Assembly. All that is needed is authorization to move empty and full carts between the two Lean Cells to control production. The single card which will be used in this example is an authorization to move a full bin of parts at Station K-1 to Station K, or an empty box from Station K to Station K-1. We will call this card an M-card (Move card).

Suppose that a workday begins with one empty box and one full box at Workstation K-I, and two full boxes at Workstation K (Figure 6.2). Both boxes at Cell K have a **M**ove card attached to them (M-Kanban card) and both boxes at Workstation K-1 have no **M**ove cards attached to either box. A full box at Workstation K-1 cannot be moved to Workstation K without M-card authorization. This is accomplished as follows. Workstation K-1 and Workstation K begin Manufacturing at the same time. Workstation K-1 will begin to fill up the empty box, and Workstation K will begin to deplete the first full box. The fill rate at Workstation K-1 and the usage rate at Workstation K should be about the same, since both are driven by Takt time in final assembly. At some point, Workstation K will use up all of the parts or subassemblies in the previously full box. When this box is empty, the operator responsible for Workstation Input will move the empty box and its M-card to a Material Handling station and turn on a red light. This is a signal that an empty box at Workstation K needs to be moved to Workstation K-1. The Material Handler picks up the empty box at Workstation K, turns the Red light from red to green and begins his trip to Workstation K-1. The M-card will specify where the empty cart is to be delivered.

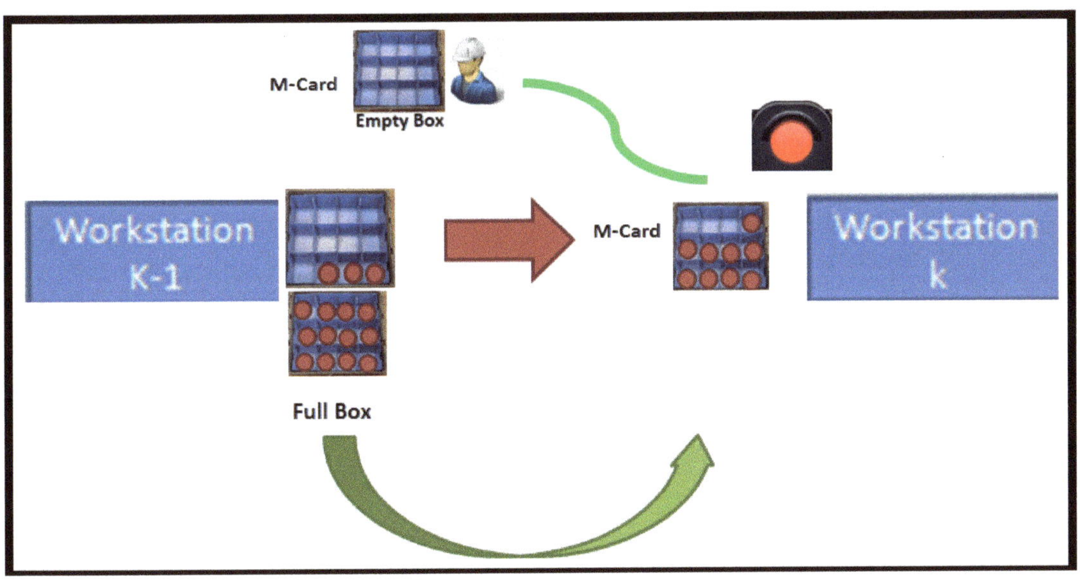

The Material Handler will replace the *Full* Box at Station K-1 with the *empty* box from Workstation K (1) He will remove the M-card from the empty box and attach it to the full box to authorize its move to Station K, and then (2) Move the full box to Workstation K. The destination of the *full* cart is also given on the M-card. During the cycle of time when the empty box is removed from Workstation K and the full box is delivered from Workstation K-1, some parts might be manufactured and placed into the empty box at Workstation K-1. In like manner, some parts might be removed from the new empty cart at Workstation K. This may not always be the case.

The 2 Workstation- linked System should now be in its initial state, and the procedure is repeated .

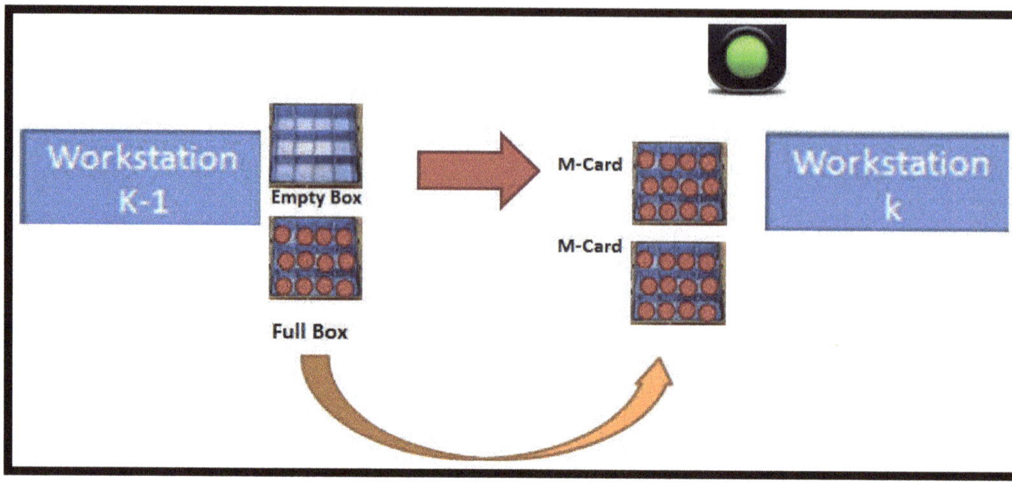

The one card system is a paradigm of simplicity, but is recommended for stable production where Workstation K and Workstation K-1 are not a great distance apart. As usual, no defective parts can be produced and machines must not fail at any time. The production rate at each Workstation is determined as usual by the customer Takt time. No finished parts at Workstation K-1 can be moved anywhere without a M-card. This system has a very tight control on WIP, which in this case can newer exceed the amount of parts or subassemblies in three full carts, plus the amount of work in either stand alone machines or in Lean Cells. Overproduction or underproduction cannot occur.

6.5.1 Numerical Examples

A customer has ordered 450 items with a 30-Day delivery date. The necessary raw materials will take 3 days to deliver, and the order must be shipped within 2 days after the last part is completed. This leaves 25 days to manufacture all of the parts. The plant operates 5 days per week , 10 hours a day. It takes 30 minutes to start up every day and there is a 30 minute lunch break. This results in 45 hours of production time each week. The production line must make 18 parts per day or 2 parts per hour. Takt time for each Machining Center or Manufacturing cell is 1/2 hour per part. The Lean Engineer decides to use a box to move parts or subassemblies between Cells that will hold two parts or subassemblies and a 1-card Kanban system to schedule and control part production and movement between cells.

Production will begin and Workstation K-1 will be initially authorized to produce 3 full boxes of parts or subassemblies, one for Workstation K-1 and two for Workstation K. Assume that A production rate of 6 pieces per hour is required by Final Assembly at both Workstation K-1 and Workstation K. Takt time for both workstations is 10 minutes per part. From when Workstation K requests another full cart from Workstation K-1 to when it arrives is 15 minutes and that each Material Handling box holds 5 units. Workstation K will deplete a full box of 6 units in about one hour and will request a Material Handler by turning on a Red light. Fifteen minutes later a full box of parts will arrive with an M-card attached. This sequence of events will repeat itself until all production requirements are satisfied.

6.5.2 A Two-Card Kanban System

The most popular Kanban Control is to use a two-card system. One card is used to authorize production (P-Card) and the other card is to authorize bin movement (M-Card), either an empty box from Cell K upstream to Cell K-1 or a full box downstream to Cell K from Cell K-1. Consider the following diagram of two Lean, U -shaped cells in series. Workstation K-1 is producing parts for use at Workstation K. The following diagram shows the output of Cell K-1 and the input of Cell K. There are two Kanban *Mailboxes*, one at Cell K and the other at cell K-1. Each Workstation or cell has both an input and output station. and both each workstation will have 2 carts. Workstation K will have two full carts at its input station (Green box) and Cell K-1 will have one empty (Red box) cart and one full cart (Green bux). No production can begin at Cell K-1 until it is authorized with a production P-card. No full container at Station K-1 can be moved from Cell K-1 to Cell k unless authorized with an M-card. The following diagram illustrates the initial state of a two station system. Station K-1 is making parts or subassemblies for Station K. Both Station K and Station K-1 are actually Lean U-shaped cells and both are producing to Takt time determined by final assembly.

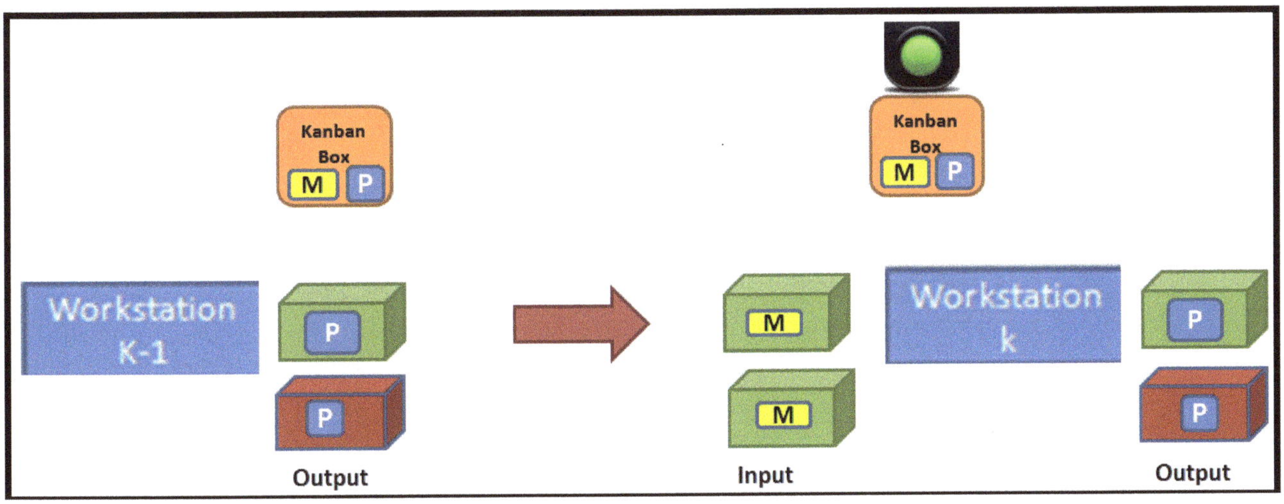

Step 1: When Workstation 2 begins work on a full box he removes the M-card from the full box and places it on the K station Kanban board. This Kanban card specifies the amount of parts or subassemblies needed and what previous workstation is supplying the parts. At approximately the same time (both workstations both operate using Takt time) Workstation K-1 has an empty box on its output side; the operator removes the P-card from the empty box and places it on the Workstation K-1 Kanban board.

Step 2: At some point in time the first full box at Cell K is depleted and a red light at the Kanban Box is turned on to request a Material Handler. When the Material Handler arrives, he picks up the empty box, goes to the Kanban board, attaches the M-card to the empty box, turns the red light off and a green one on and moves both to Workstation K-1. At Workstation K-1, he removes the P-card from a full box, attaches it to the empty box, and replaces it with the M-card on the empty box.. The Kanban M-card specifies the destination and location of the Workstation that needs the parts.

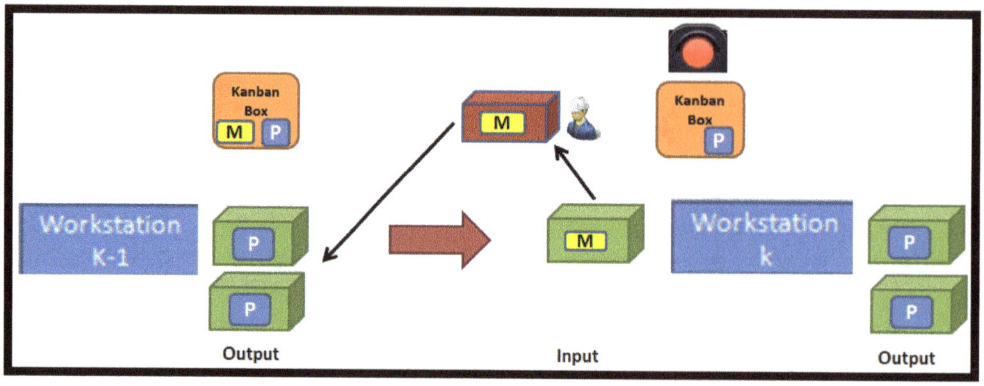

He then moves the full box to Workstation K and the procedure is repeated. Workstation K is controlled by the same Kanban control system with Workstation K+1 .

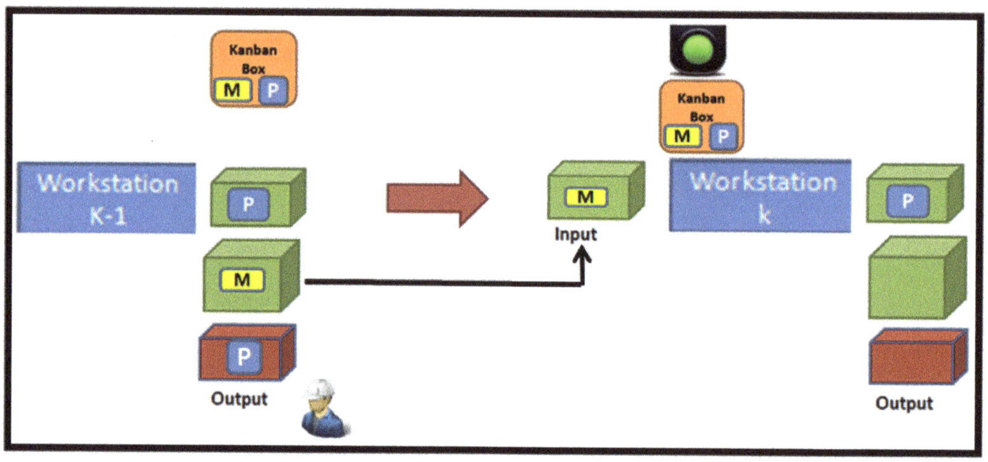

The entire production cycle is scheduled and controlled by Kanban cards. There can never be any production unless authorized by a downstream Workstation and there can never be raw materials moved from an upstream workstation to a downstream workstation unless authorized. If anything happens at any one workstation to shut the station down, everything automatically comes to a halt. WIP is tightly controlled by a WIP-Cap and Throughput Time is automatically reduced by Little's Law.

7.0 CONWIP: An Alternate Methodology to Cap WIP and Reduce Cycle Time

It should be clear from previous discussions that there is a fundamental difference between how Push and Pull systems are designed and how they operate. To emphasize these differences we offer the following definitions of a Push system versus a Pull system..

Push systems try to schedule production according to demand (Forecasted or actual), and execute order release to the manufacturing system based upon a Master Production Schedule which attempts to schedule production based upon a break-back structure with appropriate manufacturing, move and slack time components. Slack times are used to buffer uncertainty. Push systems specifically try to dictate throughput and observe the resulting WIP levels.

Pull systems schedule to intermediate or final assemblies based upon customer demands (required production rate or its inverse, called Takt time) and strategic Make-to-Stock policies. All pull system manufacturing or move activities are authorized by coordinated signals such as Kanban cards. Pull systems strictly enforce a WIP Cap, and control throughput by adjusting the WIP cap and Manufacturing Cell labor requirements to satisfy demand (Takt time) requirements.

Push systems typically release work to the shop floor in large batches. Order, production and move batches are often the same. Material movement occurs by pushing work in progress through the system as an opportunity arises, as quickly as possible. Most of the problems associated with push systems are twofold. *First*, all operations are scheduled in advance and will be executed as soon as possible. If an order gets behind schedule, the order may be split, reprioritized, *red-tagged* or expedited. The second major problem is that any of these remedies create variance components which corrupt every operation; decrease throughput; and increase WIP. Conversely, in a pull production system there are no pre-determined move schedules. Product is pulled from Workstation to Workstation based upon resource availability and system status.

The key issues are WIP control and WIP levels. By Little's Law, for a designated input rate, an increase in WIP will increase throughput time. The ***Magic*** of a pull system is the ***WIP cap*** which is strictly enforced by Kanban. The amount of WIP in any pull system can never exceed the amount of WIP allowed by the amount of product authorized by Kanban. A typical push system like MRP/MRP II will attempt to release work to the shop floor based upon a set schedule and assumed shop floor capacity. If too much work is released and/or capacity changes, a WIP explosion will occur with associated increase in cycle time. If too little work is released, then starvation can occur at bottleneck Workstations. A properly configured and managed pull system cannot exceed a WIP cap, and work will not be scheduled to either be moved or to start processing unless specifically authorized by shop floor personnel. There is simply no scheduling required except at final assembly. Sequencing should follow predetermined MMP strategies if there are multiple products; workload leveling, Workstation balancing, predetermined product mix sequencing and a FIFO discipline should always be strictly observed.

CONWIP is a term coined by Spearman. CONWIP attempts to leverage the best properties of a push system with those of a pull system. CONWIP is based upon what we would call a *Closed Product Flow Network* (CPFN). A CPFN is a manufacturing cell, system or subsystem which is only authorized to hold a specific amount of WIP, and as with all pull systems that level can be changed but cannot be violated. A CONWIP CPFN is shown below.

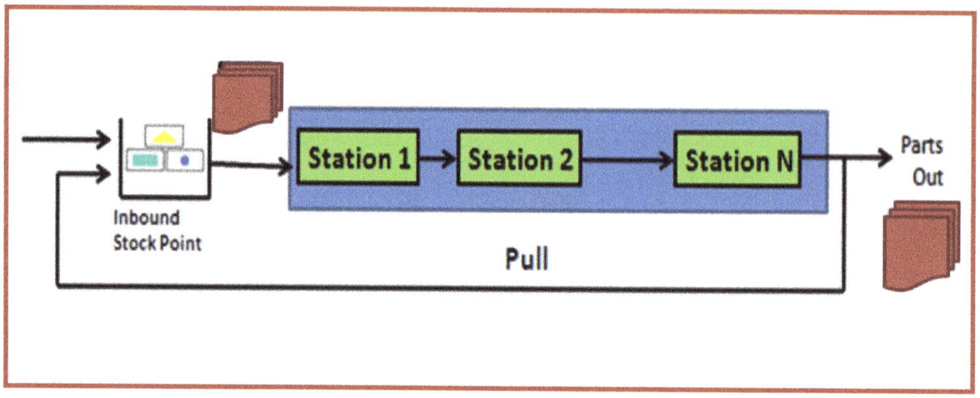

The role of the Kanban cards and the way product arrives to the system serve to define whether the system is operating as a classic Push system or as a classic CONWIP system. A push system would assume that there are an infinite number of Kanban cards, and that parts arrive for processing according to some order release strategy. If the arrival stream is random and Poisson, and each Workstation service time density function is Exponential, then we have a classical three Workstation serial push Markovian network. Usually, an MRP driven order release system follows non-Markovian behavior. To fill outstanding orders, the raw materials and the customer order are usually released to the system and production is executed on large lot sizes. The lots travel from one workstation to another as soon as possible. Large amount of WIP and excessive throughput time is a direct result. Orders that have been released to the system and fall behind in expected flow times are often reprioritized, creating excessive setups and changeovers, and introducing variance into the system. Note that in any push or pull system ,assuming that no product is lost to poor quality, the system output rate is always the same as the system input rate. A CONWIP system is a hybrid push and pull system that attempts to take advantage of the good properties of each system. Consider the following N-Station assembly line. The amount of WIP allowed in the entire line is *capped* by restricting the WIP to a maximum allowed level specified and controlled by WIP cards. Assume that engineering has decided to WIP Cap the entire assembly line to W units of work. W cards are posted at the input station, and each part which enters the production line is assigned a card. When that part finishes, the card is removed...sent to the beginning of the line....and assigned to another part. Clearly, no more than W parts can be in the line at any one time. This is a card-controlled Pull System. Once a part enters into the line, it moves from station to station in a Push System. There will be one station in the line which controls line production, and that is the *Bottleneck Work Station*. This station will continually be busy and will control the output rate. Work in Progress will not all pile up at the bottleneck workstation because of the WIP Cap...It will distribute itself among the remaining N-1 workstations.

A CONWIP system behaves in an entirely different way than either a pure Push system or a pure Pull System. To reiterate, a typical CONWIP system might operate as follows. Assume that there are only **w** unit release Kanban cards available. A part cannot enter the system unless authorized by a Kanban card. As long as cards are available, a part can seize one and immediately enter the system. When there is no Kanban card available, no part can enter the system. When a finished part leaves the system, a Kanban card is released back to the inbound stock point where it is available to authorize another part to enter the system. Once a part enters the system at Workstation 1, it is pushed through Workstations 1-3. If there are **w** Kanban cards authorizing unit entry to the system, then system WIP is bounded by **w**. A relevant question is; *where will WIP be in this system*? After a short period of operation the work-in-process will tend to gravitate or spread to each of the Workstations according to system service parameters, bottleneck processes and other interactive parameters.

Hopp and Spearman in their book *Factory Physics* show that a CONWIP system is more *efficient* than a corresponding pure push system. Being efficient simply means the ability to consistently produce product with shorter throughput time and at a higher release rate than a push system with the same number and type of workstations. System WIP is determined by the capacity to produce and the differences in maximum output rate at each Workstation. Note that in a pure push system, parts arrive to each Workstation at a rate which is difficult to control once an order is released. In a CONWIP strategy, parts are *pulled* through the system at a rate which is set by demand and available working time (Takt time). If one is to compare these two systems, the concept of operational efficiency can be formalized as follows.

> ***For a given push system configuration with an observed output rate, the same output rate can be achieved with a lower WIP level using a properly configured CONWIP system.***

Hopp and Spearman have shown that a CONWIP system with a specified WIP cap will always dominate a corresponding Push system with infinite WIP provided that the CONWIP system is never starved. This is easily accomplished by providing some work- in- process at the front of the CONWIP line. In addition, increasing WIP by either fiat or by push philosophies beyond Critical WIP will result in both unnecessary work in progress and a corresponding increase in system throughput time. This is a killing blow to any form of MRP scheduling system that release orders to the factory in a manner that exceeds Critical WIP. This is another example of the *Magic of Pull*…it is not so much the pull system itself as it is the *WIP Cap*. Finally, as previously observed, a WIP cap will obviously result in fewer requirements for expensive floor space, and less capital investment in inventory and raw materials. As already stated, another advantage is that a closed loop Pull system will by design exhibit less variance than a corresponding push system. This will be reflected in the squared coefficient of variation for time between part departures for the CONWIP system.

CONWIP systems are easier to implement and adjust, since only one set of system cards is used in any one production line to manage system WIP. Hopp and Spearman also state that, in general, CONWIP systems exhibit less variability in Throughput times, which is a persistent goal of Lean Engineering. This benefit is a direct result of negative correlation between WIP levels at each Workstation.

8.0 What is a Pull System Mathematically ?

Now that we have characterized, discussed and defined two different types of Pull Systems, what is it that is so special about this methodology? By construction, a typical pull production line is nothing more than a variation of what academics refer to as a *closed queuing system* with finite WIP, as opposed to an *open queuing system* with infinite WIP. Why would the Toyota system strictly limit the amount of WIP? The answer is clear if we again appeal to Little's Law. Little's law can be expressed as follows.

$$WIP_{sys} = \text{Throughput Rate} * \text{Time in System}$$

In a stable system , Input always equals output and the Throughput Rate is the required Production Rate. Takt time is 1/(Throughput Rate). Since the Throughput Rate is fixed, Time in the System is directly related to WIP. In other words, if you want to decrease the time in the system, decrease WIP. Taiichi Ohno did not use Little's Law when he installed the Toyota Production System, but he was intuitively aware of its theoretical implications. Lean production Systems and the use of Kanban Cards to limit WIP and control production are not only sound operationally, but have a solid theoretical basis.

To fully understand why a Kanban -based Pull system is always superior to an MPR -based Push system in terms of Production rates and Throughput Time, we again state that ideally nothing is allowed to be produced or moved without authorization, and the amount of material in the system cannot be more than the sum of all the material specified on all Kanban cards. We build on Hopp and Spearman to list several benefits of a finite WIP system.

(1) Lower WIP and shorter cycle times are achieved by limiting the total amount of items in the system. By Little's Law, for a fixed (demand) rate this translates into shorter cycle times.
(2) By strictly controlling the amount of WIP by product and material type, a degree of production smoothing is achieved and excess inventory is minimized.
(3) A system with fewer parts in process, generally promotes visual control and higher quality levels. Large levels of WIP often bury parts ready to be moved or processed.
(4) High WIP always translates directly into longer throughput times and usually promotes expediting and reprioritization.
(5) Less WIP and shorter cycle times directly translate into higher profits. By any Economic Lot Size formula, the cost of raw materials, storage space and carrying cost all go down. Of course, this will be directly reflected in lower lot sizes. Note that to achieve this goal, lower set up times are necessary.
(6) Less WIP and smaller lot sizes are usually reflected in more predictable Material Handling times
(7) Excessive WIP usually results in the shop floor personnel working *faster* not *smarte*r because of a perception that *hurry up andcatch up* is needed.
(8) Traditional monolithic monsters called "ASRS" systems are more easily controlled and managed at lower inventory levels
(9) Less WIP means less congestion and more working space

One might argue that these benefits can also be achieved with push systems if time and effort is applied. This is fundamentally true, but the very nature of push systems makes it difficult to achieve the previous benefits. The real secret is in the one fundamental difference in operating strategies.

A Kanban pull system which strictly limits the amount of product in a system automatically provides a "WIP Cap", while the traditional push system does not

In short, a WIP Cap *prevents a WIP explosion.* In its simplest terms, a Lean Pull system will *not* authorize production or material movement until it is asked by upstream or downstream processes. Hence, WIP is precisely maintained at predetermined levels throughout the entire manufacturing system. Further, the concept of *pacing the production process* at a rate called Takt time regulates production in response to customer demands. Lean Production means doing things at the right pace, at the right time in the right way. Having said this, one might infer that production in a Lean system is strictly *make to customer order*, but in reality this is almost impossible except in dedicated production lines with stable demand requirements. In actual practice, the modern Lean Engineer has learned that even the best Lean systems sometimes need to *build to stock* or *make to forecast* to some degree.

9.0 Observations on Variability and its Influence

The key issues of variance reduction, continuous improvement and JIT should continue to dominate Lean Thinking. Pull is only a part of the larger scheme of things, but a big part. Finally, we directly quote Hopp and Spearman concerning variability.

"*Regardless of (its nature, magnitude) or source, all variability will be buffered. A fundamental principle (of the manufacturing rules) of Factory Physics is that there are three types of **variability buffers**, (1) **Inventory** (2) **Capacity** and (3) **Time**. For example, safety stock represents inventory buffering against variability in either demand requirements or production capability. While the exact mix and content of buffers is basically a management decision, the decision of whether or not buffering is **required**, in the face of uncertainty or variability is not. If variability exists, it **will be buffered**. If you will not or refuse to invest significant time and resources into variability reduction, you **will** pay for it in one or more of the following ways.*

- *Long cycle times*
- *High WIP levels*
- *Wasted and underutilized capacity*
- *Lower throughput capabilities*
- *Loss of customer satisfaction*
- *Lost revenue*

10.0 Waste Must Be Eliminated

The philosophy of complete elimination of waste means that we aggressively and continuously attack waste whenever and wherever it is found. Ohno had a nasty reputation of wandering around on the shop floor in search of waste, and anytime he spotted waste of any kind he would rant and rave uncontrollably. This was probably just a contrived conception of Ohno, but the principle is sound. Applied to Lean Manufacturing, it is both necessary and critical to manage inventory levels and prevent waste of any time. This is the essence of Lean Thinking.

Below is a simplified version of the six rules for an effective Kanban system as defined by Toyota:

- Never ignore of fail to address the "why" of defective products
- Make only the type and quantity requested
- Level the production sequence for mixed model production
- WIP for a particular part type is held in identical delivery boxes
- Work must be moved immediately upon request
- A Kanban Card controls all movement of parts from Raw Material Procurement to Customer Delivery

11.0 The Pervasive Impact of Kanban

A complete Lean Manufacturing System Controlled by Kanban, WIP caps and drives the entire Manufacturing Enterprise.

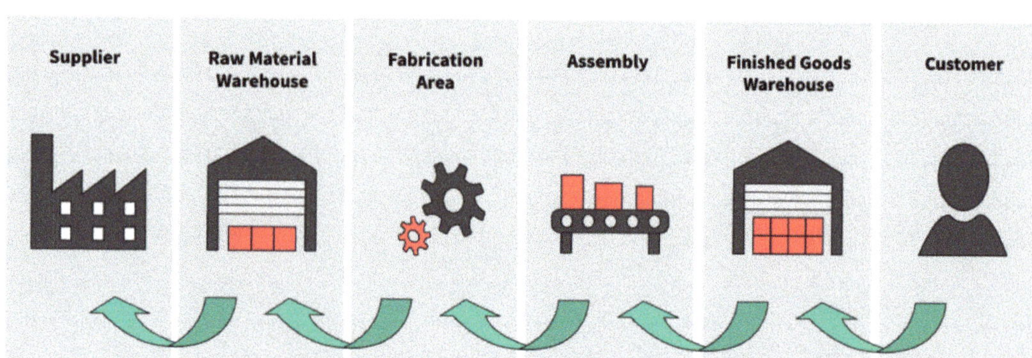

Kanban system for a manufacturing process
https://kanbanzone.com/2019/step-by-step-guide-kanban-system/

Kanban is a highly visual method for managing workflow, inventory, and more. Kanban encourages the use of visual indicators to improve the flow and efficiency of production, making it an ideal method for manufacturers of physical goods

12.0 Summary and Conclusions

In this Chapter, we build a case for transforming push systems into pull systems, and explored the difference in the two operational philosophies by using either the Toyota Production System controlled by Kanban Cards or a CONWIP strategy. CONWIP is just one of many possible pull strategies, but the results which were given were typical of those expected as one transitions from a push to pull operating strategy. CONWIP was chosen for this study because it is easy to implement and is essentially self-regulating in a manufacturing cell. A major contribution to our discussion was to recognize that a pure pull system is always superior to a pure push system in terms of required throughput time and WIP. This fact was fully exploited and explored in Chapters 1-5,which were concerned solely with the design and operation of U-Shaped manufacturing and assembly cells. In this Chapter we have also pointed out that most operational systems have both push and pull embedded in their operation and that almost all implementations have some form of a Push-Pull interface. Finally, we stated that the *Magic of Pull* is not a set of operating rules, system configuration or operating principles; but the simple fact is that a Kanban pull system of any type puts a *WIP Cap* on the amount of inventory in the system.

Push systems typically release work to the shop floor in large batches. Order, production and move batches are often the same. Material movement occurs by *pushing* work in progress through the system as an opportunity arises, as quickly as possible. Most of the problems associated with push systems are twofold. First, all operations are scheduled in advance and will be executed as soon as possible. If an order gets behind schedule, the order may be split, reprioritized, *red-tagged* or expedited. The second major problem is that any of these remedies create variance components which corrupt every operation; decrease throughput; and increase WIP. Conversely, in a pull production system there are no pre-determined move schedules. Product is pulled from Workstation to Workstation based upon resource availability and system status.

The key issues are WIP control and WIP levels. By Little's Law, for a designated input rate, an increase in WIP will increase throughput time. Once again it must be emphasized: **The *Magic* of a pull system is the *WIP cap* which is strictly enforced by Kanban.** The amount of WIP in any pull system can never exceed the amount of WIP allowed by Kanban. A typical push system like MRP/MRP II will attempt to release work to the shop floor based upon a set schedule and assumed shop floor capacity. If too much work is released and/or capacity changes, a WIP explosion will occur with associated increase in cycle time. If too little work is released, then starvation can occur at bottleneck Workstations. A properly configured and managed pull system cannot exceed a WIP cap, and work will not be scheduled to either be moved or to start processing unless specifically authorized by shop floor personnel. There is simply no scheduling required except at final assembly. Sequencing should follow predetermined MMP strategies if there are multiple products; workload leveling, Workstation balancing, predetermined product mix sequencing and a FIFO discipline should always be strictly observed.

The role of the Kanban cards and the way product arrives to the system serve to define whether the system is operating as a classic Push system or as a Pull system. In terms of the amount of WIP allowed, a pure push system would assume that there are an infinite number of Kanban cards, and that parts arrive for processing according to some order release strategy. A pure pull system would be defined and controlled by the number items allowed in the system by Kanban cards. The rate of production is dictated by customer demand and enforced by Mixed- Model final assembly.

Review Questions

1. Write out your own definition of a Lean Engineer. Provide a definition of *Lean Thinking.* Contrast Lean Engineering to Lean Thinking.
2. If a company wanted to hire a Lean Engineer, provide a description of what he/she would do, what expectations and what sort of projects.
3. Explain the difference in *push* production control systems and *pull* systems.
4. What was the most important thing(s) that Henry Ford accomplished? What important operating principles and management philosophies eventually caused problems and why?
5. Contrast the manufacturing and corporate objectives of Henry Ford to William Durant. What were the commonalities?
6. List and discuss three major problems and three good things about the original concepts and implementation of MRP.
7. Discuss the major differences in MRP and MRP II.
8. Why are MRP/MRP II Production Planning & Control Systems often referred to as "push" systems?
9. Discuss the fundamental differences in the Toyota Production System and MRP/MRP II.
10. What is the fundamental difference in a "push" versus a "pull" system?
11. Using textual material and the World Wide Web, discus and contrast the difference between a "two card pull" system and a "one card pull" system.
12. List 5 major benefits in using a pull system versus a push system.
13. Consider a major United States retail store such as KMART or Sears. . Discuss how such a store represents a typical "push-pull system" Where are the pull components, the push components and the push-pull interface? How could you improve that system?
14. Discuss a general CONWIP system which places a WIP Cap on the total amount of work-in-progress for a serial, three Workstation production line. What prevents all of the product from building up at the bottleneck workstation?
15. For the following four Workstation production line: calculate (1) System WIP, (2) Workstation WIP, (3) System Throughput time and (4) Individual Workstation WIP and Throughput times. (5) Verify that Little's law holds at the system level. Each Workstation has an infinite WIP holding area and the system arrival rate is λ=10 parts/hr.

Workstation ID	Service Rate(Parts/hr)	SCV
1	10.8	1.25
2	5.5	0.85
3	3.4	1.50
4	11.0	1.0

16. What two things must be true for an "N" Workstation serial system to achieve "optimum" system performance?
17. Discuss the concept of *Critical WIP*.
18. Consider the following four Workstation production line with no variance in the processing rates.

Workstation ID	Number of Servers	Service Time
1	2	5
2	4	7
3	1	1.5
4	3	10.0

(1) Calculate the raw throughput time for this system.
(2) What is the Bottleneck Rate?
(3) What is the "Critical WIP"?
(4) What is the expected WIP level at each Workstation?

19. Explain the following statement: *Variability in a Manufacturing System will always be buffered.* How will it be buffered? Give three examples illustrating this principle.

Chapter 7
Preventive Maintenance, Reliability and Continuous Improvement

Money can always be gained; lost time can never be recovered
Matshona Dhliwayo

7.1 Introduction

Time must be allotted in any Manufacturing System for executing Preventive Maintenance, Reliability and Quality Assurance. Unless all the elements in the manufacturing system are properly designed, maintained and monitored; breakdowns will occur and disrupt the flow of products. Some manufacturing processes need to run continuously, particularly those that are environmentally sensitive (plating, cleaning, etc.). Usually the problem with continuous processes is that they are very expensive and difficult to shut down. All the maintenance except for emergency service is performed during a scheduled shutdown period. The other class of Manufacturing resources (machines) which must run continuously during normal usage are the Bottleneck Machines or production processes. It has been accurately said that any time lost during a Bottleneck Machining operation cannot ever be recovered. The people and equipment in a Lean Manufacturing system must be ready to produce what is needed when needed. This may not be possible in a tightly balanced system if slack time is not provided for catching up or for maintenance. The solution is to develop a less-than-full-capacity schedule that includes time for maintenance. Repairs made under pressure may not be done well, leading to further downtime for re-repair, tinkering, and adjusting.

Reliability is primarily a function of engineering Design, but designed Reliability can be seriously compromised by using any machine under stress or pushing range of usage. For example, drilling a hole in a metal sheet can result in broken tools if too much downward pressure is exerted. All machine operators must be properly trained and not allowed to abuse any machine. Of course, Reliability and Maintenance are not mutually independent. Lack of lubricant, the use of incorrect oil and grease, operator non-standard practices and a plethora of other factors can seriously affect machine Reliability. One of the most critical is routine, timely Preventive Maintenance.

Quality Assurance problems are less dependent upon machine design unless a machine is operator is using tools that were not designed to be used on a particular machine or raw materials that cannot be used on a machine. Quality of final products can ruin a company's reputation and destroy product sales. In Lean Manufacturing a proactive approach is taken to detect quality problems before they occur. This is called *Poka-Yoke*. quality Assurance becomes a shopfloor objective and a great deal of responsibility is placed upon shop-floor workers and less upon corporate, ivory-tower organization.

With *integrated preventive maintenance (IPM)* in Lean Manufacturing Systems, operators are required to become more aware of the behavior of their equipment and its routine problems and to execute Preventive Maintenance as part of their job requirements when time permits. IPM is clearly linked to Integrated quality control (IQC). Quality Control problems are sometimes incorrectly diagnosed as a Maintenance problem. Getting the process centered and then maintaining the aim or accuracy of the process is different from reducing the variance and then maintaining the process spread (variability or precision). Finally, keep this thought in mind – if you are not maintaining or improving a manufacturing process or system, then it *will* degrade. (layman's restatement of the second law of thermodynamics).

7.2 Integrated Preventive Maintenance

Integrated preventive maintenance includes the maintenance of machine tools, work holding devices, cutting tools, and personnel. This function is integrated into the daily regimen of people on the plant floor. Preventive maintenance responsibility is shifted from a corporate maintenance department to the manufacturing personnel. This is not to say that a corporate Maintenance Staff is not required. It does imply a shift in how maintenance is executed and how. An example of this characterization is that a Maintenance Engineer will develop specific maintenance requirements, and delegate responsibilities to the shop floor. The important result of such a strategy is to be *proactive* and not *reactive*. The operators are involved in the preparation and use of maintenance tasks and checklists, much like the checklists pilots use to check out the aircraft before takeoff. The goal is to prevent any machines from failing during a scheduled eight-hour shift. Workers are responsible for most of the routine machine tool maintenance. The Maintenance Department does major machine overhauls and takes the lead role in the event of major breakdowns.

For a lean factory to operate effectively, the entire plant is best run on a 4-8-4-8 schedule. The four-hour time blocks between the two eight-hour shifts allow for maintenance or unavoidable long setups. In addition, the eight-hour shift can begin early or run over when needed without disrupting the next shift. Therefore, if possible Lean manufacturing systems are designed to be run at less-than-full-capacity (LTFC) so that there is breathing room to keep everything and everybody up and running 100% of scheduled time.

Machine life and tool life are further enhanced by operating the equipment at reduced speeds or reduced production rates. This concept is totally foreign to most American factories. The idea of less-than-full capacity operation suggests less-than-100-percent utilization of equipment. What we should do is to worry about effective people utilization and let the machine utilization be whatever is needed to meet the demand. The main culprit that blocks this type of thinking is the corporate accounting function and capital expenditures group. the "bean counters" are focused to running machines at 100 % utilization as much as possible. This to amortize machine costs as fast as possible. Corporate Capital Investment people are usually only concerned about minimum procurement costs and often follow a "low bid" policy. Low bid does not always translate into superior performance.

7.3 Scheduling Major Preventative Maintenance

Scheduling Major Preventive Maintenance (PM) is a task typically assigned to the Lean Engineer. Production supervisors often fail to accept that their manufacturing processes should not be shut down for an extended period of time so that major PM can be performed. The main cause of this attitude is slipped schedules and a terrible practice called "piece rate" incentives. To overcome resistance, PM should be between eight-hour shifts in the four-hour time blocks, or on weekends if necessary. This can present some inconvenience for the maintenance engineers and specialists. It can also result in ordinary shift workers participating in these events. When Maintenancc is rcquired during scheduled work periods, Lean Cell design can help considerably.

Unused or underutilized machines can be used on an alternate basis when overhaul maintenance or technology upgrades are required. One machine can often be removed from the cell and replaced with another, so that the necessary maintenance can be performed.

The pace of the manufacturing system in the entire plant is synchronized to Takt time, and production rate (cycle time, etc.) for a specific piece of equipment or cell are determined by worker allocation in each cell. The machines in each cell are not run at 100% utilization. Furthermore, the entire eight-hour shift is not scheduled unless that is required to meet the daily production needs. Extra (slack) time at the start and end of each shift can be allotted for routine repair and maintenance. If the entire eight-hour shift is scheduled, this represents a problem for worker-executed maintenance, and steps should be taken to correct this situation. By executing proper maintenance practice, the equipment will last longer and exhibit higher reliability. The idea is not to overtax or stress the machine tools, the people, or the cutting tools. People are less likely to make mistakes and machines less likely to break down if they are always running at full capacity.

Suppose you prepared a race car to run the Indy 500 but instead of running the car at 200 mph, you operate it consistently at 150 mph. The car will run much longer before it breaks down. The long distance race is won by the steady and consistent performer. Taiichi Ohno once said: *The slower but consistent tortoise causes less waste and is much more desirable than the speedy hare.*

7.4 What PM Actually Means

Preventive maintenance (PM) is designed to preserve and enhance equipment reliability. A correctly integrated PM program will provide a significant increase in production capability throughout the entire production system. The ideal PM program will prevent failure of most equipment before it occurs. The operator is trained and empowered to maintain his machine(s), schedule the routine maintenance and keep the area clean and neat.

7.5 Value of Preventive Maintenance

People not exposed to preventive maintenance question its value. They believe that it costs more for regular downtime and maintenance than it costs to operate equipment until repair is absolutely necessary. However, one should compare not only the costs but the long-term benefits and savings associated with PM. Without PM, the following costs are likely to be incurred:

Costs of Poor Preventive Maintenance

- Lost production time resulting from unscheduled equipment breakdown
- Variation in the quality of products due to deteriorating equipment performance
- Decrease in the service life of the equipment
- Safety-related accidents due to equipment malfunction
- Major equipment repair and lost production time
- Missed schedules and late deliveries
- Increased WIP in push system
- Idle skilled workers

Long-term benefits

- If routine maintenance is a shop-floor responsibility there are direct benefits. Operators are more familiar with the equipment, the way it operates, and its chronic problems.
- Quality, flexibility, safety, reliability, and production capability are improved.
- Reliable equipment permits reduction in inventory

Long-term effects and cost comparisons undoubtedly favor preventive maintenance. A carefully designed and properly integrated program requires a positive managerial attitude that will set the pace for a successful program.

Machine operators should be trained to observe their equipment and to respond to their observations. If a piece of equipment need special attention and the operator cannot perform· the level of maintenance necessary, he or she should see that the proper maintenance specialist comes to the machine. Though preventive maintenance, production operators become more conscious of their performance and take pride in their work.

Operators need to realize the importance of an orderly workplace and clean equipment. This is not just cosmetic, studies have shown that orderly and organized people are generally more productive than sloppy and disorganized people. Routine. If a person is given responsibility for a piece of equipment, he/she will take pride in its operation. Process drift (loss of stability or loss of aim or accuracy) is a chief problem that operators need to continuously monitor. This task is closely linked to IQC (Integrated Quality control), and is something that an operator should be able to identify. Housekeeping should become a ritual in their everyday job performance. Routine cleaning familiarizes the operator with the machine, and results in visual inspection. When operators have a sense of responsibility rather than just a duty to perform a specific task, they become enthusiastic about their jobs.

Integrated Preventative Maintenance (IPM) emphasizes the significance of executing the correct procedures needed to operate all equipment in a manufacturing cell. When an operator is responsible for multiple machines in a Lean Cell he/she becomes aware of their linkage and impact on one another. Any irregularities in the processes will clearly stand out. Many problems are readily observed and can be prevented as operators Lo-Co-Moo around a Lean Cell

7.6 The Role of Maintenance

In large companies there will always be a need for designated maintenance mechanics but their role will change in a Lean factory. If operators are doing a good job of trying to avoid emergency shut downs, and regularly performing routine tasks that have usually been assigned to corporate maintenance, maintenance personnel should be able to concentrate on more constructive tasks. There are some tasks like welding or repair of complicated equipment that operators could not be expected to learn. If the maintenance employees have been reduced significantly due to a company's dedication to lean manufacturing, then the maintenance personnel who are left should be able to devote themselves to the role of troubleshooting and major corrective tasks sophisticated and modern diagnostic tools. Maintenance personnel should be multi-skilled. They must be able to perform electric and mechanical tasks very well as well as be trained in some operating skills. They must be able to be productive at all times

7.7 Lean Engineering and Maintenance

Lean Manufacturing engineers are responsible for the design, build, test, and implementation of manufacturing equipment. Maintenance should be considered in the design or selection of purchased equipment. Simple, reliable equipment that can be easily maintained should be specified. In general, dedicated equipment can be built in-house better than can be purchased. Many companies understand that it is not good strategy simply to imitate or copy the manufacturing technology from another company and then expect to make superior product using the same technology as the competitor. The company must perform research and development on manufacturing technologies as well as manufacturing systems in order to produce effective and cost efficient products. The competitive Lean Company should continually survey their competition to assess their market strategy and position. An effective, cost efficient manufacturing system makes research and development (R&D) in manufacturing technology pay off. Once you have implemented the L-CMS you can begin to design, build and install your own machines into the manufacturing cells. Those who have taken this step into lean manufacturing find the following advantages.

- The cell (and the machines within it) is designed for the system so the cell operates on a MO-CO-MOO basis and the machines can meet the cycle time, where CT =Takt time for the system.
- It is flexible, allowing rapid changeover for existing products and rapid modification for new products or model changes
- It has unique processing capabilities, unique machine tools
- It has maintainability/reliability/durability
- Equipment and methods are designed to prevent accidents (safety).
- It is easy to operate, has a fail-safe design (ergonomics).

The equipment is designed and developed with both the external and internal customer in mind. Although many plants lack the expertise to build machines from scratch, most have the expertise to modify equipment to give it unique capabilities. Modification of the equipment to prevent reoccurrence of a breakdown requires that management assign the highest priority to this task. The most skilled maintenance personnel must be given this task so that even subtle problems in the Linked-Cell Manufacturing System (L-CMS) are eventually eliminated.

7.8 Total Productive Maintenance (TPM)

The Japanese imported preventive maintenance from the United States over 50 years ago. Since that time they have improved and extended it into what is now labeled *Total Productive Maintenance* (TPM). Seiichi Nakajima traced the development of TPM as follows:

1950s: Preventive Maintenance -established maintenance functions
1960s: Productive Maintenance -recognizing the importance of reliability, maintenance, and economic efficiency in the plant design.

1970s: Total Productive Maintenance-achieving productive maintenance efficiency through a comprehensive system based on respect for individuals and total employee participation (Nakajima, 1988).

1980s: Total Productive Maintenance (TPM) a concept developed by the Japanese in the 1960s from maintenance practices they learned from the United States during the 1950s. TPM was institutionalized as an equipment management strategy that involves all personnel in a plant or facility. TPM is to productivity as flexibility is to competitiveness. Without it, corporate survival is questionable. It is a critical step in the lean strategy.

1990s: TPM became a key component of lean manufacturing system design (Black, 1990). TPM involves the following five elements, often called the *Five Pillars of TPM* by Nakajima.

7.8.1 Nakajima's 5 Pillars of Preventive Maintenance

Eliminate the "Three big losses" and thereby improve the efficiency of the equipment.

Uptime Losses

1. Equipment failure - breakdowns of machine tools, material handling devices, fixtures, etc.
2. Setup and adjustment - from exchange of dies in molding machines and presses, and tooling exchanges in machine tools.

Speed losses*:*

3. Idling and minor stoppages- due to the abnormal operation of sensors, blockage of work on chutes, and the like
4. Reduced speed- due to discrepancies between specified and actual speed of equipment

Losses due to defects:

5. Process defects -due to scrap and rework
6. Reduced yield-from machine start-up to stable production.

7.8.2 Develop an Autonomous Maintenance Program.

This demands that the operators are involved in the daily maintenance of the equipment. The *Seven Steps of Autonomous Maintenance* are:

1. *Cleaning*: Clean to eliminate dust and dirt mainly on the body of the equipment: lubricate and tighten: discover problems, and eliminate their causes.
2. *Countermeasures at the source of problems*: Prevent the causes of dust, dirt, and spattering of liquid, improve those parts of equipment that are hard to clean and lubricate: reduced the time required for cleaning and lubricating.
3. *Lubrication and Cleaning standards*: Establish standards that reduce the time spent cleaning, lubricating, and tightening (specify daily and periodic tasks)."

4. *General and Efficient inspection Procedures*: Follow the instructions in the inspection manual. Use Quality Circle members to discover and correct equipment malfunctions.
5. *Autonomous inspection*: Develop and use autonomous inspection check sheets."
6. *Orderliness and Tidiness*: Everything has its own place and is easily found. Implement Shadow Boxes for tools.
7. *Document maintenance problems and their solutions*: Develop a company policy and goals for maintenance; increase the regularity of improvement activities. Record the mean time between failures (MTBF), analyze the results, and design countermeasures.

These steps are based on the *Five Basic Principles of Operations Management*. In the Japanese literature, they are known as ***The Five S's**: **S**eri, **S**eiton, **S**eiso, **S**eiketsu*, and ***S**hitsuke*. A rough translation of the five S's means organization, tidiness, purity, cleanliness, and discipline.

- Develop a scheduled maintenance program working with the maintenance department. This is usually done in cooperation with Industrial Engineering and the Lean Engineer .A fixed and regular maintenance schedule greatly improves any corporate maintenance program.
- Increase the skill of operators and the maintenance personnel. The operators should work with the maintenance people at the time PM work is done on their equipment, discussing problems and solutions. Part of the operator's job is to keep records on the performance of the equipment, so the operators must learn to be observant.
- Develop an equipment management program. A record of the use of machines and tools denoting how much they were used and who used them.

7.9 Zero Downtime

The goal of preventive maintenance is simply to eliminate all machine breakdowns and troubles. In short, zero down time and 100% on-demand utilization. Preventive maintenance and its key benefit – zero machine downtime - cannot be realized without the efforts of the entire organization -operators, management, maintenance crews, and shop floor workers. Operator involvement is critical. Zero down time, or 100% on-demand utilization, can only be assured by preventing breakdowns and problems before they happen, not by the firefighting mode of a corporate maintenance group. A key problem is that in many cases maintenance crews are busy fixing sudden breakdowns, while operators think that machine breakdowns are not their responsibility. However, it can be argued that equally important is that Maintenance is considered to be a corporate function and not a shop floor problem. In too many cases, maintenance methodologies and schemes are berthed by education and not by operation. Even two seemingly identical machines will not operate exactly the same and these differences are best understood and experienced by the skilled labor that operates each machine. Ohno came from the shop floor and he clearly understood the difference between ivory tower maintenance and shop floor maintenance. He insisted that the shop floor operator must be involved in day to day maintenance. Failure to achieve zero defects is often directly related to the individual characteristics of each machine and is easily recognized by the Machine operators. There are many factors that can lead to poor machine performance. The following list is only meant to be representative and instructive .

Factors leading to machine breakdown

Machine factors:

- Dirty and oily machine
- Filled oil pan
- Leaks around the machine
- Overheating motor
- Vibration of motor
- Vibration of machine
- Inspections not routinely performed

Operator factors:

- Lack of concern for dirty machines
- Lack of training to conduct simple maintenance
- Lack of knowledgeof the machine
- Not asking for help when problem is anticipated
- Considering production more important than maintenance of machine
- Lack of knowledge of when or how inspection occur

Repair crew factors:

- Not questioning "why" when problems occur to find root cause
- Failure to coach operators on simple maintenance
- Concerned only for urgent, major machine troubles
- Accepts that machines will eventually break down
- Looks only for improvement to be made through new machines

7.9.1 Major Causes of Machine Downtime

- Basic machine requirements -housekeeping, tightening bolts, oiling, etc. -not performed
- Incorrect operating conditions - speeds, pressure, torque, temperature, etc.
- Employee skills maintenance crews, operators, etc. - not performed
- Deterioration --tooling, fixtures, gears, bearing, etc.
- Design – material, tooling, speeds ,feeds, etc.

7.10 Integrated Preventive Maintenance (IPM) in Lean Manufacturing

In the lean system, preventive maintenance includes another, broader concept: ***T****otal* ***P****reventive* ***M****aintenance* (TPM). *The goal of TPM is to achieve preventative maintenance through employee participation in maintenance activities.* TPM is closely related to ***I****ntegrated* ***Q****uality* ***C****ontrol* (IQC), where employee involvement is also a key to success in developing quality operators to produce quality products. It seems necessary at this point to challenge traditional Quality Control tools which are still being taught in most Industrial Engineering Departments. Quality Control tools include Statistical Quality Control Charts, Acceptance Sampling Plans and often Military Single, double and Multiple Inspection Plans. All of these tools have played a major role in establishing Quality Assurance as a major and important part of the Manufacturing Enterprise. These methodologies should now be abandoned in Lean Manufacturing. The problem is that all are *Reactive* tools and not *Proactive* tools. Each should be modified appropriately to predict Quality Control problems and not signal that one has occurred. This will take an entirely new approach to Quality Control.

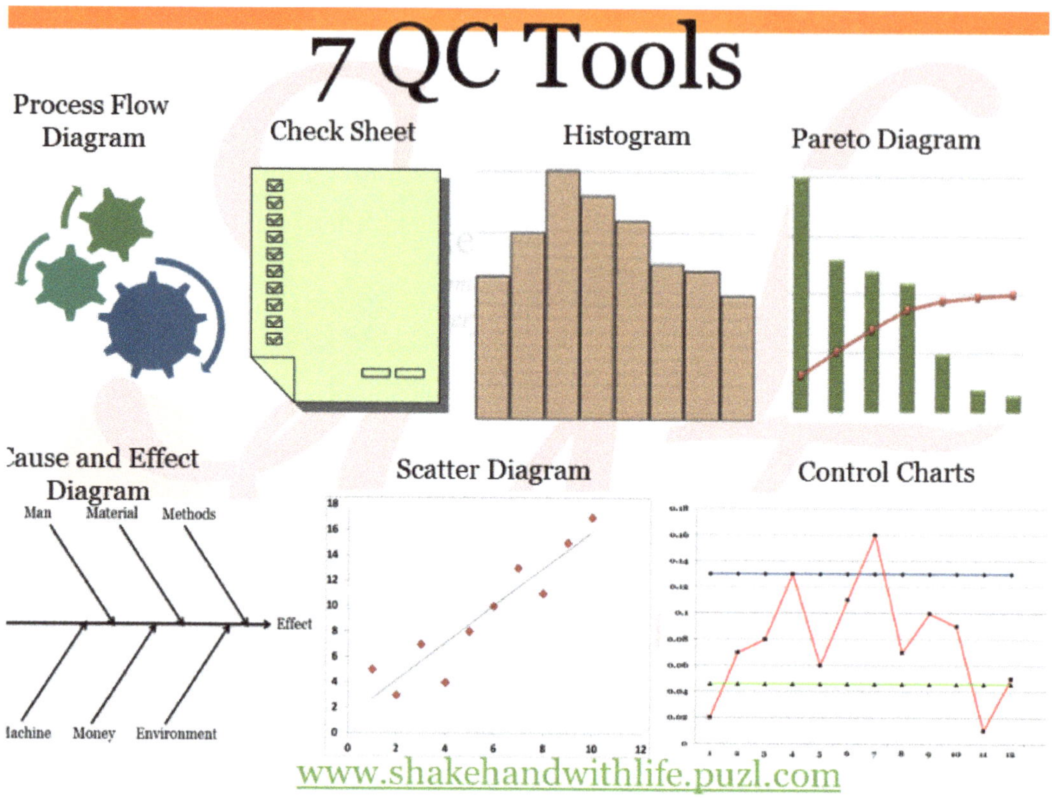

Lean production builds high quality into the production process itself, rather than achieving it later through inspections and repairs. Preventive maintenance is directed to preventing trouble at the source, rather them letting things go wrong and then fixing each problem as it occurs. At the very foundation of TPM is partnership. TPM focuses on partnership between the manufacturing people, the maintenance department, manufacturing and design engineering and technical services to improve the overall equipment effectiveness.

7.11 Benchmarking Your Company

Prior to the development of an IPM plan, benchmarking or recording the current status of your organization is a good idea. This evaluation will include more than just the maintenance organization. It should embrace all parts of the organization involved in the manufacturing maintenance, design, and purchasing of the assets of the company. The most important benchmark for TPM is the current state of the equipment as measured by the percentage uptime.

Benchmarking provides the starting point from which to calculate improvement. As such, it also provides the means to demonstrate to your management that you are making progress, saving money, and improving productivity. Benchmarking also shows the production/maintenance teams that they are making progress which keeps interest and enthusiasm levels high.

Benchmarking involves documenting as much current data on equipment as possible. From the current status you can show management where the equipment is in terms of productivity, where it is, and where the trend patterns are pointing.

One of the first steps in benchmarking should be an analysis of amount of additional capacity the company could get from their machinery if they improved its uptime from where it was at baseline. Such data could include quality information of the equipment, time data, maintenance downtime histories, setup times, and procedures. Other documentation might include photos showing the physical condition of the equipment at program start to compare with similar photos after TPM implementation.

7.12 Selecting a Pilot Program

Selecting a pilot area is an important step in the TPM sequence. Rather than initially attempting to implement TPM plant wide most lean engineers would start with selected machines that have historically caused problems.

7.13 Predictive Maintenance

We have previously presented a case for implementing predictive and proactive maintenance rather than reactive maintenance programs. Implementing a predictive maintenance program is the next step in establishing real-time methods for monitoring equipment and ensuring that wear trends are documented. In this way, equipment can be overhauled, with worn components changed, before a failure occurs. Table 16 .1 shows a typical inspection form for a hydraulic and pneumatic system. Such forms must be developed for belts, chains, and general equipment, respectively.

Table 16.1
A Typical Inspection Check Form for a Hydraulic and Pneumatic System

Hydraulic Inspections	Item	OK	Needs Repair
Hydraulic Pump	Proper oil flow?		
	Proper pressure?		
	Excessive noise?		
	Vibration?		
	Proper mounting?		
	Excessive heat?		
Intake filter	Clean?		
	Free oil flow?		
Directional Control Valves	Easy movement?		
	Proper oil flow?		
Relief Valve	Proper pressure?		
	Excessive heat?		
Lines	Properly mounted?		
	Oil leaks?		
	Loose fittings?		
	Damaged piping?		

Pneumatic Inspections	Item	OK	Needs Repair
Compressor	Proper oil flow?		
	Proper pressure?		
	Excessive noise?		
	Vibration?		
	Proper lubrication?		
Inlet filter	Clean?		
	Free air flow?		
Directional Control Valves	Easy Movement?		
	Proper air flow?		
Muffler	Proper air flow?		
	Proper noise reduction?		
Lines	Properly mounted?		
	Air leaks?		
	Loose fittings?		
	Damaged piping?		

Real -time Maintenance Programs demand the need for a computerized database for tracking and trending equipment histories, correlation to different types of problems, cause and effect analysis, use of spare parts, the impact of training, etc. Systems commonly used for this are called computerized maintenance management systems (CMMS), and they are all characterized by a comprehensive relational database accessible to the entire organization. Any way it is measured, the impact of machine downtime and defective product is reflected in reduced profit, slipped product due dates, increased overtime or subcontracting and a degradation of employee morale. As a TPM program matures, the company will see how much maintenance is required by areas, cells and individual machines. The result is that maintenance is not just a required activity, but one of the keys to a more efficient and defect free manufacturing program. Lean manufacturing implementation is not just improved, but defect free Lean Cells CANNOT operate properly if delays due to machine problems and rejected work-in-process continue to interrupt production. Every piece of equipment must run when it is scheduled and produce at the rate demanded by Takt time. TQC is not just a job, but a critical activity.

7.14 Continuous Improvement

Any company that does not continuously improve machine performance and seek new and innovative ways to achieve the same level of production at lower cost will not survive in a competitive market. The only exception might be a monopoly, but even then if increased production costs result in higher product costs, the increased price might result in two disastrous scenarios. The consumer may just decide that the price is too high, or competitive manufacturing might find a new way to produce the same product at a lower cost. Continuous improvement is not just a watchword, but is a necessity in any competitive environment.

7.15 Standardization and Continuous Improvement

Henry ford said that without standards, improvement is impossible. He was obsessed with establishing standard procedures and standard times for all processes, and then using these standards as a benchmark for continuous improvement. Standard times and standard work practices have traditionally been conducted by Industrial Engineering, and best practice work has been the bread and butter of Human Factors engineering. Unfortunately, In recent years Industrial Engineering has been obsessed with sophisticated Optimization techniques based upon Linear and Nonlinear Programming. The practice of setting both engineered time standards or work sampling time standards needs to be revived...particularly with the emergence of Lean Engineering.

The objective of standardization is to develop daily routines and methodologies that lead to continuous improvement

Without standards, the internal customers have no goals by which to judge their work. Without standards, operations and methods will fall back to the old ways. The result will be continual firefighting, not continual improvement. Standards must be set and followed. For every deviation from the standard, the problem must be identified and must be eliminated. If possible, improve the standard so the same problem does not recur. In addition, as more jobs become standardized, there will be less confusion. Training of new operators becomes easier- which is extremely important where people rotate jobs often and work on multiple machines. In short, standards make jobs easier and more cost effective.

Standard work or standard operations is a way to achieve maximum performance with minimum waste. Standard work must involve three key elements.

1. *Cycle time* - The time between completion of the last component or product and completion of the next product. (CT = 1/PR)
2. *Work sequence*- The sequence of work demanded and dictated by the internal customer.
3. *Stock-on-hand* (SOH) - The standard amount of work that is currently in process necessary to conduct smooth operations in the cell.

Standard work is not simply something management demands and operators comply with.
It is implemented to involve management and operators in the development process.
Finally, anyone should be able to follow the instructions of standard work. The goal is that a new operator should be able to master the standard work after three days of training on a process.

All operation and machining cycle times should determined by using sound Industrial Engineering Methodologies of setting *time standards*. Time standards can be set by using what is called *Engineered Standards* or *Stop Watch Standards*. Engineered standards are based upon standard practices and predetermined standard times for each element of work. Stop watch standards are set by direct observation using an average worker under average working conditions (Niebel)

7.15.1 Steps to Standardizing Work Practices

Standard work can be pursued through the following steps:

1. Standardize current work practices with assistance of engineers, team leaders, and operators.
2. Locate problem areas.
3. Solve the problem and develop new improved methods.
4. Implement the new methods.
5. If the new methods prove satisfactory, develop new standard work.
6. Require operators to write out the standard work combination tables with the assistance of supervisors.
7. Start the process over again.

In most cases, to the degree every person practices standards and improvement activities to eliminate waste; to that degree will the factory's competitive edge be improved.

The development and improvement of standard work involves everyone, especially the operators. It is said that Toyota will strive to improve their systems to reduce operator work time by as little as half a second !

7.16 Continuous Improvement and Employee Involvement

Machine failures are often caused by the way people think and act. Operators, maintenance crews, and other support people should be trained to understand how their roles interact and what they must do to support one another. Machine tools and equipment should never be asked to perform tasks beyond their design intention or capacity. To do so is simply asking for machine/tool failures and unreliable processes.

Operators should learn how to perform routine maintenance on machines, be instructed in proper operating procedures, and develop an awareness of the signs associated with early machine deterioration. Maintenance people should learn how to assist production people with routine activities, readjustments, taking corrective actions, and increasing maintenance skills.

By involving every employee, zero machine failures and defect-free-production can often be routinely achieved. Employee involvement assures early detection of problems as product is made, and early correction of abnormal or deteriorating machine conditions. The clear implication of continuous improvement is that continual skill development and training is essential to increase the abilities of all involved.

7.17 Continuous Improvement and the *Five S* Principles

Toyota developed a set of 5 basic principles to facilitate continuous improvement.

1. **Seiri ("Say-ree")-** (Sifting)-Organization

Keep only the minimum of what is needed for a task and discard anything else.

Analyze what is available for the task, determine what is required to complete the task, and discard what is unnecessary. Anything extra is wasteful. For example, having extra tools, materials, pencils, and paper is waste and should be eliminated.

2. **Seiton ("Say-ton") -**(Sorting) - Arrangement

Once the minimum requirement is determined, there must be "a place for everything and everything in its place."

Assign a location for all essential items, often using shadow boards. Make the work place self-explanatory so everyone knows what goes where. Thus, eliminate the confusion and the lost time associated with hunting for items out of their proper place. A simple example of *Seiton* is installing what is called a *Shadow Box* to keep necessary and required tools in place. Lean Cell workers regularly change their routes and assigned machines. Rather than require each worker to buy specialized or expensive tools to be used at each location, provide shared tools in a common Shadowbox.

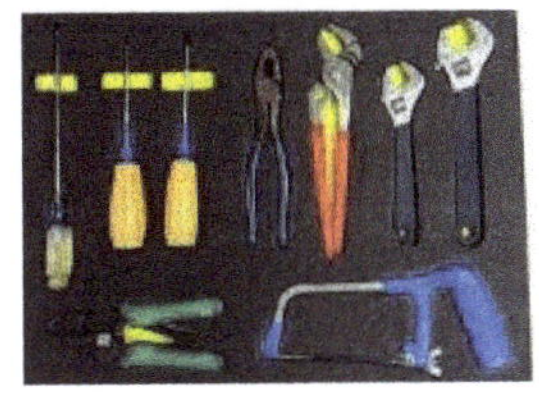

3. **Seiso ("Say-so")-** ***(Sweeping)- Cleaning***

Every tool should be clean and in proper working condition.

Once the work site is organized and arrangements are completed, tools must be kept clean so they can be easily obtained and used with no lost time. If something goes wrong, a backup tool should always be available, in proper working condition, and stored exactly where it can be readily found (Seiton).

4. **Seiketsu ("Say-ka-sue")** - (Spic-and-Span) - Hygiene

The working environment should be as clean as possible.

Hygiene usually complements the other aspects of detailed housekeeping. Effective organization and work arrangement is reinforced by keeping the entire area as clean as possible.

5. **Shitsuke ("Shay-sue-ka"}** - (Strict) - Discipline

All operational activities should be pursued with strict discipline.

Discipline is a direct by-product of standardization. No one should be permitted to "do their own thing". If a worker thinks that "his way is best" present it to the Lean Engineer as a productivity suggestion. It is not a bad idea to recognize such improvements with a Plaque and monetary reward.

These 5 general principles must be followed and become part of the daily routine. From observation, it seems the *fourth S* has been the hardest one to follow. Things start out organized, arranged and clean, but over time the workplace becomes messy. This is the second law of thermodynamics in action on the factory floor. All systems degrade and become unclean with time unless maintained. Daily discipline will greatly enhance the *Five S* tools. As we have said in different ways, the mark of good organization and effective work site arrangement is that nothing is hidden and everyonc knows where to find or place items. The Five S's are basic principles for identifying problem areas and waste. But lean production depends on everyone's active involvement. Thus, every member of the factory must follow the Five S principles before results will be noticed and sustained on a daily basis.

7.18 Kaizen and Kaizen Blitz Activities

Kaizen ("KY-ZAN") means continuous improvement.

Kaizen is a continuous, ongoing effort to improve quality, save time and money and resolve minor problems. Kaizen is the responsibility of everyone in involved in producing a finished product.

A Kaizen Blitz is a more intensive, directed continuous investigation of a major or systemic problem.

A Kaizen Blitz is scheduled at some specific point in time and utilizes the talents of several people who function as a Kaizen team. A Kaizen Blitz is conducted continuously over a designated period of time. It usually has one specific objective. Here are some specifics of Kaizen

1. The Concept

Kaizen is the responsibility and duty of every Plant Employee. Every shop floor worker is expected to spend his or her idle time thinking about new and ingenious ways to improve every facet of production. A Kaizen event involves both shop floor workers, Lean Engineers, production supervisors and often domain-specific outside consultants. In either Kaizen category, a team concept is encouraged . In continuing Kaizen, multiple workers in a Lean cell might get together either informally or formally to have brainstorming sessions. The Kaizen blitz is a more formal teaming effort. The *team* is a key vehicle through which continuous improvement is achieved. Anyone can make a suggestion for improvement. The team leader for each area is a designated *moderator*. He or she has veto power over any suggestion. If problems are of a greater scale, the team calls on other teams or departments, the team may employ the use of the Seven Tools of Quality-discussed earlier to gather and present facts about the problem. Team meeting usually occur before or after the shift with overtime paid.

2. Monthly Engineering Meetings

Monthly engineering meetings are another Kaizen activity. Engineers and shop floor workers meet on a regular basis to discuss both old and new problems that are causing waste. The 6 types of waste previously discussed provide a framework for focused discussions. After a time of open discussion, the problems discussed should lead to a brainstorming discussion to suggest future course of action. Through this exercise, the engineers and others in the group learn how to become problem solvers. an important by-product of open and frank discussion is that shop floor employees feel that they are "being heard and not ignored". Corporate engineers often discover that college and "ivory tower" education can fall quite short of actual problem solutions. This process is not intended to improve a person's presentation skills or writing skills. The idea is to empower everyone involved in manufacturing the ability and opportunity to resolve and solve problems.

The Lean Engineer must demonstrate the ability to resolve problems every day. Most young college graduates fail to realize that a college education and a college degree imparts basic knowledge to anyone in a specific field, but the main value in a college education is to teach students to be creative and innovative with a problem solving mentality. They learn to define, address and solve problems in a systematic way. As a result, the corporate engineer becomes a "real person" with "real solutions" to "real problems". Taiichi Ohno wore the same clothes as people on the shop floor...he ate with them.... and he often joined them in daily work tasks. He wasn't just a genius but he was a *people engineer*. He coined the term *autonomation.* Automation implies usage of machines, technology etc. to make processes run on their own without human intervention. Autonomation is automation with a human touch. Kaizen events are always recorded and archived. They are called *story boards* or *A3 reports* and they include the various implementations for improvement, projected benefits, problems solved and the current status of the situation. As with the other elements of lean production, it takes practice and discipline for this process to become a daily problem solving habit. The goal, however, is not simply to go through the motions of Lean Manufacturing but to use and pioneer new, more effective, innovative and challenging ways of thinking and acting.

3. The Suggestion System

The suggestion system also involves everyone in the factory. It is another mechanism to involve everyone in the production process ..... and to send the message that employees contribute through their minds, not just their hands. If suggestions are not given through a team, they can be given through a formal cost reduction program. Ideas are submitted to a cost reduction coordinator who in turn presents the idea to a cost reduction team. If the suggestion can be handled by a production team, it will be given to them for review and implementation. If a detailed study is required, the cost reduction team will recommend a Kaizen event to address the problem.

The cost reduction team, monthly meetings and, indeed all Kaizen activities encourage continuous improvement through employee involvement It is a system designed to let improvements happen quickly without bureaucratic red tape. It is a system which underscores the principle that employees should be given the responsibility to suggest improvement and the authority to have their voice heard. Ohno repeatedly said that the common man was a company's greatest asset...not machines.

7.19 The Concept of *RAM*

RAM is a term which was popularized several years ago but has not received the recognition that it deserves. RAM stands for the integrated concepts of **R**eliability, **A**vailability and **M**aintainability. The reality of RAM is completely understood by shop floor workers and some Industrial Engineers, but is often ignored or misunderstood by Maintenance Engineering.

Reliability

Reliability - ***The probability of a system or device continuing to function over a given period of time and under specified conditions***

An essential element in any lean manufacturing system is system reliability. Lean systems have serial, linked processes and by definition usually do not have standby or machines which simply "stand by". Machines and machinery are often expensive, highly automated and complicated pieces of equipment. Therefore, every component must operate properly every time and must be very reliable. In order to accomplish this goal of high reliability, a great deal of time and effort is usually required. This cannot be avoided...it is a fact of manufacturing "life".

Availability

> *Availability in manufacturing is the ability of a machine or process to operate as designed at any time that it is asked to do so*

Reliability and Availability are often confused with one another. Reliability is actually the capability of an operating critical resource to execute its assigned function from start to finish without failing. Availability is defined as the ability to immediately function as designed when called upon to do so. The difference in the two is best explained by a real-world example. When one of the authors was a young man he worked in a chemical plant which required a great deal of heat. That heat was continuously supplied by a large dowtherm heating unit. It was very old but functioned continuously 24 hours a day. Periodically an alarm would go off signaling that it had unexpectedly gone off-line. An old man that iI worked with would run to the heating unit and sparks would be flying everywhere. Rushing through the door, he kicked to h__l out of the heating unit with his #12 boot. The unit would immediately return to operation and function flawlessly for days until it failed again. This heating unit was not very reliable, but since it could be immediately restored to service it had very high availability.

The third leg of RAM is **M**aintainability.

> *Maintainability is the ease with which maintenance activities can be performed on an asset or piece of equipment*

The heating unit just described had low reliability but high availability. Maintenance of that unit was costly and difficult. If that heating unit is not running continuously all distillation columns in the plant would not operate. Small maintenance operations (low oil, cheap switches, etc) could easily be replaced, but a major, extensive maintenance was not easily done.

Each of these components are usually viewed as things that cannot be controlled without major equipment acquisition or major overhaul, but this is not the answer. Most RAM problems can be mitigated if not eliminated in the design phase. The corrupting influence of RAM in Lean Manufacturing must be addressed since Machine failure cannot be tolerated. In-line failure of Manufacturing processes (broken tools, component failure, etc) not only disrupt production and affect due-dates, but they usually result in product failure and unusable items.

Each of these concepts can be studied and attacked independently, but the greatest improvements are realized when it is realized that all three are highly correlated but can be functionally independent.

Another way to improve machine tool reliability and extend tool life is by recognizing the physics of tool wear and performance. For example, the life of a cutting tool (as well as surface finish) is greatly influenced by the speed and feed of the tool. Any materials engineer will recognize that there is a fundamental equation that relates these parameters to cutting conditions.

$$V T^n = C$$

V is the operating speed of the tool. T is the tool life, and both n and C are constants related to the particular tool material. From the equation, it is clear that as cutting speed of the tool (V) is decreased, the tool life (T) will increase since the other variables are constant for each tool material/ work material combination. As tool life increases, the time between failures will decrease, thus improving reliability. Lower speeds and feeds will also result in a better surface finish; improving quality. This is an example of a solution that requires almost no additional monetary investment but often represents significant monetary savings.

It should be pointed out that historically, manufacturing engineers have been trained to run machines at near 100% capacity, and run cutting tools as fast as possible to speed up production. The Lean engineer rejects this philosophy in light of the previous discussion. It should be recognized that the opposite is true in Lean manufacturing. Lean cells are designed and built on the premise that low utilization is expected, which actually enable preventive maintenance and superior machine performance. For example, reducing the cutting speed will increase machining time, but will not change the Cycle Time for the cell so long as CT > MT. Therefore, a second way to improve hardware reliability is to avoid putting stress and strain upon machine processes.. One way to do this is to decrease the operating speed of a cutting tool.

A third way to improve equipment reliability is through the implementation of successive checks following each step of a process. Inspecting at each station will allow identification and correction of the causes of errors before they become masked by other processes.

In order to improve hardware systems, the operators must have detailed knowledge of how the equipment works and be focused on the reliability of the process. Lean manufacturers are consistently better than their competitors when it comes to hardware reliability. A lean manufacturer allows the operator, someone with the required level of knowledge, to be directly involved in design development and improvement of the hardware. This is in contrast to the manufacturer that has a *specialized* or *focused* corporate Maintenance organization Increased reliability requires that the operator be involved in the maintenance of the equipment .Operators do not necessarily need to be able to fix every problem, but need to personally be able to call upon an advanced technology when a more complex problem develops. Any corporate Maintenance should work with the operators to return the machine to its original condition. Remember that the goal should always be to restore the equipment to a like-new condition, rather than just restoration of operational capability. Remember this: computers are unreliable but humans are even more unreliable. Any system that depends on humans is unreliable (a Murphy's Law).

7.19 Workforce Reliability

Workforce reliability refers to those personnel working in the manufacturing process and their task of keeping the process flowing continuously. A reliable workforce is a goal of almost every industry, frequently studied and theorized about the optimal strategies to attain this goal. Every manager has their own theory on how to keep people alert and productive. How did the Japanese rebuild its manufacturing from the ashes of WWII to become a world leader? Was it simply by inventing better manufacturing process technology or was it the lean manufacturing system? Actually...it was neither. They excelled out of necessity and effort. The lean manufacturing system design simply provided the right environment for success. Because the Japanese Industrial sector did not universally have the extensive computer resources that was possessed by the United States, and could not in general afford millions dollars of investment in new machines, they retrofitted what they had and turned to their labor force to get the job done (Ohno comments). To fully exploit and maximize labor involvement they defined *5 P's for the people in manufacturing systems.*

1. **Purpose** - *Do my efforts always complement the task?*
   - ➢ Make certain that everyone in the organization knows intuitively what the purpose is.
   - ➢ Be the best in the world at what we do.
   - ➢ Have a design, cost and quality that permits us to be market competetive.
2. **Passion** - *How much am I really, really holding back?*
   - ➢ Have a sincere, deep and unrelenting desire to pursue our purpose.
   - ➢ Elevate the intensity, hours and planning - amount and method
   - ➢ Hire only those willing to exert maximum effort
   - ➢ Be accountable for motivating and requiring our people to do the same.
3. **Pragmatism** - *Could I explain it to my grandmother?*
   - ➢ Learn to use brains and not money
   - ➢ Eliminate waste- don't plan it forever, start today
   - ➢ Refuse to delegate work that does not add value
   - ➢ Stop useless reporting
4. **Participation** - *Should I ask my subordinates for advice?*
   - ➢ Ultimately, people are our most valuable asset, with them we can do anything. Without them we can do nothing. One individual's efforts are always limited by the contribution of others.
   - ➢ People must be educated, trained, respected, and fairly paid.
   - ➢ We must arrange responsibilities so that it gives an opportunity to make a meaningful contribution to the organization's success. Without this our self-respect suffers. Without self-respect we cannot serve others and they cannot serve us.
   - ➢ Use every employee in a humane and challenging manner. Everyone is paid to think and their thoughts are valuable. They understand this even if we do not.
5. **Principles** - *Is our daily routine predictable?*(See Table 16.3)
   - ➢ Standardize all activities as much as possible.
   - ➢ Train all involved personnel in those standard practices
   - ➢ All rules are made to be broken - rule of last and first
   - ➢ Only break rules after good participative investigation
   - ➢ Institutionalize continuous improvement

Guiding Principle

Our life is more important than our work.

We improve our work in order to add value to and enhance our life, not to escape from it. Try applying the 5 P's to civilian private life and see how we are changed to be more fulfilled, more creative and more usefully to ourselves and others.

Author's note: Ed Adams birthed these principles and Japan adopted them. Adams was a manufacturing engineer who designed machine tools, work holders, decouplers, Poka-yokes and more for manufacturing cells and lean manufacturing vendors.

7.20 Implementation of Integrated Productive Maintenance and RAM

Here are the 12 steps involved in developing and implementing productive RAM programs (Nakajima).

Step 1:

- Announce top management's decision to introduce IPM.
- State IPM objectives in company newsletter/emails.
- Place articles on IPM in company newspaper/emails.

Step 2:

- Launch an educational campaign.
- For managers, offer seminars/retreats according to level.
- For internal customers, provide slide presentations.

Step 3:

- Create organizations to promote IPM.
- Form special committees at every level to promote IPM.
- Establish central headquarters and assign staff.

Step 4:

- Establish basic IPM policies and goals.
- Analyze existing conditions.
- Set goals
- Predict results

Step 5:

- Formulate master plan for development.
- Prepare detailed implementation plans for the five foundational activities

Step 6:

- Hold a TPM kickoff event.
- Invite external customers, affiliated and subcontracting companies

Step 7:

- Improve effectiveness of each piece of equipment.
- Select model equipment.
- Form project teams

Step 8:

- Develop an autonomous maintenance program.
- Promote the Seven Steps
- Build diagnostic skills and establish worker procedures for certification

Step 9:

- Develop a scheduled maintenance program for the maintenance department.
- Include periodic and predictive maintenance.
- Include management of spare parts, tools, blueprints, and schedules

Step 10:

- Conduct training to improve operation and maintenance skills.
- Train leaders together·
- Have leaders share information with group members.

Step 11:

- Develop initial equipment management program.
- Use MP design (maintenance prevention).
- Use start-up equipment maintenance.
- Use life cycle cost analysis.

Step 12:

- Perfect IPM implementation and raise IPM levels.
- Evaluate for PM prize.
- Set higher goals.

7.21 Summary

Increase Reliability in the two main components of Lean manufacturing

- Hardware -the machine tools that are necessary to produce the product Software- the data and logic systems that control the machine tools and collect and distribute data essential to the control of the manufacturing system
- Workforce Reliability- the people required to actually operate the machinery within the manufacturing system

Promote Reliability from 3 different viewpoints

- Project Managers.- up time vs. down time
- Plant Managers- relative frequency and severity of unanticipated events
- Equipment Manufacturer- financial cost, engineering effort, and reputation costs associated with a new technology

Recognize the basic Do's and Don'ts of Reliability

Do .....

1. Understand failures caused by interactions
2. Be aware of program timing
3. Understand what customers expect
4. Use common systems and methods
5. Communicate with other engineers
6. Track vendor performance
7. Design for easy assembly
8. Consider external influences
9. Regard reliability as a moving target

Don't .....

1. Invent during product development process
2. Focus too narrowly and forget your component interactions
3. Think your job is complete if everything is running smoothly
4.

Think about *faults* rather than *failures* when considering the reliability of a system. Faults can be wrong information, disconnected lines, operator induced crash, or just inappropriate responses. *(From the International Journal of Quality and Reliability Management, Volume 14).*

Standardize Support Functions

- In the lean production factory system, all support functions should reinforce the core principles of stability and continuous improvement through employee involvement.
- Standardization of both machines and work routines lends stability to the workplace and production process.
- Preventive maintenance – Regular planned activities which prevent break downs in machinery and assures a stable, predictable production environment.
- The *Five S Principles*, for keeping work places clean, orderly and always ready at the command of the worker smoothes the road to effective work.
- *Kaizen* activities, *Quality Circles* and an employee *Incentive Program* draw the minds and energies of the workforce into the production process. Using active teams, employee suggestions and a systematic approach to problem solving, stability allows continuous improvement through people.

Over 50 years ago, the Japanese adopted preventive maintenance from the United States. They have since developed and improve the preventive maintenance program into what they presently call Total Productive maintenance. The objective of integrated preventive maintenance is to prevent failure of all equipment before it actually occurs. The objective of total productive maintenance is to increase productivity, improve efficiency, increase the percentage of time equipment is able to operate, and minimize the number of required steps by keeping it simple. The program requires heavy involvement with production workers. This is shared responsibility.

Review Questions

1. Why is the Lean Manufacturing system designed to run at Less-Than-Full-Capacity (LTFC)?
2. What are the benefits of Preventive Maintenance
3. Discuss five pillars of IPM.
4. What are the factors that lead to machine breakdowns? List a few causes.
5. What is the importance of piloting in IPM?
6. What are the three elements of standard work, why is SOH so important in standard work?
7. What are the functions of standard work charts?
8. List five principles of Continuous Improvement and explain.
9. Explain the “Bath Tub” plot of failures.

10. "Any system that depends on humans is unreliable!" Explain.
11. List Ed Adam's five P's for the people in the manufacturing systems.
12. What are Nakajima's 12 steps for developing and implementing an integrated maintenance program ?

Reliability....Availability......Maintainability

Chapter 8
Balancing and Leveling Production

The slower but consistent tortoise causes less waste and is much more desirable than the speedy hare that races ahead and then stops occasionally to doze. The Toyota Production System (TPS) can be realized only when all the workers become tortoises *Taiichi Ohno, 1988*

8.1 Introduction

In addition to implementing the Lean practices of Just-In-Time, pull production, and minimum WIP manufacturing cells, a fundamental requirement of smooth, synchronized Lean Manufacturing is to balance, smooth and level production requirements in *Mixed Model Manufacturing*. It is important to create and maintain even, consistent flow of work-in-process, including raw materials. Production requirements need to be allocated to Workstations in fixed order release quantities and then synchronized by using small batch sizes. It is also critical to minimize setup and changeover times to create smooth, uninterrupted flow of products through an assembly or manufacturing sequence. Taiichi Oho called this SMED (**S**ingle **M**inute **E**xchange of **D**ies) or QC (**Q**uick **C**hangeover) SMED was a term that was initially applied to reducing the time required from changing dies in a drill press from hours to minutes. We will categorically state that a balanced and leveled production system is a necessary prerequisite to Lean Transformation. This fundamental principle was recognized very early in the Toyota Production System (TPS).

"In general, when you try to apply TPS, the first thing you have to do is to even out or level production. That is the responsibility primarily of production control or production management. Leveling the production schedule may require some front-loading of SOME order release quantities, or postponing SOME shipments, and you may have to ask SOME customers to wait for a short period of time. Once the production level is more or less the same or constant every month, you will be able to apply the principles of pull systems and balance the assembly lines. But if production varies from day to day, there is no sense in trying to apply other (Lean Engineering) systems because you simply cannot establish standardized work under such circumstances"

Fujjo Cho, President, Toyota Motor Corporation

There are a couple of remarkable things to be noted about Mr. Cho's comments. *First*, this is not a Manufacturing Engineer or a Production Foreman making these comments.....this is the President of the company. Just as Deming observed and stated concerning Total Quality Management, there can be no real transformation unless higher management has completely bought into the program and supports a paradigm shift in manufacturing philosophies. *Second*, it completely dismantles the piecemeal approach to implementing Lean Manufacturing by starting with Kaizen Events and haphazardly implementing Pull Production. If Lean Manufacturing is going to be implemented, it MUST start at the top and require the involvement of every company employee. It MUST be preceded by a total and complete examination of the Manufacturing Enterprise, and it MUST be accompanied by analysis and modeling exercises to help determine the correct place to start. Piecemeal implementation may sometimes help, but our observations are that unless coupled to overall system objectives, it will only hinder or destroy the journey. It is our opinion that Lean Implementation must be preceded by a comprehensive Systems Analysis, and then followed by a roadmap and plan of integrated change. The primary focus of most companies on *Muda, Kaizen Events* and *Kaizen Blitzes* is both appealing and intuitive, but sometimes extremely short sighted. It is relatively easy to *recognize* waste, but a more difficult and correct path is to aggressively pursue waste reduction by executing *root cause analysis* and killing the source of the problem…not the symptom of the problem.

The subject of this Chapter is to address the difficult problems of balancing and smoothing workflow to support a Lean transformation. In the limited space available, we can do little more than to emphasize the importance and magnitude of this problem, and to demonstrate representative solution techniques. The process of leveling and smoothing product flow will be called *Heijunka*, and we will broaden the use of this term to include single or multiple product flows, mixed model production and the visual procedures of *Andon*. Andon refers to any mechanism which immediately signals that problem has occurred on the shop floor which will disrupt or stop production. It carries with it the authority of any shop floor worker to shut down any machine or process in the company .

Historically, it is probably fair to say that production control, accounting functions and even shop floor foremen have been opposed to the Heijunka practices of mixed model sequencing and small lot sizes. This fact stems from the persistent use of MRP *push* order release systems; the accounting philosophy that large lot sizes are needed to amortize equipment procurement costs; and the manufacturing philosophy that long production runs spread setup and changeover times across many production units. This *strategy* is also consistent with the erroneous concept that material handling should follow two primary rules; move things in large quantities that usually match production runs, and move product when something has completed a material transformation at a particular machine or work station, and not when it is needed. Heijunka can properly be defined as *creating level or constant order release to every Workstation on the factory floor such that volume and product mix combine to create a consistent flow of work.* ***Lean flow*** primarily means the avoidance of large lot sizes, while ***consistency*** means that every work Workstation will be working on the correct product, in the correct quantity, in the correct sequence, at the correct point in time. Large lot sizes contradict and defeat this logic. The Toyota Production System developed ways to effectively move and process individual units or small lot sizes. These practices were largely enabled and preceded by *SMED.* SMED is directed to minimize setup and changeover times and support the philosophy of Just-In-Time. The term Heijunka relates to two separate but interrelated principles. The first is to level production by *tasks,* the second is to level production by *volume.* Successful Heijunka has as its goal the coordinated movement of work-in-process from order release to order completion, such that all materials flow in a uniform fashion and arrive at final assembly or final production in the right *quantities*, at the right *time* and in the right *order*. Expediting and priority scheduling are to be avoided at all times. There are two key concepts associated with this definition of Heijunka which we will address in this Chapter. The first is *balance* and the second is *synchronization.* Balancing workload implies equalizing the amount of time spent at every Workstation, consistent with demand or Takt time. Synchronization is determining the correct product mix delivered to each Workstation such that variance is minimized and production resources equally utilized. The key issues are to ***avoid*** (1) Congesting and slugging the system with large lot sizes;

(2) Starving the system due to poor order release practices, and poorly sized transport lot sizes; (3)Workstation imbalance and (4) Whiplashing the system due to poor order mix policies in Mixed Model Production (MMP). If these basic operating procedures are accomplished, Just-in-time manufacturing and pull production are just a step away.

8.2 Balancing Workload

Balancing workload in either single model or multiple model production lines is a relatively old Industrial Engineering problem. Modern Lean Engineering production practices have caused this problem to re-emerge with renewed importance. Traditionally, dedicated assembly lines and Focused Factories were designed and implemented as older factories were being rebuilt. Large, cumbersome machines were placed in a common work area, and little thought was given to moving them around to different uses or locations. All of that changed with the advent of flexible and versatile manufacturing, particularly within the Lean Manufacturing context. Modern production management is driving toward flexibility not only by multiple products but variety within products. In any case, if the product mix, product demand or product design is changed; the production system usually needs to be re-sequenced, re-balanced and often re-configured. Two similar manufacturing paradigms have emerged. The first is the serial production line, which is really not new at all; dating back to the classic work of Henry Ford. The second is the U-shaped Manufacturing Cell or the U-shaped production cell. These are discussed in great detail in Chapters 2-4. In either case, any dedicated arrangement is now usually referred to as a *Focused Factory* (Burbridge), or as a *Mass Production Line*. Each are composed of Workstations dedicated to specific products and tasks manufactured in large quantities. The objective of Heijunka is to achieve the advantages of large volume production by improving the flow of a process to better match customer demand, reduce waste, and decrease or eliminate batch processing. The idea is to evenly distribute both the volume of work and the mix of work evenly across the production line. There are two major design decisions to be made, and they are not independent. The first is *how many Workstations are needed to satisfy production requirements (demand)*, and the second is *what set of tasks are to be allocated to dedicated Machine or process operators*? To address these issues, we must first decompose the production process into a linked set of individual tasks, and then allocate these tasks to workers in U-Shaped Lean Cells. For each task, it is necessary to have a standard time. In Single product production, this is determined by Work and Time standards for each part type. In Mixed Model Production, we will only be concerned with the *average time* to complete a task or processing step. There is ongoing current research directed to stochastic task times, but algorithms are currently not widely available to address this complexity. Hence, for our purposes (and in most industrial settings) we will only use an average time for each task.

Having specified a set of required tasks, and the corresponding time required to execute each task, we will now need to determine the precedence relationship between each of the tasks. Precedence relationships specify which task or set of tasks must be completed before other tasks can be started. An analogous problem is the set of precedence relations which define a PERT/CPM network. Having specified the set of required tasks, task times and precedence relationships between individual tasks; we are now ready to determine how many Workstations will be required and how much workload (tasks) to assign to each Workstation. The basic problem is that both customer demand and the amount of available working time will determine the limit on how much work/tasks can be assigned to any one Workstation. Assume that the planning horizon is one week of 5 days working 8 hours per day, and that a demand rate of 100 parts per week must be made and delivered. For this example, one part must be completed every 24 minutes. In this book we call this the required *Cycle Time* for each part. It is not the same as *Throughput time*. The inverse of Cycle Time using this definition is the *Production Rate* (parts/unit time). It is obvious that if multiple Workstations, assemblies or subassemblies are used to produce this part, each of these areas cannot take more than an *Average* time of 24 minutes between part completions. This time is called the *Takt* time. This upper bound on time between part completions must be strictly enforced for both single and mixed model part production lines. Notice that in Mixed Model Production lines some parts may take a little longer and some a little less time, but *average* production must strictly adhere to Takt time. Of course, we recognize that there should be a minimum amount of spread between the minimum and maximum Cycle Times, and so we see that our old enemy *Variance* is once more very important and must be minimized.

8.3 Balancing Single Product Production Lines

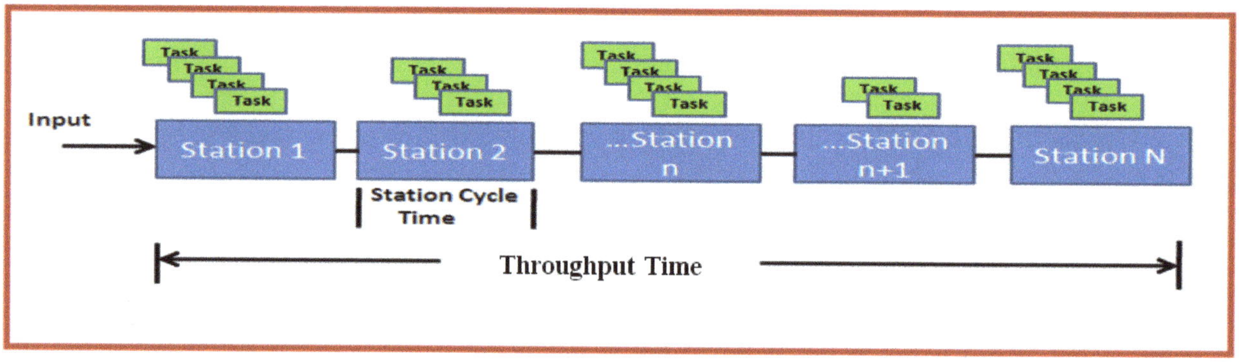

Figure 8.1
A Typical N Workstation Production Line

Figure 8.1 shows a typical serial production system or final assembly for a single product. Each Workstation will be assigned a fixed number of Tasks to execute. The number of Tasks assigned to each Workstation is constrained by both precedence relationships and by the required Takt time. The primary goal is to assign K production tasks to a series of N Workstations such that the total workload is evenly distributed across all Workstations.

The required Throughput Time to complete all tasks assigned to the line is the total time required to execute all tasks. The required time between part completions must be strictly less than or equal to the *Takt time,* and is given by: $Takt = \frac{Total\ Production\ \ Time\ Available}{Total\ Demand}$ over a specified planning period. Takt is more of a goal than an absolute, because it is rarely the case that tasks can be assigned to workstations such that each Workstation has identical throughput times due to variability in task times. An ideal balance for a production line is one where all workstations take exactly the same amount of time to execute the set of assigned tasks, and that time is as close as possible to the required Takt time. This primary goal is usually difficult to achieve because the assignment of tasks to workstations is dictated by the following conditions.

(1) Tasks must be assigned to each Workstation in a specific order that cannot be violated without design change. The order is defined by a set of *Precedence Constraints* which dictate that certain tasks must be completed before other tasks can be started.
(2) Workstation assignments should be done in such a way as to level or smooth workload. The goal is to meet Takt time.
(3) The bottleneck workstation *average* cycle time cannot exceed Takt time
(4) Variance in processing times cannot usually be totally eliminated. Since Tasks are assigned to workstations to meet Takt time, the variance of individual part cycle times must be Minimized

Assume that a production line consists of N Workstations in series such as those shown in Figure 8.1, and that there is a set of K Tasks which must be completed in this work sequence. The nth Workstation is to be assigned a set of K Tasks that will individually take an average of t_j j =1, 2.... K time units to complete. We assume that the sum of all task times allocated to a single Workstation k cannot exceed a specified Takt time and is given by $Takt_k$. The following relationship must hold for any line assignment.

$$\sum_{k=1}^{N} Takt_k = \sum_{j=1}^{K} t_j$$

The maximum workstation Takt time that can be allowed is calculated as the available production time per planning period (P_p) divided by the demand rate per planning period (D_p). In addition, the amount of work assigned to each workstation cannot violate any precedence constraints. To achieve workstation balance (Takt time - tt_n) at each workstation should be minimized. Note that the allowable Takt time is a constant across all workstations. In other words, we can define the problem as follows.

$$\text{Minimize: } \sum_{n=1}^{N}(TAKT\ time - tt_n) \qquad (7.1)$$

Subject to precedence constraints and
$tt_n \leq$ Takt Time n = 1,2,....N

Note that the minimum of Equation 8.1 can be achieved by minimizing workstation idle time or by minimizing the number of workstations used, subject to the precedence and Takt time constraints. This is true because:

$$\text{Min } \{ \sum_{n=1}^{N}(TAKT\ time - tt_n)\} = \text{Min } \{N^*\text{Takt } time + \sum_{n=1}^{N} tt_n\} \qquad (7.2)$$

Since $\sum_{n=1}^{N} tt_n$ is just a constant,

$$\text{Min } \{N^*\text{Takt } time\} = \text{Takt } time^* \text{ Min } \{N\} \qquad (7.3)$$

Hence, one can choose to assign tasks to a series of N workstations while attempting to minimize idle time with respect to Takt time, or try to assign the set of tasks to a specified number of workstations without exceeding Takt time at any individual workstation. The exact solution to either problem is very difficult, particularly for industrial problems where the number of tasks and Workstations are large. For this reason heuristic solution procedures have been proposed by a number of researchers. We will generate an initial, feasible solution to this problem by using two representative techniques; (1) A simple Assignment Rule called *The Largest Eligible Time Rule* (LETR), and (2) an effective procedure called the *Ranked Positional Weighting Technique* (RPWT). We will then discuss heuristic based techniques to improve on the initial solution. The RPWT is widely used in industry, and lends itself well to a microcomputer.

8.4 A Single Product Balancing Problem

Specialized Products, Inc. is under contract to Phillips's Sweat Shop to produce 400 Widgets per month. Phillips's shop works two 8 hours a day shifts, 5 days per week with 30 minutes off each shift for lunch. The first thing is to create a 'Precedence Diagram" which shows the precedence structure between tasks. The precedence diagram for our Widgets is shown in Figure 8.2.

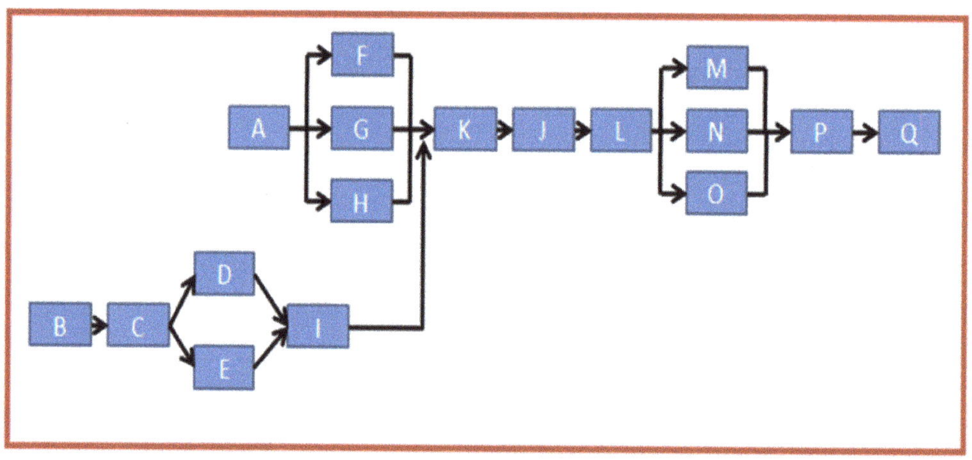

Figure 8.2
Precedence Diagram for Phillip's Sweat Shop

Black's available working time is $(8\frac{hrs}{shift}*2\frac{shifts}{day})$ - $(2\frac{shifts}{day}*0.5\frac{hrs}{shift})$ = 15 $\frac{hrs}{day}$.
Daily product demand is $\{\frac{400\ Widgets\ /mo}{20\ Days\ /mo}\}$ = 20 Parts/Day, so Takt time is $\{\frac{15\ Hrs/Day}{20 Parts\ /Day}\}$ = 0.75 Hrs/Part or = 45 Minutes/Part. This is the *drumbeat* of the production line. In other words, assigned task times to each Workstation cannot exceed 45 minutes or 20 units per month cannot be produced on time. Table 8.1 shows the average task times to be assigned to each station.

Table 8.1
Black's Sweat Shop Precedence Tasks Times (Average)

Task ID	Task Time (Min)	Task ID	Task Time (Min)	Task ID	Task Time (Min)
A	21	G	15	M	21
B	36	H	22	N	36
C	15	I	25	O	12

D	25	J	20	P	19
E	14	K	12	Q	44
F	41	L	18		

The total task times are T=396 min. Since the maximum workstation cycle time is 45 min, the minimum number of workstations required is C=$\frac{396}{45}$ = [8.8] = 9, where the brackets indicate that any fractional number should be rounded up to the next integer. This is a lower bound of the required number of workstations (see the following example) due to the differences in individual task times and the constraints imposed at each workstation. We are now ready to assign tasks to workstations. We will first use the *Largest Eligible Time Rule* (LETR).

8.4.1 The Largest Eligible Time Rule (LETR)

The LETR rule proceeds as follows. Let "j" represent the Workstation being considered.

(1) Construct a list of all tasks, task times and precedence constraints. Set j=1.

(2) Identify all tasks that can *immediately be* started (no precedence constraints) and place those tasks in a list **L**. Order list L from the task with the longest processing time to the task with the shortest processing time.

(3) Open Workstation j for assignment of tasks. The available unassigned time is initially the Takt time. From the ordered list L, choose the task with the LARGEST processing time that can be allocated to Workstation j without violating the unassigned time. If the first eligible task is too long in duration, try the second task in list L. Proceed until no task can be allocated or all tasks have been assigned. (By construction, at least one task can always be assigned). If the entire list of tasks has been assigned, terminate. Otherwise; empty set L, open Workstation j+1 and Repeat step (2).

8.4.2 The Assignment Procedure

Set j=1. Open Workstation 1. The set L is initially L = {A, B}. Unassigned time is Takt time=45. Since Task B has the largest processing time, assign to Workstation 1. Remove Task B from further consideration, and add Task C to set L. Set L is now L = {A, C}. The remaining unassigned time for Workstation 1 is [Takt time-t_B] = [45-36] = 9. The required processing times for both Task A and Task C exceed the time available at Workstation 1. Close Workstation 1. Set j=j+1=2.

Open Workstation 2. Set L is still L = {A, C}. Set the remaining unassigned time at Workstation 2 to Takt time. Choose Task A for assignment to Workstation 2 since it has the largest processing time and does not exceed the available Workstation time. Remove Task A from further consideration; add Tasks F, G, and H to list L.List L is now L={C, F, G, H}. The remaining unassigned time is [Takt-t_A] = [45-21] = 24. The largest task time which can now be assigned is Task H. Assign Task H to Workstation 2, and purge Task H from the list L. Note that Task F was passed over because it's processing time was too large for the unassigned time remaining at Workstation 2. List L is now L = {C, F, G}, and the remaining unassigned time is now [45-21-22] = 2. No further assignments can be made at Workstation 2. Close Workstation 2.

Set j=3. Open Workstation 3, and set L= {C, F, G}. Set the remaining unassigned time at Workstation 3 to Takt time. Proceed with this same methodology until all task assignments have been made. The final solution is shown in Figure 8.3.

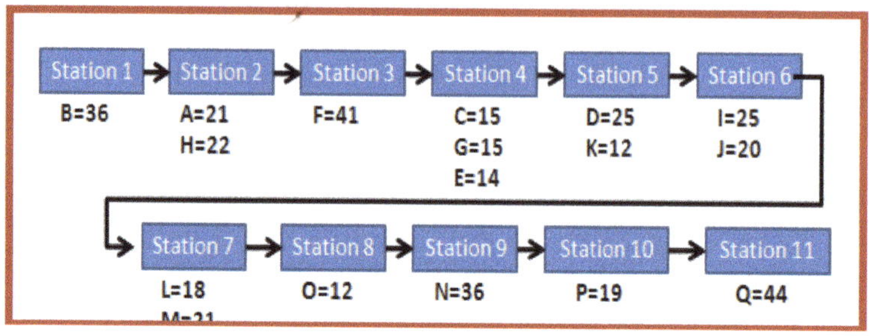

Figure 8.3
Phillips's Sweat Shop Initial Task Assignments

The task assignments result in a line with 11 workstations. The assignments all satisfy Takt time and no precedence relations are violated. However, *looking at the Workstation Utilizations we find that the* ***line efficiency*** *is:*

$$\text{Line Efficiency} = \left(\frac{\sum_{n=1}^{K} t_n}{Number\ \ of\ Stations\ *TAKT\ Time}\right) = \left[\frac{396}{(11*45)}\right]*100 = 80\%.$$

Another performance measure is *Workstation Efficiency*, which is defined as the ratio of individual Workstation usage to Takt time. For example, Workstation 11 has an efficiency rating of $E_{11} = \frac{44}{45} = 0.98$.In contrast, Workstations 8 and 10 have efficiency ratings of only 0.27 and 0.42, respectively. The worker at Workstation 6 is busy 100% of the time, while the worker at Workstation 8 is busy only 27% of the time. The Balance is poor.

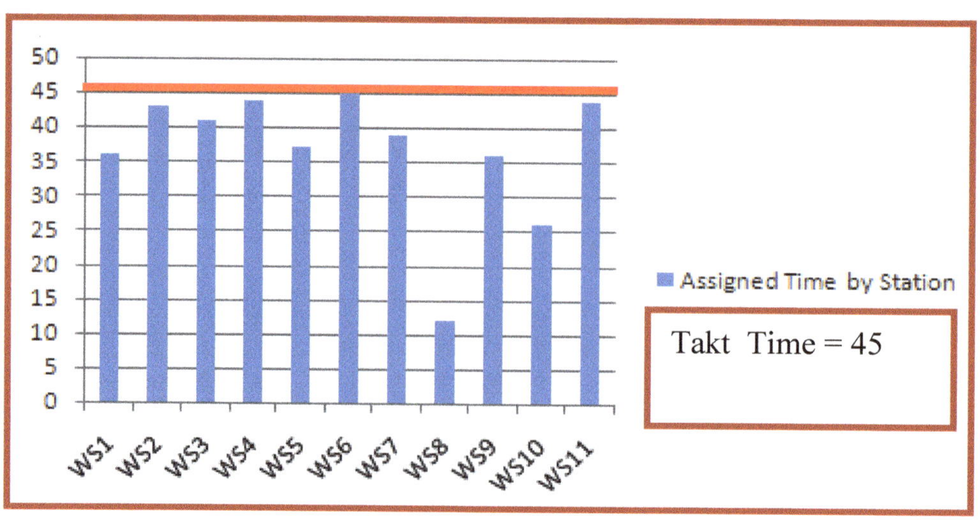

Figure 8.4
Station Time Assignments vs. Takt Time

The *Longest Time Heuristic* always produces a feasible solution, but it is rarely optimal or even near optimal. In almost every case, the initial solution can be improved by moving around feasible task assignments. For example, at Workstation 7 we could we could assign Task l to be worked first and Task O to be worked second, resulting in a Workstation cycle time of 30 minutes. Workstation 8 can be assigned Task N with a Workstation cycle time of 36 minutes, and Workstation 9 assigned Tasks M and P with a Workstation cycle time of 40 minutes. Workstation 10 and Workstation 11 remain unchanged. The new assigned Tasks and Task times by workstation are shown below. The balance is now much better, with every workstation being utilized between 80% and 100%, with the exception of Workstation 7 which is 67%. However, this could be good since note that Workstation 6 is running at 100% utilization and is the bottleneck workstation. Since Workstation 7 is right next to Workstation 6, if Workstation 6 happens to get behind, Workstation 7 might help out Workstation 6. This, of course, emphasizes the need to cross train workers at all Workstations, and promote friendly relationships. Finally, observe that there is one less workstation in this allocation scheme (10) than in the initial solution (11).

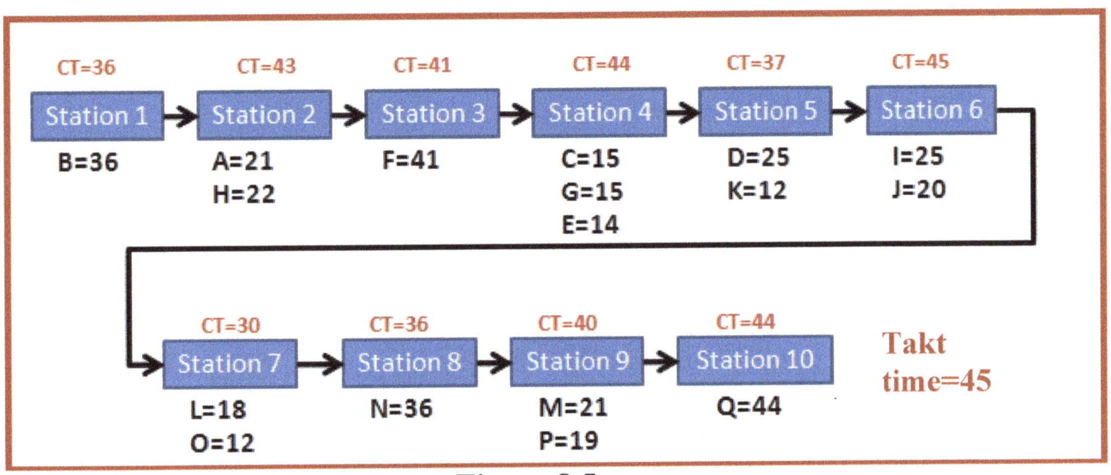

Figure 8.5
Modified Task assignments

8.4.3 The Ranked Positional Weighting Technique (RPWT)

It is intuitively clear that if a good initial solution can be generated, it can be used without further modification, or it can be used as a feasible starting solution for further refinement. The previous example was only given to present the basic principles of Line Balance. The second rule which we will use is very popular, and has been shown to usually dominate the LETR rule. This rule is called the *Ranked Positional Weighting Technique* (RPWT). The basic procedure is as follows.

(1) Identify all tasks, task times and precedence constraints. Set j=1
(2) For each task, calculate the SUM of that particular task time and ALL OTHER task times for those tasks which cannot be started until that particular task is finished. This is the Ranked Positional Weight (RPW) for that task. Open up Workstation j.
(3) Identify all tasks that can be immediately started (no precedence constraints) and place those tasks in a list "L". Order list L from the task with the largest RPT to the task with the Smallest RPT.

(4) Open Workstation j for assignment of tasks. The available unassigned time is initially the Takt time. From the ordered list L, choose the task with the LARGEST RPT that can be allocated to Workstation j without violating the unassigned time. If the first eligible task is too long in duration, try the second task in list L. Proceed until no task can be allocated or all tasks have been assigned. (By construction, at least one task can always be assigned). If the entire list of tasks has been assigned, terminate. Otherwise; empty set L, open Workstation j+1 and repeat step (3).

8.4.4 Manufacturing a CD/VHS Combo Player Unit

Consider a simplified manufacturing line for a standard VHS/CD player unit. The basic process consists of cutting a case and bottom out of a piece of sheet metal, drilling holes, cutting out user interface slots, and installing mounting brackets ,various controller cards, and screwing the unit together. The assembly flow chart is shown in Figure 8.6

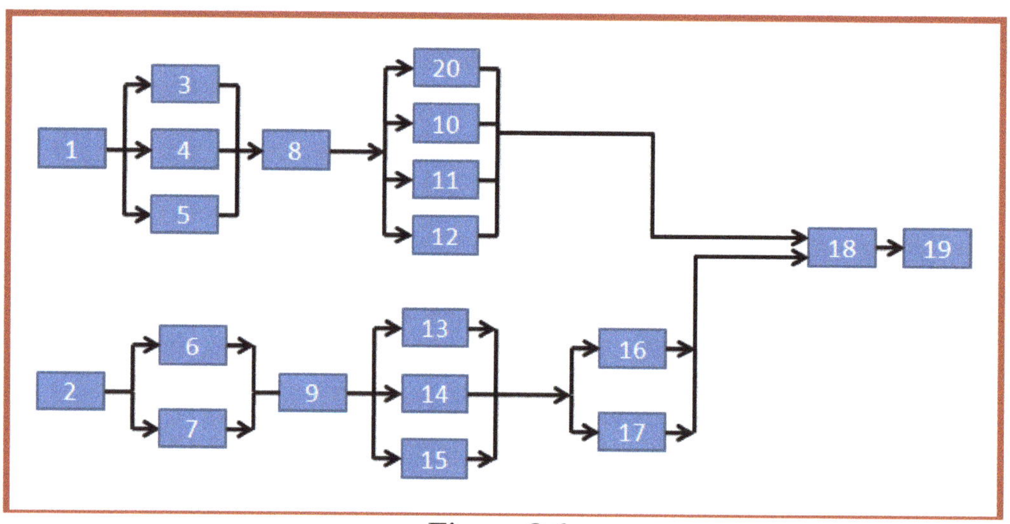

Figure 8.6
A CD/VHS Combo Player Unit

The tasks, task times and positional weights are shown in Table 8.2.

Table 8.2
RPT Example Data

Task ID	Task Time (Sec)	RPW	Task ID	Task Time (Sec)	RPW	Task ID	Task Time (Sec)	RPW	Task ID	Task Time (Sec)	RPW
1	50	331	7	42	290	13	30	143	19	15	15
2	40	363	8	20	186	14	25	138	20	42	85
3	40	226	9	45	248	15	35	148			
4	25	211	10	36	79	16	30	73			
5	30	216	11	25	68	17	40	83			
6	32	200	12	20	63	18	28	43			

The available working time is $(7.5 \frac{hrs}{day})(60 \frac{min}{hr})(60 \frac{sec}{min}) = 450 \frac{min}{day} = 27{,}000 \frac{sec}{day}$, and the daily demand is $325 \frac{unit}{day}$. The total time required to assemble a unit is 650 sec. Hence, the Takt time at any one workstation is Takt = $\frac{27{,}000}{325}$ = 83 seconds per unit. This is the calculated Takt time, which is the drumbeat of the production line. In other words, no more than 83 seconds of work can be assigned to any one workstation or 325 units per day cannot be produced .The total task times are T=650 min. Since the maximum Workstation cycle time is 83 seconds and the minimum number of workstations required is C=$\frac{650}{83}$ = [7.8] = 8. We are now ready to assign tasks to workstations.Open Workstation 1 and create the list L. In this case, L= {1, 2}. Since Task 2 has the highest RPW and its task time is less than the available remaining time, assign this task to Workstation 1. The remaining time available at Workstation 1 is [83-40] = 43. Remove Task 1 from the L list, and add Workstations 6 and 7. L= {1, 6, 7}. Task 1 is the next preferred task to assign, but requires 50 seconds. Remove this task from L. Now L= {6, 7}. Both tasks can be assigned, but Task 7 has the highest RPW (290). Assign Task 7, and then remove Task 7 from the L list. The L list is now L= {1, 6}. The remaining time available at Workstation 1 is [83-40-42] = 1. The L list is now L= {1, 6}, but both require too much time to be assigned to Workstation 1. Close Workstation 1, clear the list L, Open Workstation 2 and repopulate L with L= {1,6} . Since Task 1 has the largest RPW, we assign it to Workstation 2. Remove Task 1 from the L list, and add tasks 3, 4 and 5. The L list is now L= {6, 3, 4, 5}. The remaining unassigned time at Workstation 2 is [83-50] = 33. Of these four eligible tasks, Task 6 is preferred with RPT=280. The required Task 6 time is 32 minutes. Assign Task 6 to Workstation 2, purge Task 6 from the L list, and add Task 9. Now L= (3, 4, 5, 9}. The remaining unassigned time at Workstation 2 is now [83-50-32] = 1. No tasks can be assigned. Open Workstation 3, reconstitute the L list, and proceed as before to assign tasks to Workstations until all tasks are assigned to some Workstation. The results are shown in Figure 8.7 and the line balance in Figure 8.8

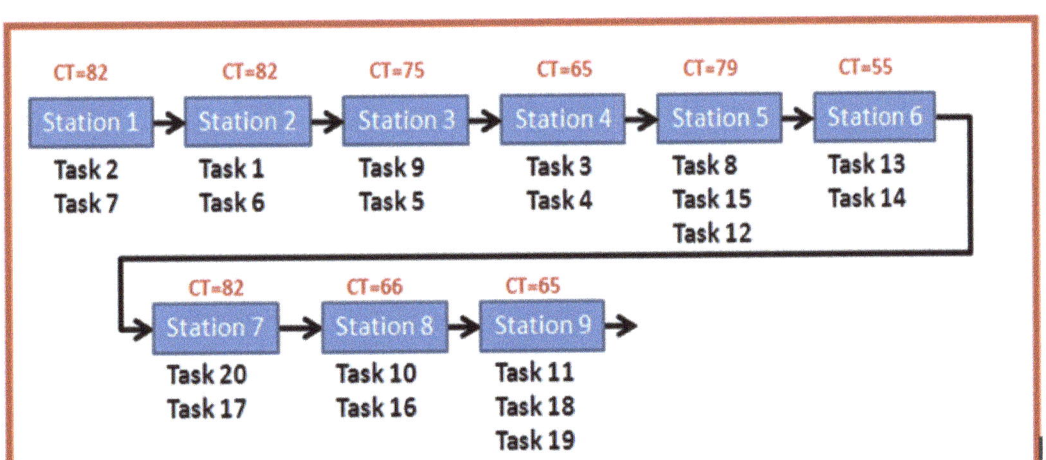

Figure 8.7
Combo Manufacturing Initial Task Assignments

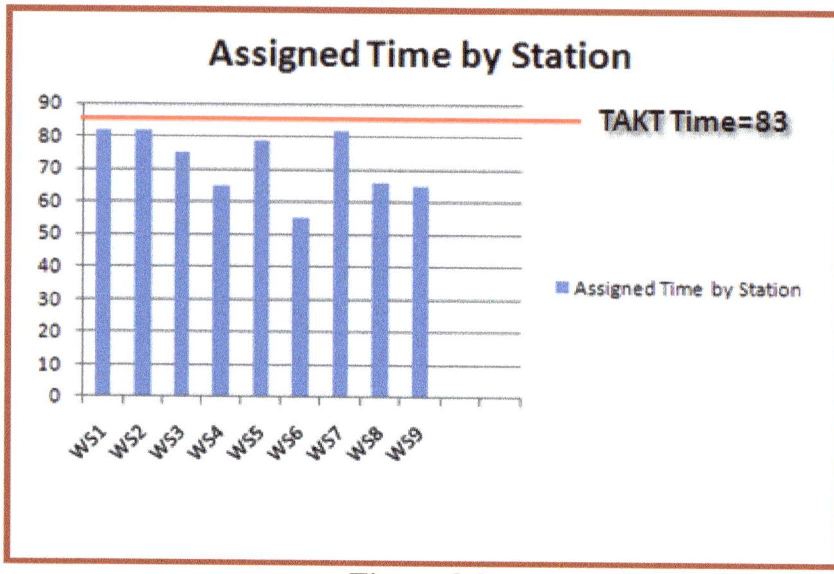

Figure 8.8
Station Line Balance

The solution is fairly well balanced, with Balance Efficiency = ($\frac{650-103}{650}$)*100 = 84.2%. However, this balance wastes 15.8 % of capacity. Workstation 6 is only utilized 65.5%, and Workstations 1, 2 and 7 are working 97.7% of the time. Workstations 1, 2 and 7 are all bottleneck workstations with Workstation CT_{WS}=82. This allows for some slippage since Takt = 83, but the margin is slim.

Again, improvements can be made starting with this solution. The reader can verify that the following steps are feasible.

(1) Move Task 6 at Workstation 2 to Workstation 1.
(2) Move Task 7 at Workstation 1 to Workstation 3.
(3) Move Task 5 at Workstation 3 to Workstation 2.
(4) Move Task 17 to Workstation 6 and Task 16 to Workstation 7.

The new assignments are shown in Figure 8.9

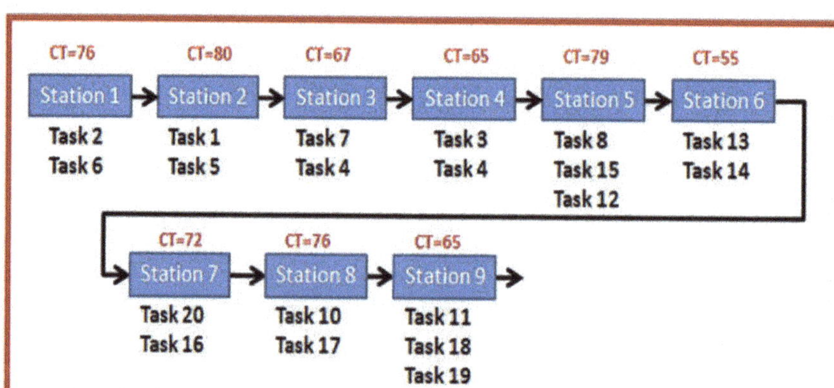

Figure 8.9
Final Combo Task Assignments

The results are significantly better. The maximum Workstation Cycle Time is now 80 at Workstation 2. The line balance is good except for the idle time at Workstation 6. The Line Balance Efficiency is now = $[\frac{(650-104)}{650}]*100 = 84\%$.

8.5 Mixed Model Production

It is not uncommon to find serial Production Lines in almost any type of manufacturing or production facility, and Final Assembly in a Lean Manufacturing Facility. Whether or not facilities can be dedicated to a single product rest upon many business factors; including volume of production, market share, inflexible production lines that are difficult to set up and the degree of automation desired. Dedicated lines often result in Batch production or long production runs, and we have already discussed how this inflates both WIP and throughput times. Modern Lean Engineering encourages flexible, versatile and agile production capabilities. As already noted, this necessitates minimum set up and changeover times, a cross-trained workforce, flexible and versatile production equipment and Just in Time material handling. Conceptually, the Balance of a Mixed Model Production Line (MMPL) is no harder than Single Model Production Lines (SMPL), but in practice it is more difficult to control, sequence and tie multiple production lines together. We will now demonstrate how to balance MMPL's, and then point out how to improve the initial balance as we did for SMPL's.

8.5.1 Bottleneck Processes and Pacing Parameters

Inherent in all Balancing models is the principle of *Workstation Capacity*. It is always assumed that adequate capacity is available to process the offered workload. This is tantamount to saying that all processing Workstations have or will have enough processing capacity to meet the required demand. This is also assumed in Line Balancing for Lean production facilities, but it is enforced in a different way. When balancing an assembly line for either single or multiple products, the available production time over a specified planning period is divided by the demand rate(s) to calculate what we call Takt time. Every Workstation produces according to this Takt time. In a typical push system, final assembly is dictated by the rate at which subassemblies or work in progress arrives to final assembly. In a Lean Pull system, the output rate is dictated by what we will call the *Final Assembly Schedule* or FAS. Pull systems have already been discussed in Chapter 6. The fundamental assumptions underlying a FAS driven system is that each workstation has enough processing capacity to meet the FAS schedule; there are only minor deviations from the planned processing Workstation times; quality is high (no rework); equipment has 100% availability; and the product resources required arrive just in time, just at the right place and just in the right quantities. However, we have observed that line throughput is strictly dependent upon the slowest Workstation in the line, and we have called that the *Bottleneck Process*. The bottleneck process can be anywhere in the production line, and in actual practice the bottleneck process will periodically change positions in the line due to variations in demand and product mix. This phenomenon is called the *Shifting Bottleneck Problem*. Loading a production line being constrained by a bottleneck process is inherent in all assembly line balancing.

In a Lean production line the cycle time at any process step, including the bottleneck process, cannot exceed Takt time. Takt time is the *pacing parameter*, and line output cannot be more than the rate at which the bottleneck process produces. This is called the *Bottleneck Rate*. Note that the converse is *NOT* true. Goldratt called this the *drumbeat* of the production line. For mixed model production lines , keeping the production line and all of its individual Workstations in sync with the required Takt time is difficult to enforce. Note that without some form of production control, every assembly line, subassembly line and manufacturing cell would have its own drumbeat. In an ideal *synchronized* Lean production facility, every component production cell, subassembly cell and assembly line is controlled by the Takt time set by Final Assembly. In an ideal pull system, the *only* production component that is not driven by subsequent operations is the final assembly or shipping line, which sets the Takt time for the *entire system*. Ideally, the final assembly line will be the Bottleneck operation for the entire manufacturing facility, and every other manufacturing component follows the drumbeat of the FAS.

For single product production facilities, the manufacturing times at every workstation or process are simply the time to process each individual item, and Takt time is determined by the demand for that single product. However, in multiproduct production, each product type will have its own processing time and its own demand schedule. This is called *Mixed Model Production* (MMP) and necessitates the concepts of *average production time* and *average demand.*
The usual approach is to *aggregate* and calculate *weighted* average service times to get a composite service time using the appropriate demand rate for each product type. This procedure will be used when balancing multiproduct Lean production lines. This is usually referred to as balancing *Mixed Model Production* (MMP), and has two primary goals; (1) Balance the production line by assigning mixed model production requirements to each Workstation without exceeding an average Takt time, subject to any precedence constraints and (2) Establishing a Mixed Model Production Sequence (MMPS) which will minimize batching and provide smooth, uninterrupted product flow to each manufacturing operation.

8.5.2 Balancing Mixed Model Production Lines: Calculating a Weighted Average

We will demonstrate how to calculate a weighted average of processing times at each workstation by using a highly simplified example. The basic approach is to use a Weighted Average to determine the average Takt time and workstation processing time requirements across all products. In other words, we will balance to an average processing strategy. In this case, it is acceptable to have any one product exceed the Takt time requirement at any one workstation, as long as the *average* workstation cycle time across all products does not exceed Takt time. To demonstrate the basic procedure, we follow an example first published by Nicholas (18) in his excellent textbook *Competitive Manufacturing Management*. Consider the production of four different products with the demand, cycle time and processing time requirements shown in Table 8.3.

Table 8.3

A Four Product MMP Example

Product	Demand (per week)	Task 1 Time	Task 2 Time	Task 3 Time	Task 4 Time
A	500	1.40	0.60	***	1.20
B	250	1.55	2.20	1.75	3.20
C	300	2.60	***	3.22	2.70
D	200	1.35	2.75	1.25	1.00

Totals	1250	6.90	5.55	6.22	8.10
Weighted Average		1.710	1.12	1.323	1.93
	(Units)	(Min)	(Min)	(Min)	(Min)

We start by calculating the proportion of total production allocated to each of the four products. Define $p_k = (\frac{D_k}{\sum_{j=1}^{K} D_j})$, where D_k = demand for product k=1, 2.... K products. Hence, $p_1 = \frac{500}{1250} = 0.40 p_2 = \frac{250}{1250} = 0.20 p_3 = \frac{300}{1250} = 0.240$ and $p_4 = \frac{200}{1250} = 0.160$. The weighted average production time for Task 1 is given by (1.4*0.40) + (1.55*0.20) + (2.60*0.240) + (1.35*0.160) = 1.710 minutes. Assume the number of production minutes available each week = 3300 minutes. Hence:

$$\text{Takt} = \frac{3300\ Min/Wk}{1250\ Units/Wk} = 2.64 \text{ min/unit}$$

The Takt time for this MMP is given by 2.64 Min/Unit. This implies that one unit must be produced on the average every 2.64 Minutes, or the demand cannot be met. Note that this not imply that the time required to process each individual part must be less than or equal to 2.64 minutes. Some parts may exceed this number, and some parts will be less than this number. The only precedence constraints that will be enforced are that Task1 must precede Task 2 and that Task 2 must precede Task 3 for all products In general and more realistic problems with multiple products, each product will have routes with both common and unique Workstations. The goal is to assign product dependent tasks to a set of workstations such that Takt time of 2.64 min/part is not violated. We will use *the Longest Processing Rule*. The average Takt times are: Task1 = 1.710, Task 2=1.12, Task 3=1.323 and Task 4=1.93.Hence, the Assignments would be Task 1 to Workstation 1, Tasks 2 and 3 to Workstation 2 and Task 4 to Workstation 3. Three workstations are required. The result is shown in Figure 8.10.

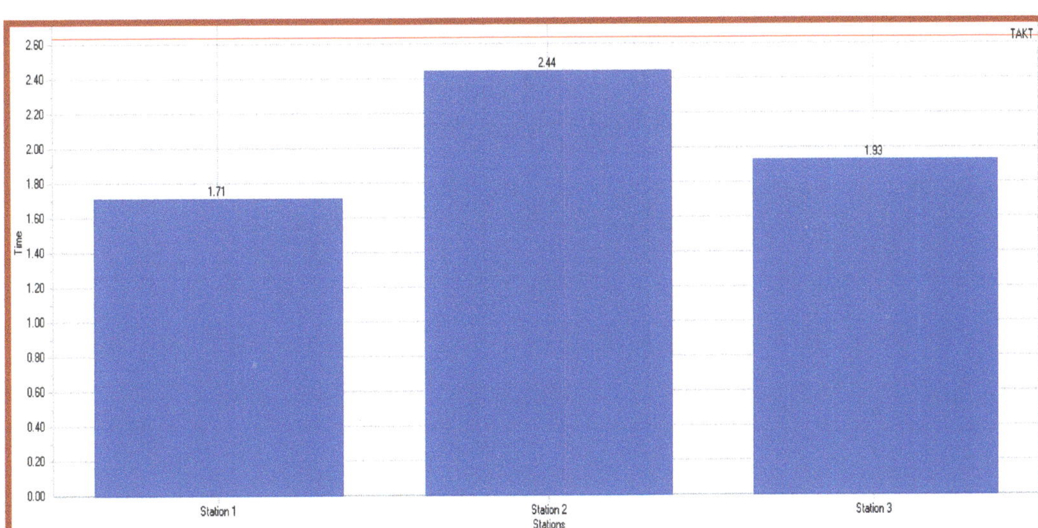

Figure 8.10
Initial Line Balance: LPT rule

For now, we will note that there are two basic procedures which can be used to balance a line. The *first* is to Minimize the number of Workstations required given a set of products, associated tasks, precedence relationships and Workstation cycle time (Takt). The *second* is to Minimize Takt time given a fixed number of Workstations. Both objectives can be addressed by a slight modification of either the Longest Time Rule or the Ranked Positional Weighting Technique previously presented. Both rules will provide a feasible solution, which will almost always require improvement using some heuristic.

We would like to point out that although some optimal solution procedures have been reported in the technical literature, any solution technique used to solve real problems will likely be based upon a heuristic procedure. That means that a two phase process will probably be needed to refine and rebalance an initial, feasible solution. This being said, the goal and fundamental issues of MMLB which we will discuss are true regardless of the solution technique used. We will address the basic problem, and then conclude with a discussion of the impact of additional real world constraints and how they can be incorporated.

The solution procedure will be based upon the RWPT heuristic presented in Section 8.4.3, and an initial balance will be found using *average* production requirements. Tasks will then be assigned such that the number of Workstations will be minimized subject to Takt time and predecessor constraints. After the assignments are made, we will suggest a *Mixed Model Sequence* which will smooth and level product flow.

8.6 A Mixed Model Line Balancing Example

To demonstrate Mixed Model Line Balancing (MMLB) we will assume that three different products (A, B and C) are to be manufactured. There are 17 tasks that must be accomplished, each with a product specific task time. Not every task is required for all of the products A, B and C. Table 8.4 shows the processing time requirements for each product by task, and the Weighted Average for each task. . The product mix is 50% product A, 25% product B and 25% product C.

Table 8.4
MMLB Processing Times

Part	T1	T2	T3	T4	T5	T6	T7	T8	T9	T10	T11	T12	T13	T14	T15	T16	T17
A	8	8	16	*	*	8	8	*	*	10	16	*	*	*	10	12	10
B	10	6	*	*	*	12	*	10	16	12	*	*	14	12	*	8	12
C	*	*	*	6	18	12	18	*	*	10	*	8	*	16	*	6	10
Wt Avg.	6.5	5.5	8	1.5	4.5	10	8.5	2.5	4	11.5	8	2	3.5	7	5	9.5	10.5

An asterisk indicates that a specific task is not required for that particular product. The weekly demands for products A, B and C are A=20 units, B= 10 units and C= 10 units. The assembly line is expected to work 960 minutes per day. Hence, the Takt time for this line is Takt =

$\left(\frac{960\frac{Min}{Day}}{40Day}\right)$ = 12 Min/Unit. In other words, a part needs to be produced on the average every 12 Minutes to meet demand. This is the *drumbeat* of the production process. The objective is to assign tasks to assembly workstations to meet Takt requirements The strategy that we will use is to once again use the *Ranked Positional Weighting Technique* (RPWT) to generate an initial line balance, and then improve that initial solution. For Mixed Model Production, the RPWT will use the *average weighted* Workstation time. The weighted service time will use the product demand percentages and the product dependent task times given in Table 8.4. The Weighted Average task times (WATT) are shown in the last row. For example, Task 6 WATT is given as 10. This is calculated as 0.50(8.0) + 0.25(12.0) + 0.25(12.0) = 10. The RPWT will now be applied as before to the set of ***average*** task times. The required tasks are numbered 1-17. The red letters of the alphabet (A, B or C) next to each task correspond to the products which must be processed through that task. For example, only product B requires Task 8, while Task 10 is required of all products. The precedence diagram is shown in Figure 8.11.Below each task box in Figure 8.11 is the Ranked Positional Weight for each task. For example, the Ranked Positional Weight for Task 12 is given by (10.50+9.50+7.0+2.0) = 29.0.Of course, the time to execute each task by product is product dependent as shown in Table 8.4 The solution procedure is as follows.

Open up Workstation 1 for task assignments. The initial list L of eligible assignments (those with no precedence constraints) is L= {1, 4}. Tasks 1 has the largest positional weight (257), so remove Task 1 from the list L and assign it to Workstation 1. Workstation 1 now has [12-6.5] =5.5 min of unallocated time. List L now becomes L= {4, 2}. Since Task 4 has the largest positional weight (251) and only requires an average of 1.5 minutes, allocate Task 4 to Workstation 1 and Purge Task 4 from the L list. Calculate the remaining available processing time [12-6.5-1.5] = 4 Min. Add Task 5 to the L list so that L= {2, 5}. Task 2 now has the largest positional weight (250.5), but the average time required for Task 2 (5.5 Min) exceeds the remaining available time (4 Min). Pass Task 2, and consider Task 5. Task 5 requires an average of 4.5 Min of processing time, which also exceeds the available remaining time at Workstation 1. Hence, close Workstation 1; Open Workstation 2 and reconstruct L = {2, 5}. Task 2 has the largest positional weight (250.5) and can be assigned. Allocate Task 2 to Workstation 2. Calculate the remaining available time at Workstation 2 [12-5.5] =6.5.No task can be added to the L list, so that L= {5}. Task 5 requires only 4.5 minutes of processing time, and 6.5 is available, so allocate Task 5 to Workstation 2; Calculate the remaining available processing time [12-5.5-4.5]= 2.The L list is now L= {6}. Since the processing time for task 6 (10.0) is more than the remaining available time, Close Workstation 2; Open Workstation 3 for assignments; set L= {6}. Proceed to allocate Tasks to Workstations until all tasks are assigned to some Workstation, such that precedence constraints and Takt time are not violated. The final Line Balance requires 11 Workstations and is shown in Figure 8.12. The Workstation assignments are shown below the task box and the Workstation Cycle Time above the task box. The Bottleneck Workstation is Workstation 7 (CT=12). The Manufacturing Cycle Time is given by (12*11) = 132 time units. The total line idle time is given by 24 Minutes, and the line Efficiency is given by [$(\frac{132-12}{132})$*100] = 90.91 %.

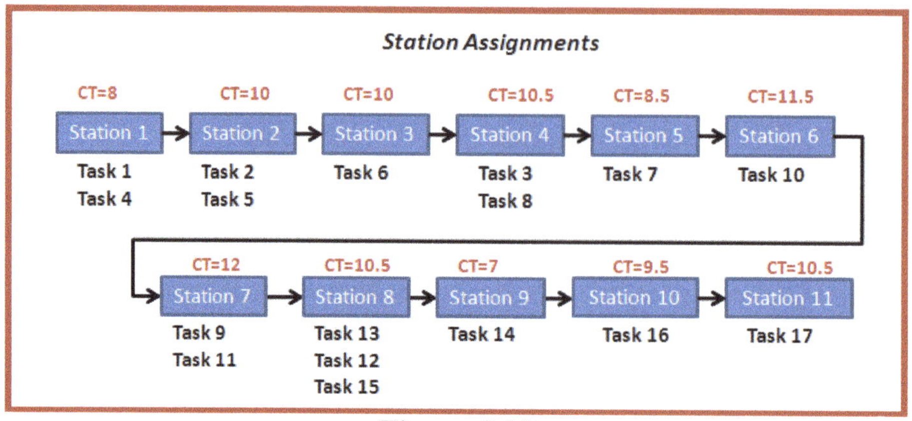

Figure 8.12
Final Station Assignments

A *Balance Graph* illustrating the required processing times for all four products and the average required time is shown in Figure 8.13. Note that some of the *individual* task times at each Workstation exceed Takt time. This is allowed, but the *Average Workstation Time* must not exceed Takt=12 Minutes.

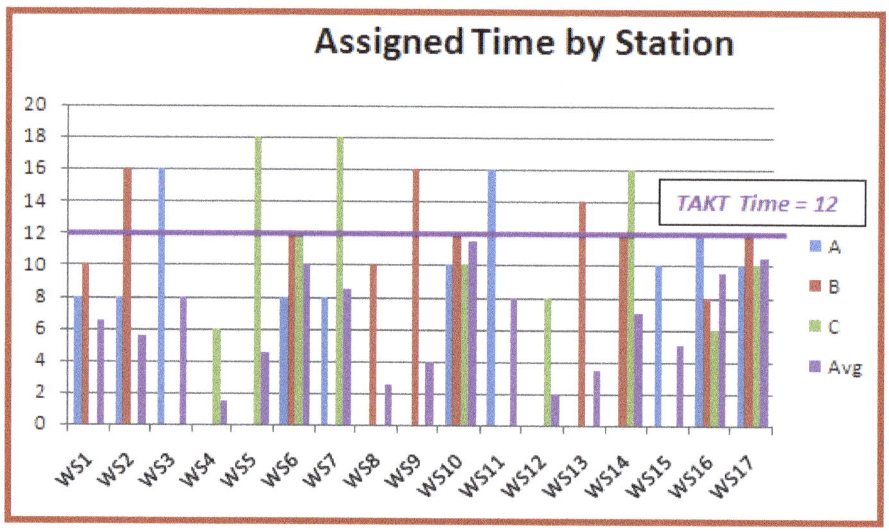

Figure 8.13
A Balance Graph

It might be interesting to ask the following question. If there was only room for 8 workstations, what would be the minimum Takt time to support this scenario? In this case, the procedure would be to increase Takt time until the minimum number of workstations required was equal to eight. The correct way to do this would be to determine a balance for a specified Takt time =13, 14, 15…..until eight workstations are all that is required. Alternatively, one could set the Takt time to some upper bound of, say, [2*Takt] and successively reduce the upper bound until a balance is found for eight workstations.

8.7 Maintaining Balance

The Lean Literature is full of philosophy as to how Balanced and Level production can be established, but there is not much on how it can be adjusted and maintained. The most recent line of thinking is to adapt flexible and versatile manufacturing capabilities such that machines and resources can be reconfigured and changed as production requirements and customer demands change. We have seen one or two companies with sophisticated production lines that can accomplish this, but this is generally very difficult to sustain. Another problem which we have not addressed is the use of *Key Workstations* that are environmentally sensitive and cannot be moved. These are called *Monumental Resources* by Dr. David Sly. It is true that line reconfiguration and relocation might be required as new products are introduced, but the realistic challenge is to adapt to normal fluxuations in demand and seasonal variations without disrupting ongoing operations. Experience has shown that with a capable, well trained workforce if demand, and hence Takt time, changes by only 10-15% over a short period of time; that this change can usually be absorbed. In any case, if demand changes over an extended period of time or demand spikes occur in excess of 10-15%, rebalancing will usually be required. In addition, when any demand change occurs, it will always be necessary to change the product mix to level production.

Assuming that Takt time needs to be maintained to keep from slipping pre-set delivery requirements, there are two fundamental ways to meet increased and changing demand. The *first* is to simply increase the amount of time allocated to production, and the *second* is to increase machining capacity. Increasing the amount of time allocated to production is perhaps the most common and easiest solution, and the common management response is to simply schedule overtime. If capacity is to be increased, this requires purchasing faster machines or adding parallel processing capabilities to the bottleneck workstations. In either case, some inventory buffering might be required to offset the fact that concurrent production rarely exists at the same time that overtime is being worked.

Overtime is often a good (short term) solution over a finite planning horizon, provided that key resources and Workstations are not being used continuously over a 24 hour day. Extra or partial shifts can often be scheduled. While management does not like overtime, most workers like it to earn extra pay. Changing the production capacity or rate of production is usually more difficult. This requires adding new or modified machines or executing dedicated Kaizen events to increase capacity. Reducing set up times, changeover time, providing quick mount jigs and fixtures and design for production can all create capacity. Minimizing batch sizes, streamlining material handling, increasing product quality and implementing a good off-line TPM program all effectively create more operating capacity. All of these solutions require dedicated Lean Engineering work and will not happen quickly. However, there are two other realistic short term solutions: Adding temporary parallel capacity by subcontracting work or reassigning experienced workers to take advantage of their superior skills on some tasks. It must be quickly pointed out that if cycle time requirements at bottleneck processes are not reduced, nothing will help.

Consider the following operational scenario. Products A, B, C and D are scheduled for production. The manufacturing line operates two 7.5 hour shifts per day six days a week. Hence, there is currently 5400 $\frac{Min}{Wk}$ available for production. Using the required production rates for each product, the Takt time is given by Takt = $(\frac{5400}{600})$ = 9.0.The product demands, Workstation Cycle Time, Product Mix and Takt time requirement is shown in Table 8.5

Product	Demand (Unit/Wk)	Required Workstation Cycle Time (Min)	Product Mix
A	200	27.0	0.330
B	200	27.0	0.330
C	100	54.0	0.167
D	100	54.0	0.167
	Total	Takt Time Requirement	Total
	600	9.00	1.00

Table 8.5
Creating Excess Capacity

Management would like to set up a single three-Workstation production line to produce all four products. The concept is to execute all Task 1 work for products A, B, C and D at Workstation 1; all Task 2 work at Workstation 2; and all Task 3 work at Workstation 3. Each product requires all three tasks in the order 1, 2, 3. The required processing times for each product at Workstation 1, and the Weighted Average Processing Time for all four products are shown in Table 8.6. For example, the product mix demands that 33% of all products passing through Workstation 1 is Product A. The time per part is 7.50 min, so the Weighted time is (0.33)*(7.5) = 2.5 minute/part.

Table 8.6
Weighted Average Processing Times

Products	Product Mix	Normal Time Required Task 1	Weighted Time Task 1	Normal Time Required Task 2	Weighted Time Task 2	Normal Time Required Task 3	Weighted Time Task 3
A	0.330	7.50	2.50	8.70	2.90	6.30	2.10
B	0.330	9.60	3.20	8.40	2.80	7.50	2.50
C	0.167	8.20	1.37	9.70	1.62	6.00	1.00
D	0.167	5.40	0.90	11.82	1.97	7.40	1.23
Average Time Required:			7.97		9.79		6.83

We observe that with Takt = 9.0, the proposed production line cannot meet demand requirements, since the Weighted Average Workstation (WAW) time at Workstation 2 (9.79) exceeds the required Takt time = 9.0 Recall that individual product task times can exceed the required Takt as long as the *average time* is not exceeded. If Takt time cannot be met, a typical management strategy would be to either increase available working time or increase the capacity to produce. Although the product demand profiles and the standard times at Workstation 2 are fixed, working overtime will temporarily fix this problem, but if this profile is expected to continue for some time, a more permanent solution may be preferred. Increasing the capacity to produce is the long run solution, but this will require capital acquisitions or some focused Lean Engineering work at Workstation 2. There is another solution which would provide both a short term and long term remedy: assign workers to tasks so that the fastest Task 2 workers (possibly more experienced workers) are assigned to Task 2. Faster workers at Workstation 2 essentially reduce the task time(s), effectively creating more capacity.

8.7.1 A Worker Assignment Strategy

Before we proceed, it might be beneficial to appreciate and understand what a *Standard Time* means. A standard time is the average time required to execute a fundamental operation or task by a qualified, experienced person under normal working conditions. Time standards can be pre-engineered standards or set by observation. If set by observation, the standard should be set using an average worker under normal operating conditions that has been fully trained to execute the task for which the standard is being set. A concept which we will define relative to a time standard is what is called *Normal Time*. Operators and working personnel assigned to specific jobs which work at the established time for that job are said to be operating at normal time or at 100%.Less trained and experienced operators may operate at some other percentage of normal time, say 110%. Experienced and seasoned veterans may take less time than the normal time, say 90%. A person who exceeds normal time on the average is typically rated between 100%-120%, while a person working less than normal time usually works between 80% -100%.One must be cautious in using these ratings. For example, if a task normal time is 50 minutes, and a worker is rated on that task at 110%, that worker will on the average complete that task in 55 minutes. A typical task time that is also sometimes adjusted with allowances, such as fatigue factors. We will not consider any predefined adjustments to normal time. The issue here is one of how to use standard times and worker task ratings to effectively increase Workstation ***capacity*** at no cost. Capacity can be gained by properly allocating skilled workers to bottleneck tasks, and we will use this example to illustrate the procedure. We recommend that every Lean Manufacturer or would-be-Lean Manufacturer maintain a worker/task rating database and follow this procedure to help in responding to short term capacity problems.

In this example, assume that such a data base exists, and for Worker 1 the Standard Time Ratings have been set by task and by product as shown in Table 8.7. For illustrative purposes, we will assume that this worker has been cross-trained and rated on all tasks and all products. This restriction will be relaxed in the next example.

Table 8.7
Worker 1 Standard Time Ratings

Products	Task 1	Task 2	Task 3
A	1.00%	88%	110%
B	95%	90%	120%
C	110%	100%	90%
D	90%	110%	85%

Using these standard time ratings, it is now possible to calculate the average times which one might expect to occur if Worker 1 is assigned to a particular task associated with a particular product. These times would be an adjustment of those times shown in Table 8.6 by Worker 1. Standard time ratings shown in Table 8.7. We will refer to these (adjusted) times as the *Task Execution Rating Matrix for Worker 1* (*TERM1*). The results are shown in Table 8.

Table 8.8
The TERM1 Matrix for Worker 1

Products	Task 1	Task 2	Task 3
A	2.50	2.552	2.310
B	3.04	2.520	3.00
C	1.507	1.620	0.90
D	0.810	2.167	0.680
TERM1	7.857	8.860	6.890

For example, the weighted normal production time listed in Table 8.6 for Product B, Task 1 is (0.33)*(9.6)= 3.20 minutes/part. From Table 8.8, worker 1 is rated at 95% on Task 1, Product B. Hence, that task can be completed on the average in (3.20)*(0.95) = 3.04 time units. All other entries in Table 6.9 are determined in the same way. The last line of Table 7.9 is the most important one to this discussion. If Worker 1 is assigned to Workstation 1, the average time to complete one set of products (A, B, C & D) is 7.97 time units. More importantly, the Cycle time for a set of Task 2 units (executed at Workstation2) is 8.860 time units, which is below the required Takt time of 9.0.In the simplest case, if workers 2 and 3 work all task times at the Normal rate, then assign worker 1 to Workstation 2, worker 2 to Workstation 1 and worker 3 to Workstation 3. This will result in an average sojourn time of 7.47 at Workstation 1; 8.860 at Workstation 2; and 6.83 at Workstation 3. This will result in a production line which can now satisfy Takt time requirements.

In general, all three workers will be rated differently when working certain tasks attached to certain products, and not all workers will be cross trained or qualified to work all tasks. We will now show haw a simple Linear Programming procedure called the *Assignment Problem* can be used to find a feasible strategy to satisfy Takt time, provided such a solution exists. Following the procedure just discussed, suppose that all three workers are task/product rated and the TERM matrix in Table 7.9-A is available from the Industrial Engineering Methods & Standards group.

Table 7.9-A
A Composite TERM Matrix for all Three Workers

	Task 1	Task 2	Task 3
W1	7.857	8.86	6.89
W2	****	7.21	8.10
W3	8.52	9.81	7.25

Notice that worker 2 has not been trained to do Task 1. Also note that Worker 3 cannot execute Task 2 for all products without exceeding Takt time. The assignment problem that we wish to solve is to assign M workers to N Workstations to Minimize the total time required to process all products at each Workstation. We can formulate the Assignment Problem as follows.

Minimize $z = \sum_{i=1}^{M} \sum_{j=1}^{N} c_{ij} x_{ij}$
Subject to: $\sum_{i=1}^{M} x_{ij} = 1$ i=1,2….N

$\sum_{j=1}^{N} x_{ij}$ = 1j=1,2….M
x_{ij} = 0 or 1all I, all j

The c_{ij} values are the ith row and j^{th} column in Table 8.9. The x_{ij} values are the solution variables. If Worker i is assigned to Task j, then x_{ij}=1. If not assigned, then x_{ij}=0. This formulation assures that only one worker will be assigned to each Workstation and that the objective function will be minimized. If a Worker cannot be assigned to a task, then add a large number to the largest c_{ij} in the Term matrix and enter that number into the corresponding cell. Note that M=N (required). The Assignment problem can be solved in a matrix form in its own space. The rules can be found in any Operations Research Textbook, such as Phillips, Ravindran and Solberg.

Step 1: Construct a 3x3 matrix from entries in the Composite TERM Matrix. Note that from Table 7.9 (a), Worker 2 cannot be assigned to Task 1. Hence, the matrix entry in Row 2 and Col 1 will be arbitrarily assigned a large number (10.0 + 9.81) = 19.81 (See table 7.9 (b)).

Table 7.9 (b)

	Task 1	Task 2	Task 3
W1	7.857	8.86	6.89
W2	19.81	7.21	8.10
W3	8.52	9.81	7.25

Step 2: Examine each row and column. (a) Subtract the ***minimum*** number in each row from every other number in that row. Next (b) from this set of new numbers, subtract the ***minimum*** number in each column from every entry in that column. Step 2 will always result in a number of cells that have zero entries ....See Table 8.9 (c) and Table 8.9 (d)

Table 8.9 C

	Task1	Task2	Task3
Worker1	0.967	1.97	0
Worker2	12.6	0	0.89
Worker3	1.27	2.56	0

Table 8.9 D

Worker1	0	1.97	0
Worker2	11.33	0	0.89
Worker3	0	2.56	0

Step 3: Since we are trying to assign a single worker to each Workstation in such a way as to minimize the total assignment *cost (*time*)*: We now try to make an assignment using matrix (d) which will result in *zero cost*. To do this, there must be only one assignment in each row and each column. By inspection, there are two solutions which exist for this problem. The time required at each Workstation will be the appropriate Task times recorded in Table 8.9 (d).

Solution 1: Assign Worker 1 to Task 1, Worker 2 to Task 2 and Worker 3 to Task 3
Solution 2: Assign Worker 3 to Task 1, Worker 2 to Task 2 and Worker 1 to Task 3

(1) Solution 1 will yield a Workstation 1 sojourn time of 7.857, a Workstation 2 sojourn time of 7.21 and a Workstation 3 sojourn time of 7.25.
(2) Solution 2 will yield a Workstation 1 sojourn time of 8.52, a Workstation 2 sojourn time of 7.21 and a Workstation 3 sojourn time of 6.89

Both solutions have resulted in all Workstations operating below the Takt time requirement of Takt = 9.0, without any extra investment or cost and without working overtime. The modeling approach is the Assignment problem and if a feasible solution cannot be immediately obtained from Steps 2 and 3, the solution procedure may require a number of extra steps other than those shown here. This methodology can also be used with more workers than machines, and will always find a solution. The reader is referred to any one of many Operations Research textbooks on the market, including Phillips, Ravindran and Solberg (20). Once Task and Worker assignments have been determined which will satisfy Takt time, it is necessary to determine a production sequence which will result in a smooth flow of products and level production downstream.

8.7.2 Sequencing the Products

Now that the task allocations and worker assignments have been determined which satisfy the Takt time requirement, it is now necessary to turn our attention to the production sequence which will be used at each workstation. A common production sequence which is sometimes used is to simply release product using production lot sizes as the demand quantities for each product. This is called a *no lot splitting policy* or *maintaining order integrity*. For this example, the production requirements are A=200, B=200, C=100 and D=100. A possible production sequence is shown in Figure 8.14.

This is not a good production sequence (although it might minimize set-up/changeover times), since Product D cannot be sent downstream until all of the previous A, B and C products have been completed. Of course, one could still prevent starvation of downstream processes if enough make-to-stock is stored as in-process WIP in front of downstream processing stations.

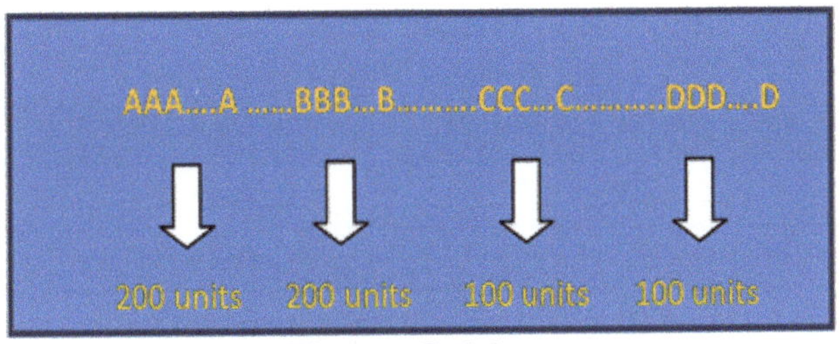

Figure 8.14
A Mixed Model Sequence

However, this is not an option to the Lean Engineer. Such practices should be avoided, since they cause excessive WIP and long throughput times. The correct Lean Engineering plan is to use a Mixed Model Sequence (MMS) which continuously produce a product mix and levels downstream production. One possible sequence is as follows. A=10, B=10, C=5 and D=5…..repeated 20 times. The astute reader will recognize that while there are many possible MMP sequences, each sequence will require a set of associated set-up and changeover (SCT) times. For example, the possible production sequence (A=200, B=200, C=100 and D=100) will require four different SCT….A to B, B to C ,C to D and D back to A. For our given demand rates, the first production sequence will require a total of (4*20)=80 different SCT .If any one setup and changeover time took between 5 and 10 minutes, between 400 minutes and 800 minutes of production time would be lost. This is the reason that Ohno put so much emphasis on SMED in the Toyota Production System. To properly frame the problem and propose a solution procedure, we need to recognize that set-up and changeover times are usually sequence dependent. For a four product sequence, there are exactly (4*3*2*1) = 24 different sequences that are possible. It is also possible that set-up and changeover times could be Workstation dependent. An optimal basic production sequence is one that will minimize total SCT. To simplify the presentation, we will assume that SCT is not Workstation dependent, although this is not a limiting assumption. For numerical purposes, assume that Table 8.10 contains the set-up and changeover times for products A, B, C and D in minutes.

Table 8.10
Product Dependent SCT for Four Products (Minutes)

	A	B	C	D
A	0	6	10	14
B	4	0	3	9
C	6	1	0	8
D	4	3	6	0

Table 8.11 lists the 24 different sequencing options for four products, and the total SCT time for each permutation.

Table 8.11
Four Product Permutations

Sequence	Time Required	Sequence	Time Required
A-B-C-D	21	C-A-B-D	27
A-B-D-C	27	C-A-D-B	26
A-C-B-D	24	C-B-A-D	25
A-C-D-B	25	C-B-D-A	24
A-D-B-C	26	C-D-B-A	25
A-D-C-B	25	C-D-A-B	21
B-A-C-D	25	D-A-B-C	22
B-A-D-C	25	D-A-C-B	23
B-C-A-D	26	D-B-A-C	25
B-C-D-A	21	D-B-C-A	26
B-D-A-C	24	D-C-A-B	27
B-D-C-A	27	D-C-B-A	25

For example, the sequence B-D-C-A yields a value of (9+6+6+6) = 27. There are three alternate optimum solutions: A-B-C-D=21, B-C-D-A=21 and C-D-A-B=21. We now need to note that since the product demand is A=200, B=200, C=100 and D=100 the three optimal solutions must follow the following ratio of lot size sequences: (A, B, C, D) = (2:2:1:1). It should be clear that one of the optimal lot size sequences is to run (B=200, C=100, D=100 and A=200). This is one of three solutions that minimizes SCT=21 minutes. However, we have already noted that such a solution is not a good Mixed Model Sequence. Here we can note that any subset of this solution that maintains production lot size ratios of (A, B, C, D) = (2:2:1:1) will also be optimal for that particular subset solution. For example, the Mixed Model Solution of (B=10, C=5, D=5 and A=10)....repeated 20 times is optimal with respect to that particular production lot size and Mixed Model Sequence. The only possible set of such solutions for this example is shown in Table 8.12

TAble 8.12
Subsets of All Optima Solutions

ID	Solution (2:1:1:2)				Reps	Time(Min)
	B	C	D	A		
1	200	100	100	200	1	21
2	100	50	50	100	2	42
3	50	25	25	50	4	84
4	20	10	10	20	10	210
5	10	5	5	10	20	420
6	4	2	2	4	50	1050
7	2	1	1	2	100	2100

The question of which solution is best depends upon many factors, but it is generally true that the preferred solution would be one with minimum lot size. However, recall that in order to satisfy customer demand, Takt time cannot be violated. That means that any feasible solution is one in which the sum of processing plus SCT time cannot exceed Takt time. For this example, the total weighted average processing time for each option was calculated in Table 8.6. Recall that the SCT times are the same at each Workstation.

We can now combine the required Processing times and the Set-up/Changeover times to get a total time required at each Workstation. These results are shown in Table 8.13
There are several interesting conclusions that can be drawn from Table 8.13.

Table 8.13
Total Average Sojourn Time

WS	PT (Hr)	Process Time+ SCT1 (Hr)	Process Time+ SCT2 (Hr)	Process Time+ SCT3 (Hr)	Process Time+ SCT4 (Hr)	Process Time+ SCT5 (Hr)	Process Time+ SCT6 (Hr)	Process Time+ SCT7 (Hr)
1	7.857	8.207	8.557	9.257	11.357	14.857	25.357	52.857
2	7.210	7.560	7.910	8.610	10.710	14.210	24.710	52.210
3	7.250	7.600	7.950	8.650	10.750	14.250	24.750	52.250

(1) Since Takt is 9.0 hours, the only feasible solutions are those shown in the blue shaded cells. The best MMP solution is to run production lot sizes of B=100, C=50, D=50 and A=100, in that order, repeated 4 times. The total time (Production + SCT) at each Workstation is Workstation 1 = 8.557 hours, Workstation 2 = 7.910 hours and Workstation 3 = 7.950 hours. The system Throughput Time is TPT=23.607 hours, and the system Output Rate=one unit every 8.557 hours. (2) There is an opportunity to use production lot sizes of B=50, C=25, D=25 and A=50 if a SMED- Kaizen study is directed at the set-up and changeover times at Workstation 1. A reduction of (Process time +SCT) =9.257 to below 9.0 would allow these production lot sizes. (3) Table 8.13 provides targets for SMED studies which could allow for other Mixed Model production sequences. (4) Recall that there were three alternative optimal solutions using different Production Sequences. A complete investigation of optimal operating policies would require the construction of two more tables similar to Table 8.13, which reflect different changeover times. It is possible that there is a better solution than that which we have obtained. The reader will be asked in a homework problem to see if this is true or not. Finally, if the SCT times are actually Workstation dependent, the above procedure is still valid but will require more Workstation specific calculations and Tables. The entire procedure is easily programmed in EXCEL or any programming language.

8.7.3 Computational Difficulties

The sequencing solution presented in Section 8.7.2 is based upon total enumeration of the different permutations (ordering) of N products. The example we used required the examination of N=4 products with 24 different permutations. For N=5, this would require 120 permutations and for N=6 it would involve 680. Since most dedicated production lines or *U-Shaped*

Manufacturing cells are dedicated to a small number of products, modern computers are capable of enumerating a fairly large number of products in a short period of time. But what if N is large, say 25-50 products? In this case a *Monte Carlo Procedure* is suggested. The procedure is to randomly generate a large number of possible sequences and evaluate the time required for each successive sequence, retaining the pair-wise least time feasible solution sets. Assume that a million solutions are randomly generated and keep only those solutions which meet a specified Takt time. Depending upon the value of N, the feasible solutions can be rank-ordered. Limited computational experiments by the authors have yielded interesting results. For N=52, the best solution obtained for several test problems was within 5%-10% of the global optimum solution. In any case, such a solution is certainly better than a guess and serves as a platform for improvement.

8.8 Set-up and Changeover Times Revisited: General Observations

For multiple product production lines, there are several batching policies that could be used to satisfy demand, each of which will directly impact the amount of production time lost to set-up and change-over. In Lean Manufacturing, if N products consisting of $n_1, n_2, n_3 \ldots n_N$ units are to be produced, and if a Kaizen event directed to SMED (Setup Time Reduction)has resulted in minimum setup and changeover times, then any product sequence can be used. Line rebalance is unnecessary in this case since the RWPT balancing heuristic previously used is based upon an average of all parts produced. For example, suppose that there are N=3 parts that are to be produced with demand n_A=500, n_B=200 and n_C=100. The required production mix is in the ratios 5:2:1. One way to do this is to produce 500 units of part A, 200 units of part B and 100 units of part C. Each production machine must be set up 3 times in a full production run. This production sequence has in fact historically been specified by production scheduling and accounting because the setup times for A, B, and C can be amortized over the maximum production sequence. However, we clearly demonstrate in Chapter 13 that this sort of batching causes variance components to significantly increase, with a corresponding increase in WIP and Throughput Time.....*the pay me now or pay me later syndrome*. Architects of the Toyota Production System (TPS) recognized this phenomenon fairly early, and spent a great deal of time and effort to minimize setup and changeover times. This effort was known as SMED. Further, emphasis was placed on using simple multipurpose machines which were designed to minimize setup time. Production and design engineers joined forces to design for minimum changeover. The results are well documented and often spectacular, reducing setup and changeover times in some cases from hours to minutes. As a result, Toyota was able to operate and balance their lines with almost any mixed model production sequence which could then be modified at any time to reflect changes in demand.

For this example, the Lean Engineer would study the downstream effects of Mixed Model Sequencing, and might prefer to execute the sequence AABACBAA three hundred times. Even with SMED exercises, this strategy may not be feasible without dedicated production work cells, multiple servers or both. In any case, the success of MMP greatly depends upon the amount of effort put into SMED, KAIZEN and TQM. Finally, recall that if t_p time units of available working time are consumed by setup and changeovers, the net effect is a significant and avoidable loss in production capacity....a fact often overlooked or ignored by manufacturing managers. For long term changes parallel or dedicated production lines may need to be introduced. To illustrate several principles, consider the weekly production requirements shown in Table 8.14. Available working time is 960 minutes per day.

Table 8.14 Weekly Production Requirements			
Product	Weekly Demand	Percent of Production	Required Cycle Time
A	400	53.3%	2.4 min
B	250	33.3%	3.84 Min
C	100	13.4%	9.6 Min
Totals	750	100%	1.28 Min

Since product demands are A= 400 B=250 and C= 100 per week, a possible mixed model production sequence is AAAABBCCBBBAAAA, repeated 50 times. Now suppose that the demand for product C increases to 200 units per week. The weekly production profile for a 960 min workday is shown in Table 8.15. The required cycle time is 1.13 min or 85 parts per day.

Table 8.15
Weekly Profile

Product	Weekly Demand	Percent of Production	Required Cycle Time
A	400	47.1%	2.4 min
B	250	29.4%	3.84 Min
C	200	23.5%	4.8 Min
Totals	**850**	**100%**	**1.13 Min**

Note that the original required cycle time decreases from an average of 1.28 min per part to 1.13 min per part. This will require a working day of 1088 min or 2.13 hours of overtime each day. This perhaps the best alternative, although the production sequence will change and in-process inventory will increase. For example, a product mix of 8 A's, 5 B's and 4 C's repeated 50 times will now satisfy demand. As usual, the way in which this production mix is sequenced will depend upon a Kaizen/SMED Blitz to reduce set up times as much as possible. Note that if the initial mixed model production sequence had been used prior to this change in demand, the sequence AAAABBCCCCBBAAAA can now be implemented with minimum disruption, and will not change the number of setups required.

Now suppose that management is not willing to schedule overtime (a common phenomenon). In that case, the current amount of production time per day will not change, and other ways must be found to accommodate the new production requirement for product C. It is clear that the production cycle time for product C must be roughly cut in half. Some time might be gained by labor reallocation, but the decrease in product C cycle time might be too much to realize. If business is booming and production requirements for all products are expected to increase, parallel processing might provide a solution. If the increase in demand for product C is expected to continue for some time and then possibly decrease to the previous level, subcontracting might provide a reasonable solution. In any case, it will probably be necessary to restructure line balance, examine the amount of in-process inventory levels to smooth production and possibly use a combination of alternatives.

8.9 Responding to Demand Changes: General Observations

There is only one thing which is certain to remain unchanged in production planning, and that is the fact that demand will constantly change. The general strategy of MRP is to try and forecast or predict future demand, and try to smooth production by averaging order release quantities across a specified planning horizon. As discussed elsewhere in this text, even MRP II systems with capacity planning modules will release work to the shop floor based upon a set of lead times and estimated processing times, often without knowledge of current operating conditions. A properly executed Lean order release strategy always depends upon production status, since it is driven by a Pull philosophy which virtually eliminates scheduling based upon lead times. In many computerized order release systems, serious imbalances in production are covered up by long lead time quotations to the customer, and usually result in expensive and inefficient make to stock or make to order inventory policies to compensate for fluxuating customer requirements. In all cases, Work in Process is always inflated. Conversely, the proper Lean Strategy is to institute pull strategies which dynamically and directly respond to customer demand by initiating upstream production only when downstream WIP or production requires more parts. This strategy prevents WIP buildup and places a *WIP Cap* on all inventory in the system. This is particularly evident in U-Shaped production and assembly lines which are paced by labor and not production capacity. Short term imbalances can usually be absorbed using the tools and techniques discussed previously, but beyond that point radical line changes may be required. In all cases, it is likely that the production mixed model sequence will need to be adjusted and modified to balance production and maintain level flow.

7.10 Solving Short Term Imbalance in Production Requirements

We simply mention three viable alternatives with selected comments.

Parallel Redundancy

A typical way to do this is by temporarily assigning another machine(s) to the bottleneck workstation. This is sometimes called *parallel redundancy*. There are several reasons why parallel redundancy might be the best option.

(1) If the parallel machine is a flexible machining platform, it might be shared between two production lines provided that set-up and changeover times are minimized. This is particularly important when dealing with bottleneck processes. Of course, some advantages might be offset by corresponding increases in variance components.

(2) Parallel machines rarely if ever fail at the same time. Parallel redundancy protects against line stoppage, and removes the need for excessive in-process inventory to protect against unreliable equipment. Of course, this in no way eliminates the need for a good preventative maintenance program.

(3) Introduction of temporary or periodic changes in line requirements can often be difficult or even impossible at bottleneck Workstations without parallel available capacity.

(4) Production line workers often work more efficiently and make fewer mistakes when not constantly pressured to maintain Takt time. The issue here is not to allow for worker idle

time, but to try and avoid work induced variance components due to excessive demand requirements and pressure to maintain Takt time.

(5) Since parallel redundancy almost always results in available capacity, the need for working overtime and frequent line rebalance is often mitigated

Working Overtime

Another common solution in dealing with temporary production imbalance and changing demand is to work overtime. For the example in Section 7.4.3 one can verify that by working one hour per day overtime at Workstation 2, the Takt time of 9.0 can be achieved. However, since Workstation 3 works only 7.5 hours per day, there will be an increase of in-process inventory at Workstation 3 during this overtime period, and will need to be worked off the next day. This would be true in any case, but particularly for a one shift operation.

Subcontracting or Outsourcing

If product demand is too great to absorb over a long period of time, an alternative strategy to increase production might be to outsource or subcontract either assemblies or subassemblies. In fact, many companies are moving to this strategy to take advantage of subcontractor efficiency and specialized knowledge, avoid capital investment costs to procure additional scarce resources and to enforce just-in-time strategies on assembly lines. One word of caution: If too much technology and expertise is exported to a second party, it is easy to lose the understanding of how to satisfy technical requirements in complex assemblies/subassemblies. This loss of intellectual capacity could prove disastrous if the subcontractors pull out.

8.11 Summary

In this Chapter we have considered the need to balance and smooth production. Several methods were presented that can be used to accomplish these goals. Both single model and mixed model production scenarios were presented and demonstrated in a series of numerical examples.

Review Questions

1. Work-in-process (WIP) has been called the *root of all evil* in manufacturing systems. Is this true? Explain your answer. Discuss the meaning of this term and the role of WIP.
2. What is *root cause analysis* and how is it related to *Kaizen events*?
3. What is *Heijunka* ?
4. Discuss (a) *process leveling* and *process balancing.*
5. What is Andon? Use the internet to provide examples of Andon.
6. Why have large production lot sizes been used historically?
7. What is consistency in Lean Manufacturing?
8. Define SMED. Why Is SMED so important to Lean Manufacturing strategies? Use the internet to find SMED success stories and report three of these.
9. Explain the difference in *balancing, synchronization and leveling.*
10. What is a Focused Factory? Under what conditions would a Focused Factory be appropriate? How would one determine which products would be produced?

11. Precisely describe the term Takt time. What is the relation between Takt time and output rate? If a Manufacturing System has multiple Subassembly cells or work cells dedicated to producing a finished product, would they all follow the same Takt time?
12. Discuss the impact of variance on Line Balancing.
13. The following data has been collected for a sequenced production system.

Task ID	Task Time (Min)	Task ID	Task Tim21e (Min25)	Task ID	Task Time (Min)
1	22	7	16	13	24
2	34	8	25	14	32
3	10	9	22	15	15
4	18	10	18	16	23
5	25	11	15	17	39
6	36	12	19	18	17

The following precedence relationships must be maintained.

Task ID	Predecessor Tasks	Task ID	Predecessor Tasks	Task ID	Predecessor Tasks
1	None	7	1	13	12
2	None	8	1	14	12
3	2	9	4,5	15	12
4	3	10	6,7,8	16	13,14,15
5	3	11	10	17	13,14,15
6	1	12	11	18	17

The production line operates 16 hours per day, 6 days per week. There is a 15 minute rest time each shift, and a 30 minute lunch break. (1) Draw the precedence diagram (2) Calculate the required Takt time. (3) Calculate the minimum number of required workstations. (4) Assign tasks to workstations using the Largest Eligible Processing Time Rule. (5) Calculate Workstation Efficiency and Line Efficiency. Is your final solution a good line balancing solution? Improve the initial solution by visual inspection. (6) Compare the improved solution to the initial solution.

14. Rework Problem 13.0 using the Ranked Positional Weighting Technique (RPWT). (1) Compare the initial RPWT solution to the final solution obtained in Problem 13.0. (2) Improve upon the RPTP initial solution. (3) Would the RWPT solution technique always be superior to the Largest Eligible Processing Time Rule? Why or why not?
15. The following Task – Time data has been collected for a proposed manufacturing assembly line.

Task ID	Task Time (Min)	Task ID	Task Time (Min)
1	12	7	8
2	5	8	6
3	6	9	2
4	4	10	5
5	4	11	6
6	13	12	8

The following precedence relationships must be maintained.

Task ID	Predecessor Tasks	Task ID	Predecessor Tasks
1	None	7	3,4
2	1	8	7
3	2	9	5
4	2	10	9,6
5	2	11	8,10
6	2	12	11

If Takt time has been set at 16 minutes; (1) Draw the precedence diagram (2) Calculate the required Takt time. (3) Calculate the minimum number of required workstations. (4) Assign tasks to workstations using the Largest Eligible Processing Time Rule. (5) Calculate Workstation Efficiency and Line Efficiency. Is your final solution a good line balancing solution? Improve the initial solution by visual inspection. (6) Compare the improved solution to the initial solution.

16. The following Task – Time data has been collected for a proposed manufacturing assembly line. (Gaither)

Task ID	Task Time (Min)	Task ID	Task Time (Min)	Task ID	Task Time (Min)
1	6	7	7	13	7
2	3	8	4	14	3
3	9	9	9	15	5
4	5	10	8	16	8
5	8	11	11		
6	10	12	12		

The following precedence relationships must be maintained.

Task ID	Predecessor Tasks	Task ID	Predecessor Tasks	Task ID	Predecessor Tasks
1	None	7	4	13	11
2	1	8	4	14	10,12
3	2	9	5,6	15	13,14
4	2	10	6,7	16	15
5	3	11	8		
6	2	12	9		

If Takt time has been set at 16 minutes; (1) Draw the precedence diagram (2) Calculate the required Takt time. (3) Calculate the minimum number of required workstations. (4) Assign tasks to workstations using the Largest Eligible Processing Time Rule. (5) Calculate Workstation Efficiency and Line Efficiency. Is your final solution a good line balancing solution? Improve the initial solution by visual inspection. (6) Compare the improved solution to the initial solution.

17. In Line Balancing problems with a fixed Takt time, if any one Task is greater than the Takt time, a common procedure is to either work overtime to meet Takt time requirements, or to add one or more machines to reduce the average task time per unit. Both Gaither and Chase have proposed a Line Balancing algorithm which automatically calculates the number of machines required as part of the heuristic solution procedure. This method is called the *Incremental Utilization Heuristic* or the IUH. This heuristic is appropriate when one or more required task times cannot meet Takt time. The IUH proceeds as follows. Assume for simplicity that there is a single start task. Create the first Workstation with that task using a single machine. If the task time is greater than Takt time, add the required number of machines to that Workstation to make Workstation utilization less than one, add tasks to that workstation in order of precedence one at a time continuing to add machine capacity until Workstation utilization is equal to 100% or ***Utilization Decreases.*** When this happens, close out this Workstation; start a new one; and repeat the process until all tasks are assigned to workstations. Using the following set of data, use the IUH heuristic to determine line balance, and how many Workstations and Machines are required. Calculate both machine and line efficiency. Takt time is 0.10 Minutes. What is the minimum number of machines required.

Task ID	Task Time (Min)	Task ID	Task Time (Min)	Task ID	Task Time (Min)
1	0.18	7	0.38	13	0.48
2	0.11	8	0.42	14	0.30
3	0.32	9	0.30	15	0.39
4	0.45	10	0.18		
5	0.51	11	0.36		
6	0.55	12	0.42		

The following precedence relationships must be maintained.

Task ID	Predecessor Tasks	Task ID	Predecessor Tasks	Task ID	Predecessor Tasks
1	None	7	6	13	11,12
2	1	8	7	14	13
3	1	9	8	15	14
4	1	10	9		
5	2,3,4	11	10		
6	5	12	10		

18. If *a single machine Workstation has a processing time which is greater than required Takt time, the processing time per unit can be reduced by 50 % by adding another identical machine.* Under what conditions will this statement be true? Under what conditions will it be false?
19. Three products (A, B and C) are to be sequenced through a single machine. If the processing times are A=5 min, B= 7 min and C=8 minutes and the set-up and change over times are as given below in minutes, determine an optimal sequence for A, B and C which minimizes total Throughput time.

Product	A	B	C
A	0	12.5	16.2
B	9.5	0	10.7
C	8.2	12.3	0

20. Using a hand simulation, determine the throughput time to make one set of products using a single machine if A=6 min, B=4 min and C=3 min. if the sequence is A then B then C. How much time can be saved by adding a second identical parallel server? What sequence would minimize throughput time with one server? Two servers

21. A Mixed Model Production Line makes Products A, B and C for the demands A=50 Units/week, B= 25 Units/week and C= 25 Units/Week involving 12 different machining tasks. The time required per task (in minutes) by product type, and the precedence relationships are shown below. For example, all parts go through Tasks T1, T2 and T3 and tasks T1, T2 and T3 must all be completed before task T4 can be started. Assume that the line runs three 8 hour shifts per day. (1) Calculate the Takt time for this line. (2) Calculate the weighted average processing time per task. (3) Use the Largest Positional Weight Technique to assign tasks to Workstations. (4) Show the final Workstation assignments (5) Calculate line efficiency (5) draw a Balance graph for the final solution.

Mixed Model Production

Part	T1	T2	T3	T4	T5	T6	T7	T8	T9	T10	T11	T12
A	10	8	*	*	11	12	10	11	*	11	*	9
B	12	11	*	11	9	9	13	14	*	10	8	12
C	8	12	7	10	12	14	12	10	12	9	13	10

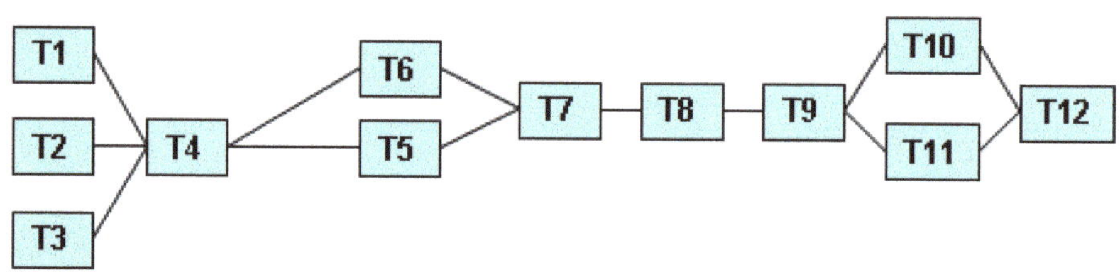

22. Assume that the production time requirements (Minutes/part) for four products (A, B, C, D, and E) assigned to three Workstations is as follows.

Products	Normal Time @ Workstation 1	Normal Time @ Workstation 2	Normal Time @ Workstation 3
A	8.25	8.75	7.76
B	7.22	7.56	9.78
C	9.10	9.22	6.98
D	7.81	6.97	7.22
E	5.32	7.23	8.55

The mix of products to be produced are A=10%, B=20%, C=40% and D=30%. (1)What is the minimum Takt time that can be achieved with this data? (2)The three workers normally assigned to each Workstation are given a one month training course to increase their efficiency. The following Table describes the new worker ratings at each Workstation by Product.

Products	Rating: WS1	Rating: WS2	Rating: WS3
A	0.98	0.91	0.95
B	0.92	0.88	0.85
C	0.79	0.98	0.88
D	1.00	0.91	0.90
E	0.88	1.0	0.89

For example, Worker 1 took 8.25 minutes per part on Product A at Workstation 1 before training, but can now accomplish this same task in (8.25 * 0.98) = 7.61 Min per part. (1) Calculate the new Product/Workstation standard times for each worker (2) determine the new minimum Takt time which can now be attained assigning Worker 1 to Workstation 1, Worker 2 to Workstation 2 and Worker 3 to Workstation 3. (3) Assume that each Worker will be assigned to a Workstation by using the Assignment algorithm presented in Section 8.7.1. What is the best Takt time which can be achieved and where will each worker be assigned?

23. Assume that a process needs to produce A=50, B=25, C=50 and D=10 units of product. The Product dependent Set- up and Changeover times in hours for these four products are as follows.

	A	B	C	D
A	0	1.67	0.87	1.26
B	1.23	0	0.76	1.23
C	1.00	0.89	0	2.21
D	1.55	1.22	1.10	0

(1) List all of the product changeover sequences that are possible and calculate the total time consumed by each. (2) What is the optimal product mix which can be used? (3) How many different subset optimal solutions can be identified? (3) Calculate the Total Average Time required for each of the subset optimal solutions. (4) Recommend a product mix and production plan that minimizes the total time required.

24. Reconsider the Line balance obtained for Black's Sweat shop. Show that a better solution can be obtained by using a lean, U-shaped cell.

Chapter 9
Lean Manufacturing and The Job Shop

Build a better mousetrap and the world will beat a path to your door   *Ralph W. Emerson*

This book has been concerned with how to convert traditional Manufacturing Systems to Lean Manufacturing. This is accomplished by converting serial production lines and Push Production systems driven by computerized MRP/ERP production planning and control to U-shaped Lean Manufacturing Cells and Pull Production systems driven by visual signals (Andon) and Kanban cards. Although Lean principles and Lean Thinking can be applied to almost any type of production or service system, it was birthed by Taiichi Ohno and Shigio Shingo in the Automotive Industry (Toyota Motor Company) and was initially applied to large volume production systems. As consumers demanded more variety and options the initial concepts rapidly expanded and were implemented. As time moved on, one area of manufacturing was virtually ignored......Traditional Job Shop Manufacturing. The general objection was that Lean Manufacturing simply would not work in a high-product-mix, low-volume environment of the typical custom fabricator or for that matter, any other job shop.

9.1 Lean Manufacturing Cells Can Replace the Job Shop

Most companies that implement Lean manufacturing ignore their existing job shop. Job shops traditionally are used to produce unique or customized finished products. Lean cells have traditionally been built to Manufacture high volume, stable products families which produce larger quantities of each type. After initial success at Toyota, Lean Principles were expanded to embrace mixed Model Manufacturing. The concepts of Lean Cell Design were implemented in large systems to support parts and subassemblies used in Mixed Model Final Assembly. Rearranging, converting and using existing equipment is a natural precursor to the implementation of mature, Lean cells. A planned stepwise conversion to Lean manufacturing and U-shaped Lean cells is good, because the company can learn how Lean cells work, how to drastically reduce setup and changeover times, enforce in-line Quality Assurance and Preventive Maintenance, implement MO-CO-MOO machine cycle time procedures and operate U-shaped Lean cells. As conversion to Lean systems mature, lean machine tools can be designed and specialized to Lean practices. The initial use of existing machines is a relatively low-cost, risk-free approach. We will call these initial Lean manufacturing cells *interim cells* to distinguish them from complete, *mature* Lean cells. The application of Lean Principles and U-Shaped Lean Cells is the last sacred cow to fall as Lean has evolved from the original Toyota Production System.

9.2 Job Shops and Lean

Consider how a traditional job shop might be used to manufacture four different parts to customer specifications. What would this small Job shop look like if Process Routing Sheets were superimposed upon the Machine Layout? Typical routing sequences for only four parts are shown in Figure 9.1.

Consider how a traditional job shop might be used to manufacture four different parts to customer specifications. What would this small Job shop look like if Process Routing Sheets were superimposed upon the Machine Layout? Typical routing sequences for only four parts are shown in Figure 9.1. Production of each part is labor and information intensive. This is typical of what might be observed in a traditional job shop. These routing patterns are often called *spaghetti flow,* which is typical of a traditional job shop operation.

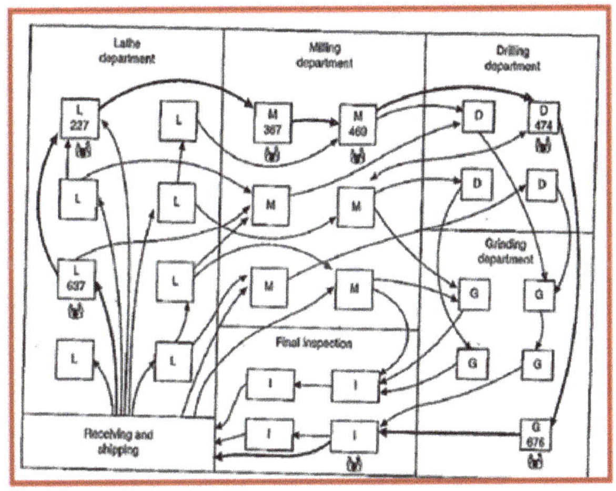

Figure 9.1
Typical Part Flow Patterns for a Job Shop Producing Four Parts

In the Job Shop, a single skilled operator would usually man each machine, and parts would move from machine to machine in tote boxes. A process plan which gives the *sequence* of operations (machines) to be used in manufacturing a part like the drive shaft is crafted by either a Manufacturing engineer or by production planning. This sheet is sometimes called a *routing sheet* or a *traveler*. The average Job Shop Foreman is intelligent, skillful and can make almost anything from a mere sketch. When asked about Lean Manufacturing, chances are you would get a blank stare or "It won't work here". In this chapter we will attempt to address both of these typical reactions

Job shops do present their own unique set of operational challenges, which include:

- Most jobs are low volume and have a high degree of variation, never allowing the plant enough time to become truly efficient at them
- Standardization due to unique design specifications are difficult
- Skilled operators simply *Know how to make this part*. Some possess a unique *Black Book* with learned trade secrets.
- Customers constantly interrupt scheduled production by emergency jobs, rush projects, and repair parts that are needed but cannot be purchased off-the-shelf
- Bottlenecks change on a daily basis according to the product mix
- Raw materials are sometimes unique and must be ordered
- Shop tooling, jigs, fixtures and available machines workstations are constantly changing
- Many of the most popular lean and six sigma tools have limited effectiveness.
- Long-range forecasting and production planning is difficult if not impossible.

A typical job shop may exist as a stand-alone company that manufactures custom products for a wide variety of *external* customers, or as a part of a larger Manufacturing system that manufactures parts for *internal* customers. Lean implementation of U-shaped cells, line balancing, load leveling, dedicated machines, and Kanban controlled part flow can be difficult if not impossible to achieve in a traditional Job Shop.. The success of Lean Concepts has changed if not revolutionized the landscape of Mass manufacturing, and even if many Lean Manufacturing tools cannot be directly applied to a Job Shop the concepts of Lean Thinking can result in reduced costs and shorter throughput times. However, there are many success stories which are beginning to surface in trade literature. Conversion to Lean Manufacturing Cells is becoming a reality.

There are two key steps that are necessary if the benefits of Lean can be realized in a typical Job Shop. The *first* is to identify all sources of waste (Muda) in the Job shop environment and *second* to identify *Part Families* which are grouping parts together that have common machining operations on common machines. Job Shops have historically not done either of these two studies. Traditional Job Shops have been seen as a manufacturing system where standardizing processes in a low-volume, high-mix environment is impossible. Customization and one-at-a-time have ruled. Realizing that even in the face of non-standard low volume manufacturing Lean Thinking should and can be applied, a lot of smart companies have figured out that by focusing on operational commonalities, job shops can successfully implement many Lean methodologies.

9.3 Identify and Rigorously Eliminate Sources of Waste (Muda)

All employees in a Job shop should meet to identify and eliminate sources of waste. Machine Operators usually know sources of waste and often what causes waste of time, raw materials and operations. There many sources of Muda which can and should be eliminated. The following categories are only for example.

9.3.1 Raw Materials

Raw materials and parts must be organized and delivered to the areas when they are needed, where they are needed and in the quantity needed. If the Job Shop exists to support internal customers in a larger Manufacturing System, treat each department as if it was an external customer. Internal WIP needs to be carefully examined, and needs to be limited to only that which is needed in sequential machining operations and at Bottleneck operations.

9.3.2 People

Usually the Job Shop work force needs to be highly trained, including a huge degree of cross training to create high efficiency at any machine in the shop. The Job shop workforce should usually be the best machine operators with years of experience. It is not uncommon to have a six month or longer period before a worker can truly be efficient, so you cannot skimp on training.

9.3.3 Standards

Standard work is a very powerful tool, but it needs to be implemented in an entirely different way. You should *not* have standard work for individual products, but rather *processes*. Those processes are then arranged in different combinations to build various products. This is part of Group Technology which will be discussed later. Different types of steel, brand of sheet metal, grinding wheels etc. can require different tooling, speeds and feeds to maintain surface finish and accuracy. As in Mass Manufacturing, if possible preferred vendors should be treated as partners and held to rigid QC standards. Standards and standard times have often been ignored and it is not uncommon in industry to find that they are years old and often based upon a single observation. This cannot be tolerated in Lean Manufacturing. Henry Ford that: *No progress can be made without standards; Otherwise how will anyone know if they are good or bad?*

9.3.4 Quality

Quality control needs to be optimized just like production processes. Zero defects should be a primary goal of part production. Common and generally accepted. Quality Control Charts simply will not work in most Job Shops. Defects are more likely to occur when you are always making something new. Kaizen events should focus more on what is *likely* to cause a problem instead of what happened only one time. However, Kaizen events which grouping defects by machines, processes and raw materials used can often lead to systemic causes of poor product quality. ALWAYS involve Job Shop machine operators in such exercises.

9.3.5 Jigs, Fixtures and Tooling

Lean practice and thinking in a Job Shop should include an investigation of jigs, fixtures and tooling to see if these three manufacturing support devices can be standardized across various machining operations. Tool life should be monitored according to usage and materials being used. In general, it is difficult (perceived or actual) to standardize tools and fixtures. It is by their very nature that they are often unique to holding, aligning, guiding and positioning unique parts in any manufacturing operation. However, this may be cause a specific jig or fixture is usually designed for a particular work-piece and not much consideration has traditionally been given to design the interim part for a jig of fixture. This may be very difficult, but well worth the time to hold a Kaizen Event to explore possibilities. This is known as Design for Manufacturing and is an extensive body of knowledge within itself. Design for Manufacturing is ideally coupled with new advances in 3-D Manufacturing and Virtual Manufacturing.

9.3.6 Machine Set-Up and Changeover Times

A great deal of time is consumed at every machine when between when one part is finished and another is started in machine set-up, tooling changes, adding jigs and fixtures and loading raw materials. Taiichi Ohno and his Chief Engineer Shigeo Shingo spent an enormous amount of time with skilled shop floor workers to reduce set-up and changeover times. This particular type of Kaizen was called SMED (Single Minute Exchange of Dies). The results reported were amazing and spectacular. Over a long period of time, set-up and changeover times were reportedly reduced from hours to minutes!

9.4 Implementing Lean Manufacturing Principles in Job Shops

The pursuit of waste (Muda) reduction can often yield spectacular results in terms of time and cost savings. However, most of the benefits of Lean Manufacturing can only be realized by redesigning the Job shop to eliminate the production of all parts as unique, stand -alone products or in small lot production. The key is to identify *part families* which require basically the same type of machining operations on similar machines. This requires that all parts be examined and become members of the same family that require similar if not identical part routing sequences, machines and materials. This is not a new problem. Group Technology was popularized by John L. Burbridge who wrote a book called Group Technology in Engineering in 1979.

> *Group technology or GT is a manufacturing technique in which parts having similarities in geometry, manufacturing process and/or functions are manufactured in one location using a small number of machines or processes. Group technology is based on a general principle that many problems are similar and by grouping similar problems, a single solution can be found to a set of problems, thus saving time and effort.*

Group Technology has recently achieved more importance in manufacturing due to the fact that Job Shop operations are beginning to adopt Lean principles. Contrary to traditional thinking, Group Technology does not attempt to eliminate variety or single product production, but rather seeks to find similarities and commonality in the types of material, machines and production sequences required to produce a final product. Products with seemingly different machining and production characteristics can be identified and grouped together into what is called *Part Families.* Once multiple products are grouped into Part Families these characteristics can be used to form Manufacturing Cells using Lean Engineering Methodologies. This is a special form of Mixed Model Production, but is directed to singular and one-of-a-kind Job Shop production. If enough products can be identified to create steady manufacturing requirements across enough parts, an entity historically referred to as a *Focused Factory* can be implemented. This "factory" may in fact be realized by a Lean Manufacturing Cell or a Linked Cell Manufacturing System. We will only demonstrate one way that Group Technology Families of products can be identified. This is by using a *Clustering Algorithm.*

9.5 Converting Job shops to Lean Cells

The number of *Part Types* which are made in a job shop are determined not by capability but by customer orders. In a typical job shop customer orders can fall into two separate categories: (1) Random orders of one product or in small lots from a wide range of customers and (2) Repeat orders of small to medium size lots from repeat customers. A typical job shop will consist of M multi-purpose machines which will be scattered across a manufacturing area. Most machines are not automated but are operated by skilled workers. The key to implementing Lean Principles in a job shop is to determine the number of products that are produced in the Job Shop, the quantity produced and the number of different machines are required to manufacture a product (part). The number of parts produced ...say over one year's time...and the number of parts manufactured per year is easily obtained from sales. A *Process Flow Chart* can then be constructed for each part which is the number of machines that will be used by each part.

Part Type	Number Made
1	286
2	225
3	205
4	185
5	165
6	155
7	152
8	147
9	126
10	110
11	95
12	90
13	87
14	80
15	80
16	75
17	72
18	68
19	65
20	62
The Rest	50

For example, an Industrial Job Shop has made and sold 150 different products over a one year period of time. The number of total parts has been 2580. The following table shows the distribution of parts manufactured. It is decided that a Lean Cell would be put into place to manufacture Parts 1-8. The next step is to construct a *Process Flow Table* for each part type which show the sequence of Manufacturing Steps and the machine used at each step.

The previous table is straightforward: For example, Part 1 requires Machines 1 and 5 and Part 9 is made on Machines 2, 3 and 6. The parts now need to be grouped into Part Families or clusters of parts which are manufactured on the same machines.

	Parts								
Machines	1	2	3	4	5	6	7	8	9
1	1			1				1	
2					1				1
3			1		1				1
4		1				1			
5	1							1	
6			1						1
7		1				1	1		

9.6 A Clustering Algorithm to Determine Part Families

The method used is called *Production Flow Analysis* (PFA). PFA was popularized by Burbridge in the early 1970's. The *Clustering Algorithm* can be used directly on the Process Flow Table. The procedure is as follows.

Example 1: A Job Shop With 9 Parts and 7 Machines

Step 1: Augment the *Process Flow Table* by adding a row along the *bottom* of this M x P matrix such that each column (Parts) is assigned a decision vector composed of P values calculated by:

$$2^{(M-J)} \qquad J = M-1, M-2\ldots\ldots 1$$

	Parts									
Machines	1	2	3	4	5	6	7	8	9	
1	1			1				1		290
2					1				1	17
3			1		1				1	81
4		1				1				136
5	1							1		258
6			1						1	65
7		1				1	1			140
	256	128	64	32	16	8	4	2	1	

In this example, M=10 and the following values comprise Row 6: $2(10-j)$ $\quad j= 10, 9, 8\ldots.1$

$\text{Cell}(8,9) = 2^{(9-9)} = 1 \qquad \text{Cell}(8,8) = 2^{(9-8)} = 2 \qquad \text{Cell}(8,7) = 2^{(9-7)} = 4 \qquad$ Etc....

Step 2: Examine each row from left to right starting With Row m=1. If Machine m m=1,2,...M has a "one" in any column, add the numbers together under each column and assign that value to the corresponding row in Column P+1. In this case:

Row (m, P+1) $\quad$ m = 1,2,3......M $\quad$ For this example: Row (1, 10) = 1*256 + 1*32 + 1*2 = 290

Row (2, 10) = 1*16 + 1*1 = 17 $\quad$ Row (3, 10) = 1*64 + 1*16 +1*1 = 81 $\quad$ Etc.......

	Parts									
Machines	1	2	3	4	5	6	7	8	9	
1	1			1				1		290
2					1				1	17
3			1		1				1	81
4		1				1				136
5	1							1		258
6			1						1	65
7		1				1	1			140
	256	128	64	32	16	8	4	2	1	

Step 3: Reorder the Column labeled *Machines* in decreasing order of the numbers in Column P+ 1.

Machines	Parts 1	2	3	4	5	6	7	8	9	
1	1			1				1		64
5	1							1		32
7		1				1	1			16
4		2				1				8
3			1		1				1	4
6			1						1	2
2					1		1		1	1

Step 4: : Augment Table xxx by adding column to the right of Column P with the following values in increasing order from top to bottom.

$2^{(M\text{-}k)}$ k = M, M-1, M-2........1

Machines	Parts 1	2	3	4	5	6	7	8	9	
1	1			1				1		64
5	1							1		32
7		1				1	1			16
4		2				1				8
3			1		1				1	4
6			1						1	2
2					1		1		1	1
	96	24	6	64	5	29	17	96	7	

Step 5: Examine each Column from left to right starting With Row m=1. If Machine m m=1,2,...M has a "one" in any column, add the numbers together in the last column and assign that value to the M+1 row in Column k = 1, 2.....P. In this case:

Machines	Parts 1	2	3	4	5	6	7	8	9
1	1			1				1	
5	1							1	
7		1				1	1		
4		2				1			
3			1		1				1
6			1						1
2					1		1		1
	96	24	6	64	5	29	17	96	7

Step 6: Reorder the Columns using the calculated values in Row M+1 from left to right in decreasing order.

	Parts									
Machines	**1**	**8**	**4**	**6**	**2**	**7**	**9**	**3**	**5**	
1	1	1	1							64
5	1	1								32
7				1	1	1				16
4				1	1					8
3							1	1	1	4
6							1			2
2							1	1	1	1
	96	96	64	29	24	17	7	6	5	

The Binary Grouping of Machines 1-6 result in the following *Three Clusters.*

	Parts									
Machines	**1**	**8**	**4**	**6**	**2**	**7**	**9**	**3**	**5**	
1	1	1	1							64
5	1	1								32
7				1	1	1				16
4				1	1					8
3							1	1	1	4
6							1			2
2							1	1	1	1
	96	96	64	29	24	17	7	6	5	

What do we do now? The astute reader will now realize that Parts 1, 8 and 4 can be made in a *Mixed Model Lean Cell* with only two machines. Parts 6, 2 and 7 can be made in a Mixed Model Cell with Machines 7 and 4, and Parts 9, 3 and 5 can be made in a Mixed Model Cell with Machines 3, 6 and 2. It is at this point that the Kaizen Events which were directed to discover wast (Muda) need to be addressed. In order to fully realize the benefits of Lean Cells to manufacture parts Set-up and conversion times from part to part must be aggressively and totally addressed; Raw material Supermarkets implemented and versatile jigs, fixtures and tools designed for each particular machining step for each part in each part family. Lean Kaizen events for Quality Control (Poka-yokes), RAM and decouplers can now proceed.

In this example, a typical, traditional Job Shop was converted into three Lean Cells in which similar machines or processes were aggregated into groups of parts. Each Cell is dedicated to the production of a part family or a limited number of parts This is called *Job Shop Cellular Manufacturing* .The advantage of dedicated cells is that:

- Manufacturing cycle times are reduced
- Lead time to order delivery is reduced
- Tools, Jigs and Fixtures can be designed for easier use
- All of the advantages of *Lean Manufacturing* can be realized
- Standardization of machine set-up and changeover times (SMED) should be addressed using Kaizen events
- Quality should improve

Example 2: Consider a Job Shop which manufactures 10 parts on 5 different machines.

Construct the Process Routing Table.

Machines	Parts									
	1	2	3	4	5	6	7	8	9	10
1	1	1	1		1	1	1	1		
2	1	1		1		1		1	1	1
3	1				1		1	1		1
4	1	1	1		1	1	1	1		
5			1	1	1		1		1	1

Construct the Last Row....and then:
Calculate the Last Column

Machines	Parts										
	1	2	3	4	5	6	7	8	9	10	
1	1	1	1		1	1	1	1			835
2	1	1		1		1		1	1	1	855
3	1				1		1	1		1	567
4	1	1	1		1	1	1	1			1020
5			1	1	1		1		1	1	235
	512	256	128	64	32	16	8	4	2	1	

Rearrange Machines

Machines	Parts										
	1	2	3	4	5	6	7	8	9	10	
4	1	1	1		1	1	1	1			1020
2	1	1		1		1		1	1	1	855
1	1	1		1		1		1	1	1	835
3	1		1		1	1	1				567
5			1	1	1		1		1	1	235

Calculate and add Row M+1

Machines	Parts									
	1	2	3	4	5	6	7	8	9	10
1	1	1	1		1	1	1	1		
2	1	1		1		1		1	1	1
3	1				1		1	1		1
4	1	1	1		1	1	1	1		
5			1	1	1		1		1	1
	512	256	128	64	32	16	8	4	2	1

Then Rearrange the columns

Machines	Parts 1	6	2	8	5	7	3	4	9	10	
4	1	1	1	1	1	1	1				16
2	1	1	1	1				1	1	1	8
1	1	1	1	1				1	1	1	4
3	1	1			1	1					2
5					1	1	1	1	1	1	1
	30	30	28	28	19	19	17	13	13	13	

Identify Clusters and Form Cells

Machines	Parts 1	6	2	8	5	7	3	4	9	10
4	1	1	1	1	1	1	1			
2	1	1	1	1				1	1	1
1	1	1	1	1				1	1	1
3	1	1			1	1				
5					1	1	1	1	1	1

Cell 1 **Cell 2** **Cell 3**

9.7 Cellular Manufacturing: Revisited

Cellular manufacturing is a derivative of the principles of group technology, which were proposed by Flanders in 1925[1] and adopted in Russia by Mitrofanov in 1933. Burbidge actively promoted Group Technology, Clustering Algorithms, and Cellular Manufacturing in the 1970s.

Apparently, Japanese firms began implementing cellular manufacturing sometime in the 1970s," and in the 1980s cells migrated to the United States as an element of Just-in-Time (JIT) production.

One of the first English-language books to discuss cellular manufacturing was that of Hall in 1983. He referred to a Manufacturing Cell as a U-line because he proposed configuration of a manufacturing cell in a U-shaped arrangement. He considered that the best machine arrangement because a U- shape puts all cell processes and operatives into close proximity, affording high visibility and less walking between machines . By 1990 manufacturing cells had come to be treated as foundational practice in JIT manufacturing. Harmon and Peterson, in their book, *Reinventing the Factory*, included a section entitled: *Cell: Fundamental Factory of the Future*. Cellular manufacturing was carried forward in the 1990s, when Just-in-Time Manufacturing was renamed Lean Manufacturing. If closely examined, the term Just-in-Time Manufacturing really is not a correct title for Lean Manufacturing. Lean Manufacturing is much more comprehensive, and it evolved directly from the Toyota Production System. Lean manufacturing exhibits the characteristics of Just-in-Time Manufacturing but they are not the same. Lean Manufacturing revolves around one overriding goal...*Waste Reduction*. The secondary goals are very important: 100% quality production and RAM Technologies to eliminate part rejects, greatly reduce downtime and minimize WIP. *Lean Cell Production* was implemented to respond quickly to changing demand and facilitate versatile, agile and to minimize cycle time; hence maximizing production.

9.8 Lean Linked Cell Manufacturing and the Job Shop

Most companies that design their first Lean manufacturing cell use their existing job shop or flow shop equipment. Job shops have been traditionally used to produce unique or one-at-a-time finished goods. Lean Cells are built around part or product families for those parts which produce larger quantities of each type. Rearranging, converting and using existing equipment is a natural precursor to the implementation of mature, Lean cells. This stepwise conversion to Lean manufacturing and U-shaped Lean cells is good, because the company can learn how Lean cells work, how to drastically reduce setup and changeover times, enforce in-line Quality Assurance and Preventative Maintenance, implement MO-CO-MOO machine cycle time procedures and operate U-shaped Lean cells. As conversion to Lean systems mature, lean machine tools can be designed and specialized to Lean practices. The initial use of existing machines is a relatively low-cost, risk-free approach. The routing sheet is accompanied, preceded or augmented by machining instructions, manufacturing/move lot sizes and set-up instructions. All of this information and much more is used to produce one or more parts in a traditional job shop

Figure 9.1 is a typical job shop layout. Suppose that this job shop is being used to produce parts to support automotive assembly. There are many different types of machines, most of which are not used except when demand dictates their use. In this case, another skilled worker is usually needed. The skilled labor pool is constantly changing, creating a dichotomy of fully skilled workers and new hires. This particular company is experiencing worker inconsistencies and quality control problems. Now let us see what happens if we convert this manufacturing system over to a Lean manufacturing system. All machines in a job shop are arranged into U-shaped configurations such as those shown in Figure 9.2.

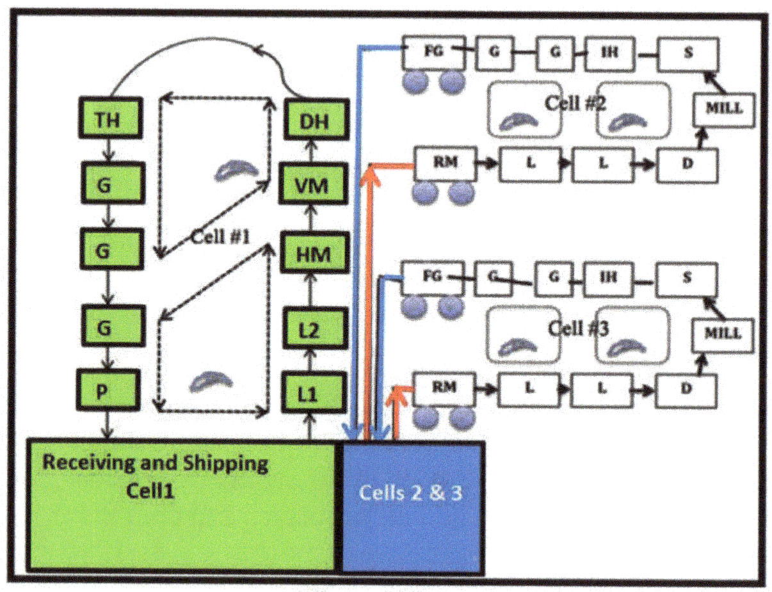

Figure 9.2
Converting a Job Shop to a Lean Shop

The Job Shop is used infrequently to manufacture a special type of manual transmission. The demand is not enough to sustain normal continuous production, but the final product is a money-maker. been used Cell 1 makes the transmission linkage. The drive shaft is manufactured in Cell 2 and Cell 3 produces a gear box. This subassembly is then moved to Final Assembly (not shown).

It should be emphatically stated that the conversion to a U-shaped manufacturing is much more than just moving machines around as many group technology gurus and value stream mapping technical papers might suggest.

A lot of work is required by the Lean Engineer to make this transition, involving everyone in the factory. For example, when designing these cells a new sequence of operations might be implemented by adding a horizontal band saw to Cell 1. The saw might be used to cut bar stock into the correct length, depending upon which of the two different transmissions is being made. This modification might eliminate the need to perform facing and cutoff operations at the first two lathe workstations and reduce the required time. Every machine should be modified to execute autonomous, single cycle operations unattended (single cycle automatics). Jigs, fixtures and cutting tools are designed or modified to drastically reduce setup and changeover from one type of drive shaft to another (SMED), and to optimize cutting parameters. Machines should all be equipped with walk away switches, safety devices, QC checks and automated loading/unloading devices. All machines are arranged according to required precedence relationships in a U-shaped layout with a narrow internal isle. Between each machine, a decoupler is added which holds only one unit to support one-piece flow through the cell. A small WIP area is designed to hold incoming parts before the first workstation, and following the last workstation in the cell, designed using Little's Law.

A complete transformation of the original job shop to the three Lean cells shown in Figure 9.2 is called a *Mature Lean Cell Design*. Any partial conversion would result in what we call an *Interim Lean Cell Design*. These three cells are designed to be flexible and versatile enough to rapidly adjust to demand fluxuations, and even changing product mix. This is done by scheduling cell production based upon the number of workers assigned to each cell and the assigned tasks for each worker.

Once the cell capacity has been reached by assigning a single worker to each machine, the cell operates as a traditional serial production line with output determined by the bottleneck operation. At this point, the cell must again be redesigned. Within each cell, cell output is set to meet Takt time requirements. In steady-state operation, every machine in each cell must obey the *First Axiomatic Design Rule:* The machining cycle time for every machine or parallel machine configuration in the cell must be strictly less that the necessary required *Takt Time* (Time between part completions). The key step is to determine (1) how many workers are needed in the cell and (2) the number of machines to be assigned to each worker (worker's loop) to meet the required Takt time for the cell. Each worker will spend a certain amount of time at each machine, determined by: (a) whether or not it is a machining or assembly operation (b) the changeover/setup time required (c) any visual inspections or QC activities, and (d) the walking times between machines. If each task is manual, the time it takes each worker to complete his/her loop of assigned processes is the sum of all these times in the assigned loop. The cell can often be manned by one or two workers. The maximum output rate is determined by the longest machining processing time. For any number of assigned workers less than the number of machines in the cell, the cell cycle time in terms of minutes/part is the *maximum of any of the assigned worker's loop times*. For a numerical example, consider the Generic10-machine Work Cell in Figure 3.3.

The Company decides to expand its options and offer two types of manual transmissions. This Lean Cell manufactures two types of transmissions; Type A and Type B. This is a simple example of *Mixed Model Production*. The single The numbers shown beside each workstation are the workstation processing times for Parts A and B. Both A and B machine (unloading, checking and loading) and that about 5 seconds is spent walking from one machine to the next: 10 seconds between Machine DH and TH, as well as Machines L1 and P (Packaging).

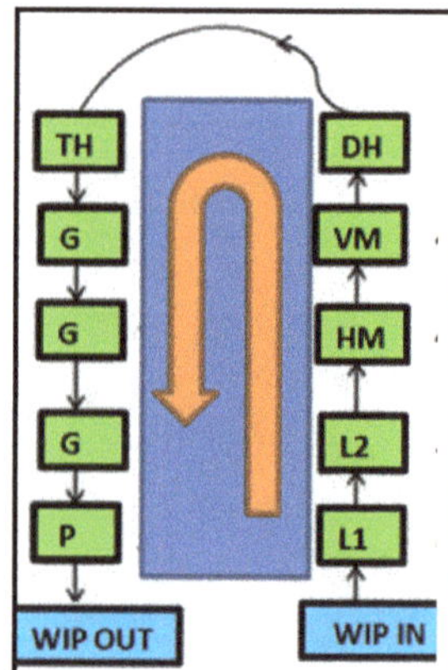

Figure 9.3
A Ten Machine Cell

Cell loading and cell unloading takes 50 seconds. At each machine, the acronym Lo-Co-Moo applies (**Load One** - **C**heck **One** - **Move One On**, and QC checks are done manually in the decoupler. Every processing step is a single cycle automatic machine except for packaging, which is a manual operation. Hence, the cell is a hybrid machining and assembly cell.

To illustrate cell balancing and how to meet Takt time requirements in *Mixed Model* production, assume that the daily demand for Transmission A is 200 units/day and Transmission B is 300 units/day. If the line runs 8 hours/day there are 489 minutes or 28,800 seconds of production time per day. The daily demand is 500 units per day, so Takt time is 57.6 minutes/part with a production rate of 0.96 parts/minute. It appears that the current system cannot meet the daily demand since the L1 operation for Part A and the packing operation for Part B both take 60 seconds. However, this conclusion is incorrect. For example, for Transmission A (Cell L1) the average processing time per part is a weighted average given by: APT_{L1} = {200*60 + 300*30}/500 = 42 seconds/unit. Similarly, For Transmission B (Cell P) APT_P = 56.0. Figure 9.4 provides the average processing times in the composite cell

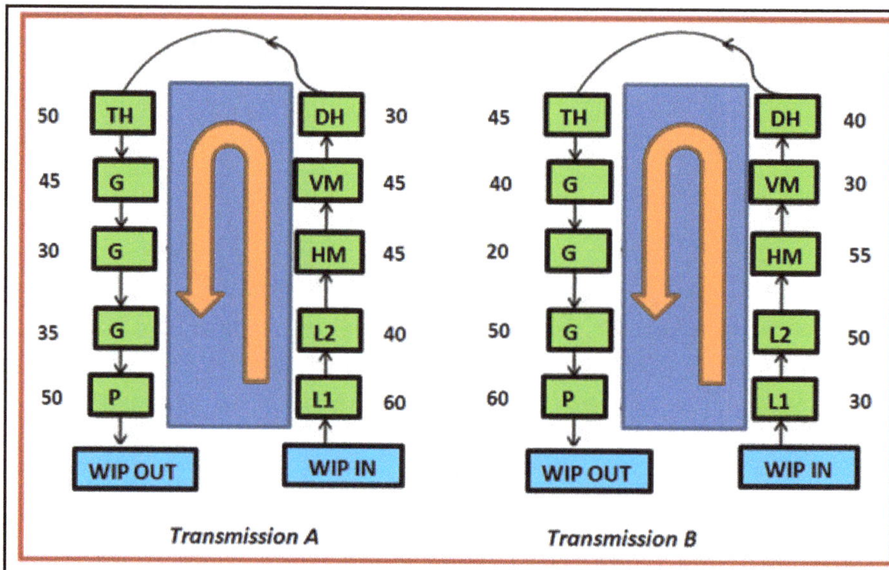

Figure 9.4
A Transmission Cell

The composite cell now adheres to the first axiomatic design principle and can be implemented. First: *What is the cell Takt time if one worker is assigned to the cell?* The worker loop time is given by:

$$Loop\ Time_{One\ worker} = (10 * 20)_{Task\ Times} + (2 * 50)_{\ Load\ /Unload} + 55_{Walk\ time} = 355 \text{ seconds.}$$

This is clearly infeasible.

Using two workers, assign Worker1 to machines 1-5 and Worker2 to machines 6-10. The two loop times are identical for this example and are given by:

$$Each\ Loop\ Time_{Two\ worker\ s} = (5 * 20)_{Task\ Times} + (50)_{\ Load\ /Unload} + 40_{Walk\ time} = 190 \text{ seconds.}$$

This is also infeasible.

Continuing to assign workers, the reader can verify that a feasible solution is given by:

- Assign Worker1 to the Cell loading task
- Assign Worker2 to the Cell unloading task
- Assign Worker3 to Machines 1&2
- Assign Worker4 to Machines 3&4
- Assign Worker5 to Machines 5&6
- Assign Worker3 to Machines 7&8
- Assign Worker3 to Machine 10& Packaging
-

Align Machines 5 & 6 so that the walking time between these two machines is 5 seconds.

The previous mixed model example was not realistic, because it ignored set-up and changeover times. In doing so, any sequencing of the 200 A's and 300 B's could be used. Let us revisit the example and assume that any changeover will take an extra 30 seconds. The effect of nonzero changeover times is to reduce available production capacity. The magnitude of lost production is totally dependent upon the manufacturing sequence used. For example, if 200 A's are run followed by 300 B's every day, only 1 minute of production time would be lost per day. However, this would result in uneven and "lumpy" delivery of A's and B's to downstream consumers. In general, small production lot sizes create less variance, reduce WIP and *smooth production* downstream. For example, another production sequence would be to run 40 A's and 60 B's; and repeat this sequence 5 times per day. In this case, 20 setups would be required and 10 minutes of production time lost. In this example, the reader will be asked in the problem set for this chapter to verify that if the sequence AAAABBBBBB is run and repeated 50 times, this will require 100 set-ups and lose 50 minutes of production time per day but the current cell configuration can still be staffed to run this pattern and meet required Takt time. Note that since the demand per day is the same, but the available production time per day has been decreased, the cell cycle time must now be shorter to meet customer delivery schedule. Overall, a part must still be passed on to final assembly every 60 seconds.

The cell cycle time and the cell throughput time are based solely upon the number of workers assigned to the cell, the operations in their individual loop(s), the time spent at each machine in their assigned loop and the walking times between machines. Here is the unique aspect of *Lean Manufacturing Cell Design*. The throughput time and production rate of the cell does not depend upon individual machine times but on the number of workers assigned to the cell, as long as there is sufficient capacity at all cell workstations to meet Takt time requirements. The machining times have been *decoupled* from the cell cycle time by the way the cell operates. In other words, *the production schedule has been made independent of the processing times at each machine.* This is the magic and genius of the TPS, linked-cell production system. *The job shop and flow lines cannot be operated in this way.* The output of any cell can be easily changed by assigning more or less workers until cell capacity is reached….then the cell must be redesigned. Generally workers assign themselves to the machines without changing any machining operations (provided capacity exists for expansion).

The number of workers and the machines operated by each worker are determined to meet two critical goals: (1) Balance the workload across all workers and (2) produce to Takt time. Balancing the workload is accomplished by assigning machines to each worker in such a way that the time required for each worker to make one cycle around his/her loop is the same for all workers. In most situations, this goal cannot be accomplished. There is usually a relatively small difference between worker cycle times. The differential can be reduced in a number of ways, including workers sharing one or more machines at the interface of their loops, but since both the number of workers and product demand will frequently change, one of the best ways to balance is to reduce long machine cycle times into multiple short, sequential operations.

Since a cell's cycle time is dictated by operator and machine assignments, and each machine will automatically cut off and wait for the operator's next arrival, the throughput time per part is not determined by the machine processing time, but the maximum loop time of all workers multiplied by the number of workstations. Hence, machine times and machine cycle times can be changed (within limits) without affecting the production rate of the cell. In traditional manufacturing lines, changing machine times would affect both the output rate and the delivery lead time. Note that the number of output cycles that any one part spends in the cell depends not only upon the maximum loop time of any worker, but also upon the amount of stock on hand in the cell and whether each machining operation is an assembly operation or a machining operation. Cell stock-on-hand (cell WIP) is limited to the number of machines in the cell. If an operation is assembly, the operator will need to be at the machine during the assembly cycle time. If the operation is machining, the operator will MO-CO-MOO, hit a walk-away switch and go to the next machine. The machine will automatically shut itself down after the machining operation is completed and wait for the operator to return.

9.9 Returning to Reality

Classic lean manufacturing principles are practically taken as gospel, but benefits can be elusive for manufacturers that produce a variety of parts in low volumes. The conversion of a typical Job Shop Operation to Lean Manufacturing Cells has been presented as a straightforward transformation, but nothing could be farther than the truth. The fact is that it is a long, difficult journey which will meet much resistance. Even if supported by all shop floor personnel and carefully planned: *Things will get worse before they get better.* This is a reality of all technological innovation. The transformation must be fully supported and funded by plant management and they must also play an active role. Training programs must be implemented to change both attitudes and work ethics of shop floor personnel.

As a champion of Lean Engineering, it is painful to admit that not all Job Shop operations should be fully converted to Lean Cells. If the Job Sop is in fact the sole business of a Manufacturing Facility and all machines must exhibit a high degree of Availability and usage, Lean transformation should be seriously investigated. If the Job shop exists primarily to support the

Manufacturing System and *internal customers* Lean Thinking will still yield tremendous benefits but may not be a candidate for full conversion or even implementation of Lean Cells. In the latter case, a support machine shop functioning as a Job Shop might by necessity have idle machines and low utilization. Every situation is unique.

Bibliography

Albin, S. Approximating a Point Process to a Renewal Process: Superposition Arrival Processes to Queues. *Operations Research, Vol. 30* , 1133-1162.

Allen, A. (1990). *Probability and Queuing Theory* (Second Edition ed.). Academic Press.

Ammer, E. D., 1985, "Rechnerunterstützte Planung Von Montageablaufstrukturen Für Erzeugnisse Der Serienfertigung." IPA-IAO Forschung Und Praxis, 81.

Askin, R., & Strandridge, C. (1993). *Modeling and Analysis of Manufacturing Systems.* John Wiley & Sons, Inc.

Bard, J. F., Shtub, A., and Joshi, S. B., 1994, "Sequencing Mixed-Model Assembly Lines to Level Parts Usage and Minimize Line Length," International Journal of Production Research, **32**(10) pp. 2431.

Bartholdi, J., & Eisenstein. (n.d.). Retrieved from http://www2.isye.gatech.edu/-jjb/bucket-brigades.html

Baudin, M. (2007). *Working with Machines.* Productivity Press.

Black, J. (1983). Cellular Manufacturing Systems Reduce Setup Time, Make Small Lot Production Economical. *Industrial Engineering* , 36-48.

Black, J. (1991). *The Design of the Factory with A Future.* McGraw-Hill.

Black, J., & Chen, J. (1994). Decoupler-improved Output of an Apparel Assembly Cell, *Journal of Appl. Manufacturing System* , 47-58.

Black, J., & Hunter, S. *Lean Manufacturing Systems and Cell Design.* Dearborn, MI: Society of Manufacturing Engineers, ISBN 2003-102-33-5.

Black, J., & Kohser, R. A. (2008). *DeGarmo's Materials & Processes in Manufacturing, 10th edition.* Wiley.

Black, J., & Schroer, B. (1988). Decouplers in Integrated Cellular Manufacturing Systems, "Jo. *Journal of Engineering for Industry, Transactions ASME, Vol. 110* , 77-85.

Black, J., & Schroer, B. (1993). Simulation of an Apparel Assembly Cell with Walking Workers and Decouplers. *Journal of Manufacturing Systems, Vol 12, No. 2* , 170-180.

Black, J., & Sipper, D. (1991). *Decouplers for Integrated Pull Manufacturing.* ASME-PED.

Black, J., Jiang, B., & Wiens, G. (1991). Design, Analysis and Control of Manufacturing Cells, PED-Vol.53. *The American Society of Mechanical Engineers* .

Blanchard, B., System Engineering Management, Fourth Edition, John Wiley and Sons, 2008.

Bourjault, A., 1984, "Contribution a Une Approche Méthodologique De L'assemblage Automatisé: Elaboration Automatique Des Séquences Opértiores." Universite De Franche-Comte, .

Boysen, N., Fliedner, M., and Scholl, A., 2007, "A Classification of Assembly Line Balancing Problems," European Journal of Operational Research, **183**(2) pp. 674-693.

Braha, D., Minai, A., & Bar-Yam, Y. (2006). *Complex Engineered Systems.* Springer.

Buffa, E., & Sarin, R. (1987). *Modern Production/ Operations Management, ISBN 0-471-81905-0.* John Wiley & Sons.

Burbridge, J. (1975). *The Introduction of Group Technology.* John Wiley & Sons.

Burgridge, J. *Production Flow Analysis for PLanning Group Technology.* Oxford, England: Clarendon Press.

Burke, P. (n.d.). The Output Process of a Stationary M/M/S Queuing System. *Annals of Math. Stat. Vol.39* , 1144-1152.

Buzacott, J., & Shanthikumar, J. (1963). *Stochastic Models of Manufacturing Systems.* Prentice-Hall.

Chen, G., Zhou, J., Cai, W., 2006, "A Framework for an Automotive Body Assembly Process Design System," Computer Aided Design, **38**pp. 531-539.

Cochran, David S. "Detaching from Management Accounting Requires System Design," Journal of Cost Management, March/April 2006, pp. 20-28.

Cochran, D. S. and Barnes, J."Enhancement of the Systems Engineering Process in the Life Cycle with Axiomatic Design," Proceedings of ICAD2013, The Seventh International Conference on Axiomatic Design, WPI, Worcester, MA, June 2013. ICAD-2013-26.

Cochran, D. S., & Dobbs, D. C. (2001/2002). Evaluating Manufacturing System Design and Performance Using the Manufacturing System Design Decomposition Approach. *Journal of Manufacturing Systems, Vol. 20 No. 6* .

Cochran, D. S., Arinez, J. F., Duda, J. W., & Linck, J. (2001/2002). A decomposition Approach for Manufacturing System Design. *Journal of Manufacturing Systems Vol 20 No. 6* .

Cochran, D. S., Eversheim, W., Kubic, G., & Sesterhenn, M. L. (2000). The Application of Axiomatic Designand Lean Management Principles in the Scope of Production System Segmentation. *International Journal of Production Research Vol. 38, No. 6* , 1377-1396.

Cochran, D., Duda, J., Linck, J., & Arinez, J. (2001/2002). The Manufacturing System DesignDecomposition. *SME Journal of Manufacturing Systems, 20* (6).

Cochran, D., Linck, J., & Neisė, P. (2001). Evaluation of the Plant Designs of Two Automotive Suppliers Using the Manufacturing System Design Decomposition (MSDD). *Transactions of NAMRI/SME* .

Cochran, D. S., Linck, J., Reinhart, G., and Mauderer, M., "Decision Support For Manufacturing System Design (MSD): Combining A Decomposition Methodology With Procedural MSD," Proceedings of The Third World Congress on Intelligent Manufacturing Processes and Systems, MIT, Cambridge, MA, June 2000, pp. 9-16.

Cochran, D. S. and edited by Pincham, W. H., "Enterprise Engineering, Creating Sustainable Systems with Collective System Design – Part I," The Journal of RMS (Reliability, Maintainability, Supportability) in Systems Engineering, Winter Journal, 2009.

Crosby, P. *Quality if Free.* New York, NY: McGraw-Hill Inc.

Curry, G., & Feldman, R. (2009). *Manufacturing Systems Modeling and Analysis, ISBN-978-3-540-88762-1.* Springer-Verlag.

Curry, G., & Phillips, D. (May 10-12, 2000). Renewal Approximations of the Departure SCV's for Batch Arrivals and Service Processes'. *International Conference on Modeling & Analysis of Semiconductor Manufacturing.* Tempe, Arizona.

Dar-El, E. M. (Dec 1973). MALB-A Heuristic Technique for Balancing Large Single-Model Assembly Lines. *IIE Transactions* .

De Fazio, T. L., and Whitney, D. E., 1987, "Simplified Generation of all Mechanical Assembly Sequences," IEEE Journal of Robotics and Automation, 4**,** pp. 640-658.

Dennis, P. (2002). *Lean Production simplified.* Productivity Press.

Drexl, A., and Kimms, A., 2001, "Sequencing JIT Mixed-Model Assembly Lines Under Station-Load and Part-Usage Constraints," Management Science, **47**(3) pp. 480-491.

Feldman, R., & Valdez-Flores, C. (1996). *Applied Probability and Stochastic Processes.* Boston: PWS.

Feldman, R., & Valdez-Florez, C. (2010). *Applied Probability and Stochastic Processes* (Second Edition ed.). Springer-Verlag.

Gaither, N. (1980). *Production and Operations Management, ISBN-0-03-074622-1.* Orlando, FL: Dryden HBJ.

Goldratt, E. (1992). *The Goal.* Croton-On Hudson, New York, N.Y.: North River Press, Inc.
Goldratt, E. (1990). *Theory of Constraints.* Croton-On Hudson, New York, N.Y.: North River Press, Inc.

Groover, M. (2001). *Automation, Production Systems and Computer Integrated Manufacturing.* Prentice-Hall.

Gross, D., & Harris, C. (1997). *Fundamentals of Queuing Theory* (Third Edition ed.). Wiley Interscience.

Gross, D., & Harris, C. (1998). *Fundamentals of Queuing Theory.* John-Wiley and sons, Inc.

Gryna, F. M., Chua, R. C., & Defeo, J. A. (2007). *Juran's Quality Planning & Analysis for Enterprise Quality, Fifth Edition.* The McGraw-Hill.

Hall, R. (1991). *Queuing Methods for Services and Manufacturing.* Prentice-Hall, Inc.

Hall, R. W. (1982). *Kawasaki U.S.A: A case Study.* Alexandria, VA: American Production Inventory Society.

Hall, R. (1983). *Zero Inventories.* Home Wood, IL: Dow Jones-Irwin.

Harmon, R. L., & Peterson, L. D. (1990). *Reinventing the Factory, Productivity Breakthroughs b Nabyfacturing Today.* Arthur Anderson & Co.

Hendryx, S. (July/Aug 1990). Manufacturing Execution: Theory and Concepts. *AT&T Technical Journal* .

Hino, S. (2006). *Inside the Mind of Toyota.* New York: Productivity Press.

Hirano, H., & Black, J. (1988). *JIT Factory Revolution.* Productivity Press.

Homen de Mello, L. S., and Sanderson, A. C., 1990, "AND/OR Graph Representation of Assembly Plans," Robotics and Automation, IEEE Transactions On, 6(2) pp. 188-199.

Hopp, W., & Spearman, M. *Factory Physics, ISBN 0-256-24795-1.* Irvin: McGraw Hill. http://en.wikipedia.org/wiki/The_purpose_of_a_system_is_what_it_does

Hunter, S. L. (2001/2002). Ergonomic Evaluation of Manufacturing System Designs, vol. 20, No. 6. *Journal of Manufacturing Systems* .

Hunter, S., & Black, J. (2007). Lean Remanufacturing: A cellular Case Study. *Journal of Advanced Manufacturing Systems, Volume 6, Number 2, World Scientific* .

Jackson, J. (1957). Networks of Waiting Lines. *Operations Research, Vol. 5* , 518-521.

Jensen, P., & Bard, J. (2003). *Operations Research: Models and Methods ISBN 0-471-38004-0.* John Wiley and Sons, Inc.

Johnson, H. T. and Kaplan, R., Relevance Lost, Harvard Business School Press, 1987.

Johnson, H. T., Relevance Regained, The Free Press, Division of Simon & Schuster, 1992.

Johnson, H. T., and Bröms, A., Profit Beyond Measure, The Free Press, 2001.

Kingman, J. (1962). On Queues in Heavy Traffic. *Journal of Royal Statistics Society, Vol. 32* , 102-110.

Kreafle, K., TMMK Quality Manager, personal conversation, Georgetown, KY.

Lenz, R. K., and Cochran, D. S., "The Application of Axiomatic Design to the Design of the Product Development Organization," Proceedings of ICAD2000, First International Conference on Axiomatic Design, Cambridge, MA, June 2000, pp. 18-25.

Liker, J. K. (2004). *Becoming Lean.* Productivity Press.

Liker, J. K. (2004). *The Toyota Way - 14 Management Principles frm the world's greatest Manufacturer.*

Little, J. A Proof for the Queuing Formula L=λW,. *Operations Research, 9*, 383-387.

Lockton, D., "The Purpose of a System is what it Does (POSIWID) and determinism in design for behavior change," Brunel University, accessed working paper, available at http://danlockton.co.uk, 2012.

Marshall, K. (n.d.). Some Inequalities in Queuing. *Operations Research, Vol. 30* , 651-655.

Martinich, J. (1997). *Production and Operations Management, ISBN 0-471-54632-1.* John Wiley & Sons, Inc

Minzu, V., Bratcu, A., and Henrioud, J. M., 1999, "Construction of the precedence graphs equivalent to a given set of assembly sequences. ," Proceedings of the 1999 IEEE International Symposium on Assembly and Task Planning, Anonymous pp. 14-19.

Monden, Y. (2001/2002). *Toyota Production System.* Industrial Engineering and Management Press, IIE.

Monden, Y. (1983). *Toyota Production System, An Integrated Approach to Just-In Time.* Institute of Industrial Engineers, second edition.

Montgomery, D. C. (2001). *Introduction to Statistical Quality Control.* Wiley.

Morgan, J. M., & Liker, J. K. (2006). *The Toyota Product Development System.* Productivity Press.

Morgan, J. M., & Liker, J. K. (2006). *The Toyota Product Development System.* Productivity Press.

Nahmias, S. (2005). *Production and Operations Analysis, ISBN 0-07-286538-7* (Fifth Edition ed.). McGraw Hill, Inc.

Nakajima, S. (1990). *Introduction to TPM.* Cambridge, MA: Productivity Press, ISBN 0-915229-23-2.

Nicholas, J. (1998). *Competitive Manufacturing Management, ISBN 0-256-21727-0.* Irvin: MCGraw Hill.

Ohno, T. (1988). *The Toyota Production System.* Productivity Press, ISBN 0-915229-14-3.

Ohno, T. (1988). *Toyota Production System: Beyond Large-Scale Production.* Productivity Press.

Ott, E. R., & Schilling, E. G. (1990). *Process Quality Control, Second Edition.* Cambridge, MA: McGraw-Hill.

Parello, B. D., and Waldo C. K., Job shop Scheduling Using Automated Reasoning: A Case Study of the Car sequencing Problem, Journal of automated Reasoning, 2, 1986, pp. 1-42

Phillips, D., Ravindran, A., & Solberg, J. (1976). *Operations Research: Principles and Practice, ISBN 0-471-68707-3.* John-Wiley and Sons, Inc.

Pound, S., & Spearman, M. (Sept 5, 2007). The Great Push vs. Pull Diversion. *Industry Week* . Pyke, D., & Cohen, M. (1990). Push and Pull in Manufacturing and Distribution Systems. *Journal of Operations Management, Vol. 1* , 24-42.

Pyzdek, T. (2003). *Six Sigma Handbook.* McGraw-Hill.

Romney, B., Godard, C., Goldwasser, M., 1995, "An efficient system for geometric assembly sequence generation and evaluation. ," Proceedings of the ASME International Computers in Engineering Conference, Boston, Massachusetts, pp. 699-712.

Ross, P. J. (1988). *Taguchi Techniques for Quality Engineering.* McGraw-Hill.

Rother, M., & Shook, J., *Learning to See.* Lean Enterprise Institute.

Rother, M. and Shook, J., "Value-Stream Mapping Workshop," Lean Enterprise Institute, Version 1.3, September 2009.
Rother, M., "Toyota Kata – Time to Retire the PDCA Wedge," Presentation, Northeast Shingo Conference, Springfield, MA, 2011.

Ruffa, S. A., & Perozziella, M. J. (2000). *Breaking the Cost Barrier.* John Wiley & Sons, Inc.

Ruiz, M., The Four Agreements, Amber-Allen Publishing, 1997.

Sakasegawa, H. (n.d.). An approximation Formula for M/G/1 Queues. *Annals of the institute of Statistical Mathematics, Vol. 29*, 67-75.

Sakasegawa, H. (1977). An approximation Formula L_(q= 〚αβ〛 ^ρ (1-ρ)). *Annals of the Institute of Statistical Mathematics Vol. 29*, 67-75.

Santos, J., Wysk, R. A., & Torres, J. M. (2006). *Improving Production with Lean Thinking.* Wiley.

Scholl, A., and Becker, C., 2006, "State-of-the-Art Exact and Heuristic Solution Procedures for Simple Assembly Line Balancing," European Journal of Operational Research, **168**(3) pp. 666-693.

Schonberger, R. J. (1982). *Japanese Manufacturing Techniques: Nine Hidden lessons in Simplicity.* The Free Press.

Schonberger, R. J. (1986). *World Class Manufacturing: The Lessons of Simplicity Applied.* The Free Press.

Schonberger, R. J. (1996). *World Class Manufacturing: The Next Decade.* The Free Press.

Schonberger, R. (1986). *World Class Manufacturing.* New York: The Free Press.

Schroer, B., & Black, J. (1993). Simulation of an Apparel Assembly Cell with Walking Workers and Decouplers. *J. Manufacturing System*, 170-180.

Sekine, K. (1990). *One-Piece Flow: Cell Design for Transforming the Production Process.* Productivity Press.

Sekine, K., & Arai, K. (1992). *Kaizen for Quick Changeover.* Productivity Press.

Shingo, S. (1985). *A Revolution in Manufacturing: The SMED System.* Cambridge, MA: Productivity Press.

Shingo, S. (1989). *A study of The Toyota Production System.* Productivity Press.

Shingo, S. (1988). *Non-Stock Production - The Shingo System for Continuous Improvement.* Productivity Press, Inc.

Shingo, S. (1986). *Zero Quality Control: Source Inspection and the Poka-Yoke System.* Productivity Press.

Shingo, S., A Study of the Toyota Production System From an Industrial Engineering Viewpoint, Productivity Press, 1989.

Shingo, S., & Translated By Dillon, A. P. (1987). *The sayings of Shigeo Shingo, Key Strategies for Plant Improvement.* Productivity Press, Inc.

Sipper, D., & Buffin, R. (1997). *Production Planning Control and Integration.* McGraw-Hill, Inc.

Skinner, W. (May/June 1974). The Focused Factory. *Harvard Business Review* , 113-121.

Sly, D. (n.d.). Retrieved from www.proplanner.com

Spear, S. and Bowen, K., Decoding The DNA of the Toyota Production System, Harvard Business Review, September – October 1999.

Spearman, M. (n.d.). Retrieved from www.factoryphysics.com

Sugimori, Y., kusunoki, K., Cho, F., & Uchikawa, S. (1977). Toyota Production System and Kanban System: Materialization of Just-in-Time and Respect-for-Human System. *International Journal of Production Research* , 553-564.

Suh, N. P. (1992). Design Axioms and Quality Control. *Robotics and CIM, Vol. 9 Number 4/A, Aug-Oct* , 367.

Suh, N. P., Cochran, D. S., & Lima, P. C. (1990). "Manufacturing System Design" CIRP Annals Manufacturing Technology. *Paris, France: Institute for Production Research. Vol. 47, No. 2* , 627-639.

Suh, N. P. (1990). *The principles of Design.* New York, NY: Oxford University Press.

Suh, N. P. Axiomatic Design: Advances and Applications, Oxford University Press, 2001.

Su, Q., 2007, "Applying Case-Based Reasoning in Assembly Sequence Planning," International Journal of Production Research, **45**pp. 29-47.

Suzaki, K. (1987). *The New Manufacturing Challenge.* The Free Press.

Swaminathan, A., and Barber, S., 1996, "An experience-based assembly sequence planner for mechanical assemblies," IEEE Transactions on Robotics and Automation, Anonymous **12,** pp. 252-267.

Taha, H. A. (2003). *Operations Research, ISBN 0-13-032374-8* (Seventh Edition ed.). Prentice Hall, Inc.

Thomopoulos, N. (Oct 1967). Line Balancing - Sequencing for Mixed Model Assembly. *Management Science.*

Whitt, W. (1983). A Queuing Network Analyzer. *Bell Systems Technical Journal Vol. 62 No. 9* , 2779-2815.

Whitt, W. (1984). Departures from a Queue with Many Servers. *Math of Operations Research Vol 9* .

Whitt, W. (1984). open and Closed Models for Networks of Queues. *AT&T Bell Laboratories Technical Journal, Vol. 63, No.9* , 1911-1979.

Whitt, W. (1994). Towards Better Multi-Class Parametric Decomposition Algorithms for Open Queuing Networks. *Annals of Operations Research, Vol. 48* , 221-248.

Winston, W. *Operations Research: Applications and Algorithms.* Thompson, Brooks and Cole, ISBN 0-534-38058-1.

Womack, J. P. (2005). *Lean Solutions.* New York, NY: Simon & Schuster.

Womack, J. P. (2006). *Value Stream Mapping.* Manufacturing Engineering, SME.

Womack, J. P., Jones, D. T., & Roos, D. (1991). *The Machine That Changed the world.* Harper Perennial.

Womack, J., & Jomes, D. (1996). *Lean Thinking.* New York: Simon & Schuster.

Yano, C. A. and Rachamadugu, R., 1991, Sequencing to Minimize Work Overload in Assembly Lines with Product Options, Management Science, **37**(5) pp. 572-586.

Glossary of Important Lean Concepts and Terms

Andon: A visual control device that indicates the status of a machine, line, or process. Frequently, audible alarms or warning messages accompany Andon status lights as a secondary method of communicating a problem has arisen.

Autonomation: Autonomation refers to the autonomous control of both defects and machine shutdowns See also Jidoka.

Axiomatic Design Principles: Used in the design of manufacturing cells and systems. It involves determining the customer needs and transforming them into functional requirements, design parameters, and process variables.

Balancing: Assigning work to each workstation such that every workstation and worker works the same amount of time. Balance is dictated by task assignments and Takt time.

Benchmarking: The process of establishing progress or comparison, particularly competitive products and competing methodologies.

Cause and Effect Diagram (a.k.a. fishbone diagram): A method of defining an occurrence of an undesirable event or problem. The effect is the fish head.Contributing factors, or causes, are the fish bones attached to a backbone and the fish head.

Cell Balancing: Assigning workers and work to Lean cells such that every cell is working to the same drumbeat or Takt time requirement.

Changeover: Process of converting a line or machine from running one product to another.

Continuous Process Improvement (CPI): Continuously improving and updating a process.

Control Chart: One of the primary techniques of statistical process control. It plots sample statistics (measurements) of a quality characteristic for the process versus time (or the sample number).

Decouplers: Decouplers physically separate (or decouple) one station or machine from another. They are placed between two workstations or sub-cells to improve capability and flexibility in a cell. Decouplers also provide the JIT pull cell with *Make One, Check one, and Move One On* capability. Decouplers can have multiple functions including part transportation and inspection.

Defect: The undesirable result of an error in a process, and one of many types of waste in a system. Defects are often expressed as either yield of good parts or as Defects per Million Opportunities (DPMO).

Deming Cycle: The Deming Cycle, also known as the Plan-Do-Check-Act (PDCA) cycle is a standardized system for making gains in quality and conducting continuous improvement.

DMAIC: Acronym for afive step problem solving process process;Define, Measure, Analyze, Improve and Control.

Downtime: A period of the time during which a device or operation is malfunctioning or inoperative causing the system to be inactive.

Error Proofing: Designing a potential failure or cause of failure out of a product or process (a.k.a. Poka-yoke)

External Customer: A person or organization outside your organization who receives the output of a process. Of all external customers, the end-user should be the most important.

Five S's: The five Japanese terms designated for maintaining an efficient and organized workspace. The terms are Seiri (sorting), Seiton (straightening or setting in order), Seiketsu (standardizing), Seiso (sweeping or shining), and Shitsuke (Sustaining).

Flexibility: A production system's ability to respond and adapt to internal and external changes, such as changes in product design or customer demand, in a cost-effective and timely manner.

Flowchart: A graphic representation using symbols to show the step-by-step sequence of operations, procedures, or activities. See Value Stream Mapping

Gemba: The place where value is being added to a product

Gemba Walk: Searching out Muda or places where improvements can be made

Heijunka Box: A visual scheduling tool that strives to level production to the mixed model final assembly.

Histogram: A a graphical representation that shows how many observations fall in certain intervals. It is used to estimate the probablity distribution function of a variable for a large set of data.

Integrated Quality Control (IQC): Everyone in an organization has a complete understanding of quality control, its methods, benefits and expectations.

Internal Customer: The people, machines, or processes being supplied with the products or parts made in preceding work areas.

Jidoka: A cornerstone of the Toyota Production System, Jidoka or autonomation prevents the production of defective products, eliminates overproduction, and preventing problem recurrence.

Jig: A work holding device used to locate a work piece with respect to the tooling during manufacture or assembly.

Job Shop: A manufacturing system where machines are functionally grouped and incoming jobs can be processed through the system using route sheets.

Just-In-Time Manufacturing: JIT is the lean production manufacturing system design which minimizes inventory and operators by having the right materials at the right place at the right time.

Kaizen: The Japanese word for improvement.

Kaizen Events: Executed by a team of workers to facilitate improvement in processes, methodologies and technology insertion

Kaizen Blitz: An intense, short term Kaizen event to execute radical or immediate change

Kanban: A pull system for production and inventory control system used in lean manufacturing, provides the links that connect cells and control inventory in the system.

Lead Time: The total time a customer must wait to receive a product after placing an order.In a lean shop, lead time is independent of the processing time for individual machines.

Lean Manufacturing/Production: A manufacturing system design to eliminate waste from all of its activities and operations. Lean produces products better, faster, and cheaper than its competitors using the economies of scope.

Leveling: Leveling or smoothing of production involves making the final assembly line into a Mixed Model subassembly or Mixed Model final assembly line to produce an even distribution of products every hour, every shift, every day.

Line Balancing: Making the amount of work, and thus the time it takes to perform the work, as equal as possible (in terms of time) at each Workstation.

Linked-Cell Manufacturing System(L-CMS): Manufacturing system design using multifunctional workers in cells that are linked by Kanban to other cells.

Little's Law: A law that shows the relationship between the work-in-progress inventory (WIP), the throughput time (TPT), and the production rate (PR) for the manufacturing system; WIP = TPT x PR

Load One-Check One-Move One On (LO –CO- MOO): Perform a step in the process; Check the product to ensure that step was done correctly; Move on to the next step.

Make-to-Stock: Intermediate product to sustain smooth, level flow or to buffer uncertainty (See WIP)

Make-to-Final Inventory: Finished product temporarily stored to satisfy on-time customer delivery.

Mass Production: Large-scale manufacturing with high-volume production and output; implies pre computer-era methods, with departmentalized operation and reliance on "economies of scale" to achieve low per-unit costs.

Materials Requirements Planning: Using software, material planning is accomplished through evaluating the Bill of Materials (BOM), Inventory Data, and the Master Schedule in order to stimulate replenishment of materials to be consumed and present purchase orders (PO's) for future materials needed.

Muda: Japanese term for any human activity that absorbs resources or time, but creates no real value.

Mura: Japanese term for *unevenness*. It refers to an uneven flow of parts in the Toyota Production System.

Muri: Japanese term for the act of overloading an area or otherwise executing unreasonable work.

One-Piece Flow Production: A production system where, through the development of a product-oriented layout and operators capable of handling multiple processes, parts are produced and transferred one piece at a time.

Pareto diagrams: Pareto diagrams styled after the Pareto Principle are a type of bar chart or histogram which displays the frequency with which a particular phenomenon is occurring relative to the occurrence of others.

Pareto's law: The law, often stated as 80/20 rule, developed by an Italian economist in the 1800s based on the principle that the vast majority of an end result (wealth, cost, quality problems, etc.) is determined by a small percentage of a group (the number of people, items, etc.).

Pokayoke (also Poka-yoke): Japanese for *mistake-proofing* or *defect prevention.* Mistake-proofing devices made by designing parts, processes, or procedures so that mistakes physically or procedurally cannot happen.

Preventive Maintenance: Activities done to insure the operating availability of a production resource and its ability to meet process specifications. They are done on a pre-scheduled basis, or based on the identification by a monitoring system of conditions that may cause future breakdowns.

Process Failure Mode and Effects Analysis (PFMEA): An analytical technique used by a manufacturing responsible engineer/team as a means to assure that, to the extent possible, potential failure modes and their associated causes/mechanisms have been considered and addressed.

Process Sheet: A set of manufacturing instructions for a specific batch, lot or run that describe the operating parameters and settings for the equipment and facilities used, and any associated tooling or supplies.

Product Life Cycle: The processes, costs and revenues associated with a product from its initial creation to its abandonment, and often categorized by the stages of introduction, growth, maturity and decline.

Product Mix: The number of individual products produced or sold by an organization. Sometimes involves the concept of *sequencing.*

Quality: The characteristic of a product or service to satisfy stated or implied needs.

Reliability: The ability of a machine, device or component to operate as it was designed to operate over a specified period of time without failure or entering a failed state.

Repeatability: Capability of a process, tool or procedure to consistently and repeatedly yield the same measureable result.

Root Cause Analysis: A problem-solving approach to reduce or eliminate recurrence of the same problem whereby the underlying cause of a problem is first identified and only then is the corrective action or solution designed.

Seven quality tools: Tools used for process improvement that are usually: Cause and Effect Diagrams, Check Sheets, Control Charts, Histograms, Pareto Charts, Scatter Plots and Run Charts.

Seven Wastes: (According to Taiichi Ohno) These wastes are: Overproduction, Idle Time, and Unnecessary Material Handling, WIP or Excess Inventory, Excess Processing, Wasted Motion Poor Quality.

Six Sigma: An improvement program using statistical analysis tools and directed Kaizen events to reduce Standard Deviation from a spread of 6-sigma to 12 –Sigma between the natural Lower Tolerance Limit and the natural Upper Tolerance Limit.

Standard Operations Routine sheets: Diagrams that show all of the tasks of a job including the sequence of operations and times, walking time, task time, the tooling needed, the stock on hand and the motion of the operator.

Taguchi Methods:A set of methods and statistical Design of Experiment tools to both enforce Design Centering (Meeting Targets) and to reduce variance (6-Sigma)

Takt Time: In a Lean Production System, Takt time is the time between production of consecutive items in a production system. Takt time is the drumbeat or heartbeat of

Lean production, and is defined as the available working time divide by demand over a specified period of time.

Throughput Time: The time needed for a product to pass through a machine, Lean cell, final assembly or an entire manufacturing system.

Type I error: The error of deciding that a process is out of control, when it is actually not making any defects. The probability of this occurring is given by Alpha (α).

Type II error: The error of deciding that a process operating in an acceptable manner, when it is actually out of control. The probability of this occurring is given by Beta (β).

Value Added Work: Any activity that actually transforms raw materials into a finished product.

Value Stream Mapping: A technique used in Lean manufacturing that maps the flow of material and data, and associated time requirements, from initial supplier to end customer for a given business process. VSM seeks to identify non-value added activities.

Variability: The extent to which the data points are spread from the mean.

Variety: Not the same as variability. Variety is a necessary component to sell finished product in different design configurations to satisfy the customer.

Withdrawal Kanban: A card used that specifies the kind and quantity of product which a manufacturing process should withdraw from a preceding process.

Work in Progress (WIP): All materials in the production process once they are withdrawn from the store until they are stored as finished goods. Wip is sometimes classified as (1) Parts waiting for value added transformation, (2) Parts waiting plus parts undergoing value added transformation, (3) Only those parts held between manufacturing areas or Lean Work cells, (4) Parts in intermediate or final storage, and (5 All of the above). Toyota called WIP *The root of all evil*.

Zero defects: The goal of a Lean system is to have no defects. Zero Defects is the utopian goal of Lean manufacturing and is necessary to operate a Lean system..

Zero Inventory: Made popular by Robert Hall's *Zero Inventories*. Hall meant target of zero WIP inventory between manufacturing areas or Lean work cells..

Appendix A

Case Study I

Case Study

Balancing, Leveling and Synchroni*zing* Lean Manufacturing Systems

There are three key concepts inherent in designing, operating and sustaining Lean Manufacturing systems. These three concepts are: (1) Balancing (2) Leveling and Synchronizing.

There are three main production facilities in a Lean production system. These are (1) Job shops (2) Serial flow lines and (3) U-shaped cells. Job shops usually produce basic components and parts. U-shaped Lean cells contain M machines which are assigned to N workers; $1 \leq N \leq M$ arranged in a U-shaped layout with all machines in close proximity to one another. Serial flow lines are M machines arranged in a straight line, at which operations are performed in sequence. Job shops usually manufacture parts or components. U-shaped Lean cells are designed to produce assemblies or subassemblies and Flow lines are typically used for Mixed Model final assembly. All three production facilities are capable of producing multiple products or families of products, although it is not uncommon to dedicate U-shaped cells to specific assemblies or subassemblies.

Balancing

Balancing is designing and operating either a flow line or a U-shaped production cell in such a way that every worker in the cell consumes exactly the same amount of time per cycle of work (Part release). If each workstation is manned by a single worker, this implies that every worker will work the same amount of time during an assigned period of work. Perfect balance is often difficult or impossible to achieve. The concept of balancing workload is applicable to either single or mixed model production.

Leveling

Leveling production has as its goal creating smooth, uninterrupted production sequences for either final assembly or U-shaped Lean cells. It only applies to mixed model production lines, except for one case. In both single and multiproduct manufacturing systems, batching can seriously inflate variance; which in turn will cause increased throughput times (time in the system) and WIP levels. The only reason to ever consider lot sizes, either production lots or movement of parts in large lots (batches), is to amortize large set up and changeover times. This is why Taiichi Ohno put so much effort into SMED. In mixed model production, the effect of batching (process batching or move batching) is most evident. For example, consider four different parts being produced in the same intermediate cell. The required production quantities are A=500, B=400, C=200 and D=100. A possible production sequence is to manufacture 500 A's, then 400 B's, then 200 C's and then 100 D's. This production sequence will cause delays, lumpy movement of products and batching of required raw materials. Both upstream and downstream workstations will be affected. A better production sequence would be to make A=10, B=8, C=4 and then D=2: This sequence would be repeated 50 times. The preferred Lean production sequence might be: AABBCCAAABBD repeated 100 times. There will be an optimal mixed model sequence and it could be determined using Linear Programming or Dynamic Programming. Clearly, the mixed model sequence used would depend upon (1) the amount of time consumed in set up and changeover activities; (2) the agility and versatility of both workers and machines; and (3) material handling efficiency and responsiveness. There are certainly other determining factors. One can discern that leveling production is a matter of sequencing and quantity, but it is not necessarily a function of typical order release or scheduling. In fact, it has been previously pointed out that the only workstation actually scheduled in Lean Manufacturing is ideally the last workstation (final assembly) in a Lean production facility. The production schedule for the entire Lean facility is determined by the Final Assembly Schedule (FAS). Work is not pushed (order release strategy) into a Lean system, it is pulled through a Lean system using Takt time as the drumbeat of the system.

Synchronizing

A properly designed and coordinated Lean manufacturing facility will be a Linked Cell Manufacturing System (LC-MS) terminating in a Mixed Model Final Assembly Line. The heartbeat of the entire production system is determined by final assembly Takt time. Every component work area, subassembly line and U-shaped Lean cell produces to final assembly; everything is produced at the correct rate, at the correct time and in the correct quantity. No one area of production either over-produces or under-produces but makes just enough to satisfy final demand. This is called *synchronization.*

Bottleneck Workstations and the Push - Pull Interface

An important concept that will always surface in any Manufacturing System is that of a *Bottleneck Workstation.* A bottleneck workstation is that workstation whose output rate is the slowest in the entire serial production sequence. The maximum output of a manufacturing line or a Lean Cell cannot exceed that of a bottleneck workstation. Bottleneck analysis is identifying and breaking the output constraint of the bottleneck workstation. It follows that the bottleneck workstation determines the output rate of the entire serial production line.

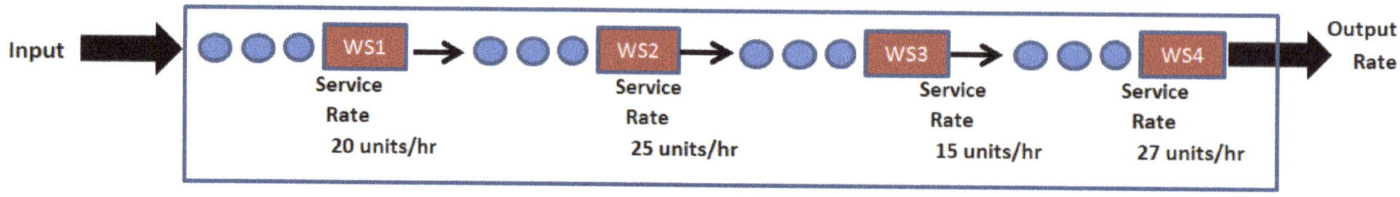

Figure 1
A Serial Production Line

The output rate of the production line shown in Figure 1 can never be more than 15 units per hour. Workstation 3 is called the *Bottleneck Workstation*. Since Workstation3 dictates product flow, it should never be idle. This means that it should never go off-line due to either lack of work or a workstation failure. In terms of Lean product flow control, Workstation 3 should ***pull*** work from Workstations 2 and 3. Since Workstations 1 and 2 can produce faster than Workstation 3, If work is not pulled from workstations one and two, WIP will build uncontrollably at Workstation3 . Conversely, Workstation 3 can *push* work to Workstation4, since Workstation4 can always produce at a faster rate than Workstation 3. Goldratt calls this a *Drum-Buffer-Rope system*. The production rate of the bottleneck workstation is called the *drumbeat* of the system. It limits the output everywhere in the system. Workstation3 must hold a limited amount of WIP at all times to prevent *starvation*. The *Rope* represents what we have called pull production control. It limits upstream production to avoid overproduction and infinite WIP buildup at Workstation3. Understanding a Drum-Buffer-Rope control strategy is key to the most efficient design of a traditional serial production system. Lean strategy using Kanban Cards at each workstation is automatically a full implementation of Drum-Buffer-Rope.

It is not commonly understood, but a major benefit of a properly designed U shaped, Lean production cell is that the concept of a bottleneck operation simply does not exist unless there is a single worker assigned to each machine. In that case, a Lean cell is identical to a CONWIP production line. In a properly designed Lean, U-shaped cell the Takt time require for each part produced in the cell is only a function of the number of workers assigned to that cell and the number of workstations assigned to each worker. The first Axiom of Lean cell design prohibits a bottleneck workstation. Of course, the cell output is determined by the cell Takt time inforced across all workers. Conceptually, the maximum worker cycle time behaves like a bottleneck workstation cycle time, but its resolution is completely different than an individual bottleneck workstation in a serial production line.

Recall that the only workstation actually scheduled in a Lean production system is *final assembly*. The final assembly schedule is based upon product demand manifested in Takt time. Ideally, the last workstation in Mixed Model Final Assembly will for all practical purposes function as a bottleneck operation, and it will pull final assembly requirements from every other upstream production cell, subassembly and component manufacturing. This will require upstream manufacturing capacity to be faster than final assembly, creating a correspondence to traditional drum-buffer-rope control with Takt time being the drumbeat and Kanban control the rope. Lean production facilities create the ability of upstream manufacturing to easily respond to increased demand and changing product mix. This in turn demands that the final assembly line must be carefully designed to be flexible and versatile enough to accommodate change and remain the bottleneck line.

A Lean Manufacturing Facility

The Phillips Production Plant has been offered a contract to produce 300 units of a product A, 200 units of B and 50 units of C. They are requesting a 60 day delivery schedule, which if met will result in guaranteed orders for 3 years. To illustrate the fundamental principles of Lean facilities design we will configure a manufacturing facility to produce the three products. The manufacturing sequence will use: (1) supply chain delivery of basic manufacturing raw materials (nuts, bolts, screws, wiring harnesses, etc.), (2) two subassembly cells and (3) a mixed model final assembly line. Figure 2 is a schematic of the proposed production facility. The design team has identified four separate final assembly steps in a Mixed Model final assembly flow line (ASM1, ASM2, ASM3 & ASM4). Two U-shaped production cells feed into final assembly; one at Assembly Station ASM1 and one at Assembly Station ASM4. All assembly stations have been designed such that all small assembly resources (nuts, bolts, etc.) will be purchased and delivered from subcontractors. These materials are planned to be stored adjacent to each workstation in storage areas called *Supermarkets*. The standard times for the 4 assembly steps for products A, B and C at Assembly Stations 1, 2, 3 and 4 are given in Table 1. Assume that a typical work day is one 8 hour shift, with 45 minutes planned for a lunch break and 15 minutes for a work break.

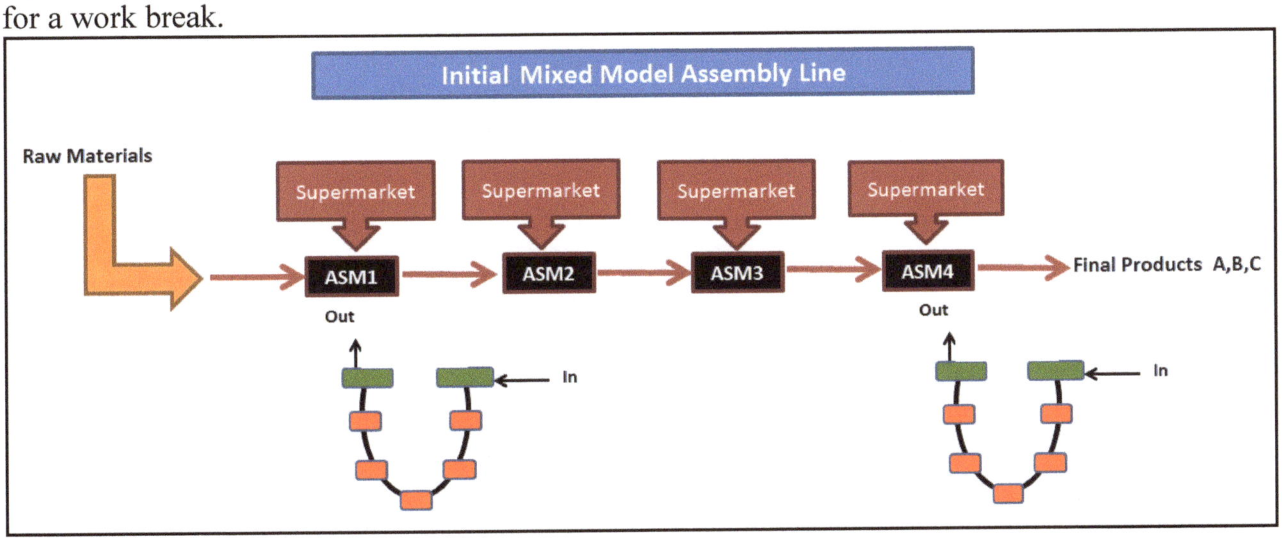

Figure 2
The Proposed Final Assembly Line

Assembly Times per Part (hr/part)

Task Times at ASM stations

		1	2	3	4
Parts	A	0.7	0.3	0.4	0.7
	B	0.65	0.5	0.35	0.75
	C	0.7	0.35	0.3	0.7
Total Time/Task		2.05	1.15	1.05	2.15

Table 1
Task Times

Balancing in Final Assembly

The first Lean Engineering activity is to determine the Takt time for final assembly. The required manufacturing cycle time for all 550 parts is 60 days. Working one shift per day will generate 420 minutes of production time per day or 25,200 minutes over a 60 day period of work, working 7 days a week. If 550 parts are to be made over this same period of time, this will require that one part be produced every (25,200/550) = 45.818 minutes. To be conservative and work with integer values, we will choose TAKT = 45 minutes/part. This will require {45 min/part*550 parts} / {60 min/hr* 7 hrs/day} = 58.93 days. Hence, it is possible to meet delivery requirements using Takt as 45 minutes. It turns out that there are 6 paid holidays per year, or one every two months. Counting one day holiday every two months, the production cycle is almost exactly 60 days (59.93 days)

To execute Line Balancing when multiple products (A, B, C) are involved, we will calculate the average processing time for each Final Assembly Workstation for the three products. A weighted average for part production time at each final assembly workstation can be calculated as follows.

A part must be produced every 45 minutes on the average. Note that some parts may take more time, and some parts may take less time, but a part must be produced on the average every 45 seconds. The average times to produce one unit of A, B and C can be calculated as follows. Using the required production rates for parts A, B and C and the weekly demands, the proportion of parts produced by type every 60 days are as follows.

Part Types	Demand over 2 Mos.	Proportion
A	300	0.5454
B	200	0.3636
C	50	0.0909

Table 2
Proportion of Required Production/Part Type

The corresponding assembly times at each ASM station by part type are:

		Assembly Times per Part (min/part) Task Times per Part (min/part)			
		1	2	3	4
Part	A	42	18	24	42
	B	39	30	21	45
	C	42	21	18	42
Total Time/Task		123	69	63	129

Table 3
Assembly Times per Task by Part Type

Using the proportions in Table 2 and the assembly times in Table 3, the weighted average task times at each ASM station can be calculated and are shown in Table 4

Tasks	Weighted Average Task Times (min)			
	1	2	3	4
	40.909	22.636	22.364	43.091

Table 4
Weighted Average Task Times per ASM Station

For example, the Weighted Average Task Time (WATT) for ASM 1 is given by:

$$WATT(1) = (42*0.5454) + (39*0.3636) + (42*.091) = 40.909$$

Table 4 shows that the weighted average task time at each workstation is less than the required Takt time, so no extra production capacity is needed.

Leveling Production in Final Assembly

The average time to execute all final assembly tasks has been determined for each of the four final assembly workstations. We now need to level production. Note that for final assembly there are many different choices as to how products A, B and C can be sequenced. As previously discussed, the demand of A=300, B=200 and C=50 can be produced by making 300 A's, then 200 B's and 50 C's. This manufacturing sequence might minimize set-up and changeover times, but it will not level production or smooth cell output. A leveled sequence is one in which A's, B's and C's are mixed in small quantities. The choice as to which sequence to use is largely determined by the magnitude of time required to set-up and change-over from one part type to another. If the following production sequence is repeated 50 times, product demand will be achieved.

AAAAAACBBBB

Note that this sequence will require 150 set-ups and changeovers; 3 for each set of 11 parts. It should now be obvious why Taiichi Ohno stressed the practice of SMED so aggressively. The set-up and changeover times are as follows.

A to B... 3 minutes

B to C... 4 minutes

C to A... 3 minutes

Applying Lean Production Principles to Final Assembly

There are four workstations in final assembly. Final assembly is usually a serial assembly line but for demonstration and pedagogical purposes we will design a U-shaped cell. Arranging these workstations in a U-shaped cell we obtain the configuration shown in Figure 3. The round circles with an "X" are decouplers which hold one unit. Load, unload, input, output and QC activities are all manual.

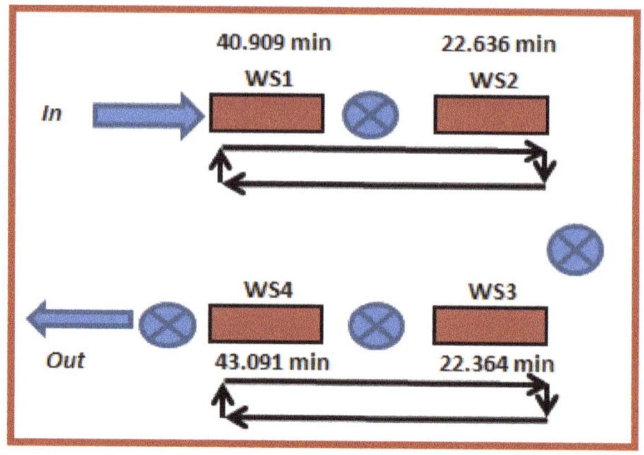

Figure 3
Mixed Model Final Assembly With Two Workers

Assume that the walking time between each workstation is 10 seconds; from WS2 to WS1 is 20 seconds; and WS4 to WS3 is 20 seconds. The time to input and output one part is 3 minutes; the time to load or unload one part at each machine is 4 minutes; and a quality control check at each workstation's decoupler takes 5 minutes.

A One Worker Cell

If the cell is manned by one worker, the following cycle time per part will be required.

CT= 60 sec (walking) + 6 min (in/out) + 32 min (load/unload) + 20 min (QC)

+ [(10 min*4)/11] (set-up per part per sequence of 11 parts) = 62.64 min

In other words, the average time that each part takes to exit final assembly using one worker is about 63 minutes. This exceeds the required cycle time of 45 minutes, so 2 workers must be assigned to the cell.

A Two Worker Cell

Assume that worker 1 is assigned to WS1 and WS2, and worker 2 is assigned to WS3 and WS4.

The worker cycle times per part are:

CT_1=20 sec (walking) + 3 min (in) + 10 min (QC) + 16 min (load/unload)

+ (10*2)/11 (setups)

CT_1 = 30.02 minutes

CT_2= 30.02 minutes

Here we must recognize the principle of a *Bottleneck Workstation.* Note that both Worker1 and Worker2 will finish their roving, working loop before the machining work cycles are completed. The bottleneck delay for each worker is as follows.

W1= (40.91-30.02) = 10.89 min

W2= (43.091-30.02) = 13.071 min

Hence, the actual Average Cycle Time per unit produced is: ACT= max {40.91, 43.091} = 43.091 minutes. Since this is less than the required Takt time of 45 minutes, this is a feasible solution. Note that both workers will have a balance Delay of about 25% idle time.

The total time required to produce a set of AAAAAABBBBC is: AT= 43.091*11 = 492.91 minutes. The entire order will take (492.91*50) / (420) = 58.68 days. This is also feasible.

Synchronizing Production to Support Mixed Model Final Assembly

Based upon product demand, the number of tasks to be completed, precedence relationships, and the time required for each product/task to be completed; a Mixed Model final assembly line was configured as a U-shaped lean cell with two workers. The mixed model assembly sequence by product was chosen as AAAAAABBBBC.

As previously discussed, each part in final assembly manufactured at Workstation 1 and at Workstation 4 require a matching subassembly. We will now look at the *subassembly manufacturing line* that feeds into *Workstation 1*. The initial final assembly line will be designed to have one *Kanban Square* located next to Workstation 1 and Workstation 4. Kanban Squares are an alternative to using Kanban cards or any other form of signal.

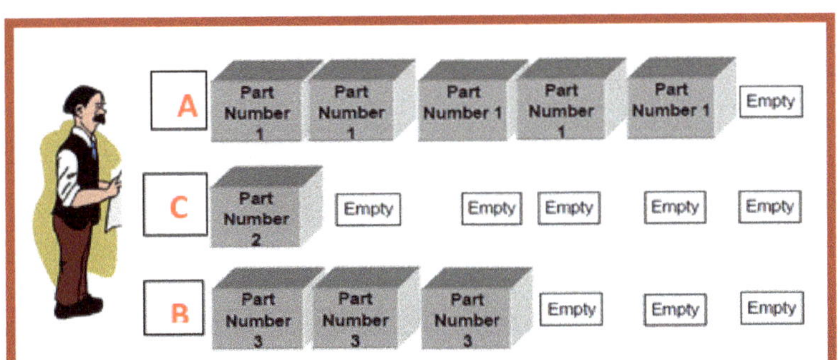

Figure 4
A Kanban Square

A Kanban square can be used very effectively for small lot production. The Kanban square is a "mini storage system" and is a form of decoupler. The Kanban square shown below holds Subassemblies for Parts A, B and C which are stored in small boxes. Subassemblies are pulled when needed. The signal for replenishing the Kanban square is when every square is empty, but the one which holds a subassembly for Part C. When the last subassembly for Part B is pulled, an order for a new batch of subassemblies is placed. This signal could be a simple red light at the output of the subassembly cell. The material handling transfer device holds 4 A's, 2 B's and 1 C.

The WS1 Subassembly Cell

From Figure 1, parts A, B and C at Workstation 1 all require a subassembly. Using Bottleneck pull scheduling, the subassembly cell can take no more than 493 minutes to produce each set (6 for Part A, 4 for Part B and 1 for Part C) of subassemblies. The Takt time per part in each subassembly is still 45 minutes. Assume for illustrative purposes, a common subassembly manufacturing sequence is used on all parts. Figure 5 shows the precedence diagram for the subassembly work area. Figure 6 shows all the standard manufacturing times at each workstation by subassemblies A, B and C. Figure 6 also gives the following data for each workstation.

- The times to input and output one part from the cell
- Load time per part at each machine
- Inspect time at each decoupler
- Set-up and change-over times between part types

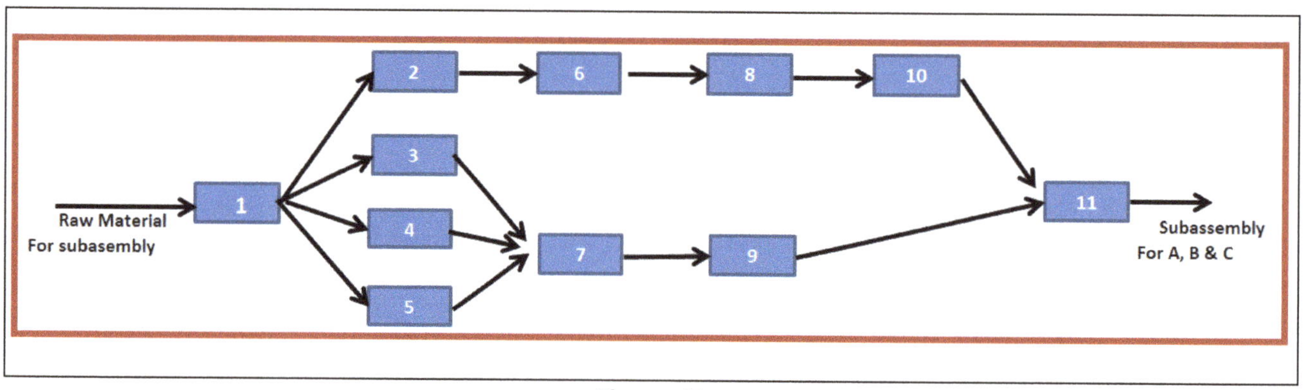

Figure 5
The Subassembly Precedence Diagram

In	2 minutes
Out	2 minutes

A to B	3 minutes
B to C	4 minutes
C to A	3 minutes

Load	2 minutes
Check	2 minutes

Task Times by Part and Task (min/part)

Task	1	2	3	4	5	6	7	8	9	10	11	
Part												Total
A	17	12	38	38	13	4	13	14	14	44	20	227.00
B	19	8	34	32	13	6	12	12	13	43	15	207.00
C	24	10	25	39	11	9	12	10	18	32	19	209.00
Total Time/Task (min)	60	30	97	109	37	19	37	36	45	119	54	643.00
Weighted Average(min)	18.364	10.364	35.364	35.909	12.818	5.182	12.545	12.909	14.000	42.545	18.091	218.09
Time for a Set (min)	202	114	389	395	141	57	138	142	154	468	199	2399.00
Inspect/Load	44	44	44	44	44	44	44	44	44	44	44	484.00
Changeover	10	10	10	10	10	10	10	10	10	10	10	110.00
In/Out	22										22	44.00
Total Time per Set	278	168	443	449	195	111	192	196	208	522	275	3037.00
Total Average Minutes per Part	25.27	15.27	40.27	40.82	17.73	10.09	17.45	17.82	18.91	47.45	25.00	276.09

Figure 6
Standard Subassembly Times for Each Task

We will refer to this as *Subassembly 1*. Figure 5 is not only a process flow chart but a *Precedence Diagram*. For simplicity, all three subassemblies (Sub A, Sub B and Sub C) follow this same production sequence. However, as in Final Assembly the times required to execute each of the 11 tasks are different for each part type. Proceeding as before, the weighted average processing time is calculated at each of the 11 workstations. These are also shown in Figure 6. Notice that the average weighted processing time at workstation 10 is 47.45 minutes. This processing time exceeds the Takt time of 45 minutes per part. Suppose we decide to install an identical machine in parallel. This will effectively yield an average processing time of 23.73 seconds per part and a station throughput time of 261 seconds, and both are now feasible. These times are reflected in Figure 7, which also gives the average time to make asset of 11 parts (AAAAAABBBBC) for each task.

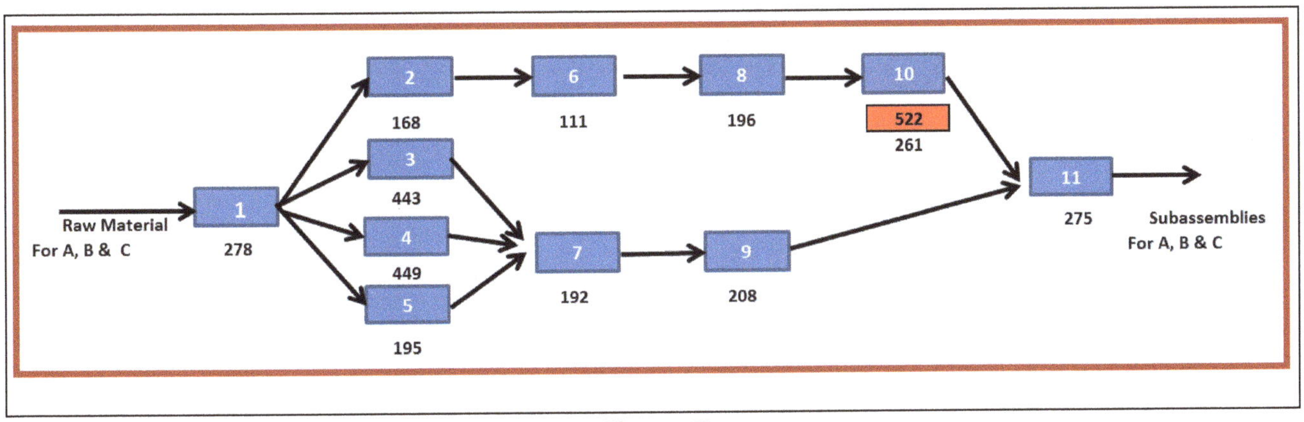

Figure 7
Station Processing Times

Traditional Line Balancing

Suppose we choose to run a batch of 11 parts (AAAAAABBBBC) at each workstation. We will now determine the number of *Work Centers* required using a traditional assembly line balancing procedure called the *Longest Eligible Processing Time* (LEPD). The solution obtained will then be compared to the design and operation of a Lean, U-shaped cell. The maximum time that can be assigned to any one workstation if two or more tasks are combined is 493 minutes. The solution procedure is as follows.

Open up Workstation 1 (WS1). Only Task1 can be assigned to WS1. Assigning Task1 to WS1 leaves (493-278) = 215 minutes. Tasks 2, 3, 4 and 5 can now be assigned; choose Task 5.This leaves (215-195) = 20 minutes. No tasks are now eligible for assignment. The time spent at Workstation 1 is 473 minutes. Close out Workstation 1.

Open up Workstation 2 (WS2). Tasks 2, 3 and 4 are all eligible for assignment. Using the LPT rule, Choose Task4 This leaves (493-449) = 44 minutes. No further tasks are eligible to assign to Workstation2. The time spent at Workstation 2 is 449 minutes. Close out Workstation 2.

Open up Workstation 3 (WS3). Tasks 2 and 3 are both eligible for assignment. Choose Task 3 using the LPT rule. This leaves (493-443) = 50 minutes. No other tasks can be assigned.

Close out Workstation 3.

Open up Workstation 4 (WS4). Task 2 and Task7 are eligible for assignment. Choose Task 7. This leaves (493-192) = 301 minutes. Task2 and Task9 are eligible for assignment. Choose Task9. This leaves (301-208) = 93 minutes. No other tasks can be assigned. Close out Workstation 4.

Open up Workstation 5 (WS5). Sequentially assign Tasks 2, 6 and 8 in that order. Assigning Task 8 will leave (493-168-111-196) = 18 minutes. Workstation time is 475 minutes. Close out Workstation 5.

Open up Workstation 6 (WS6). Assign Task 10. This leaves (493-261) = 232 minutes. Close out Workstation5. Workstation time is 261 minutes.

Open up Workstation 7. Assign Task11 to Workstation6. Workstation time is 275 minutes.

The line balancing solution is shown in Figure 8.

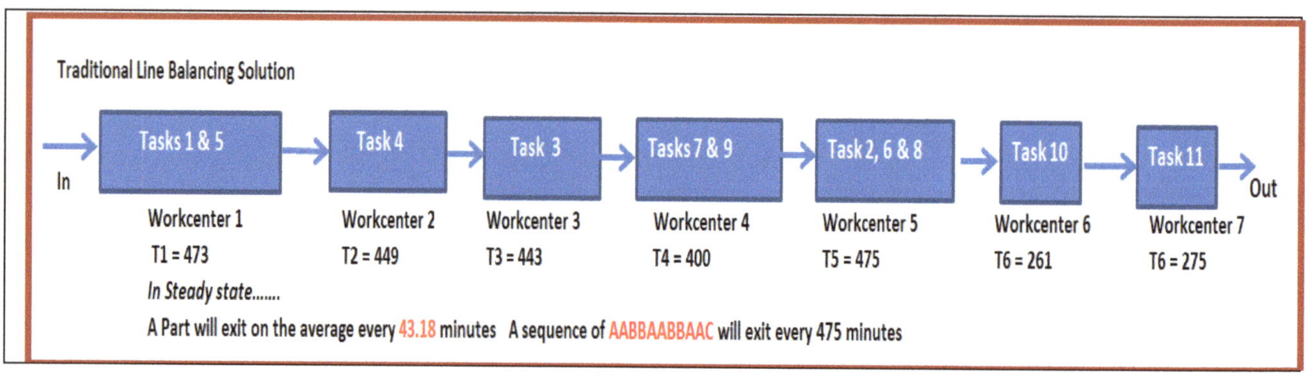

Figure 8
Traditional Line Balancing Solution

The *Bottleneck Workstation* is Workstation 5 with a throughput time of 475 minutes. Workstation 5 should pull production from Workstations 1-4 and push production to Workstations 6 and 7. It is clear that a set of AAAAAABBBBC parts will be produced every 475 seconds in steady state, or one unit every 43.18 minutes. Note that this solution is based upon a set of 11 parts being produced as a batch in each workstation. This is not an ideal solution to the Lean engineer. The most desirable solution to the Lean engineer is to produce one part at a time with a Mixed Model sequence that will smooth production. Since the sequence AAAAAABBBBC is a relatively small batch that also minimizes set-up and changeover time, we will produce to the following rule: *Produce parts one at a time using the production sequence AAAAAABBBBC repeated 50 times.* We will now design this subassembly cell using Lean principles. Figure 9 shows the Lean cell diagram. A decoupler which holds only one part is shown between workstations

Lean Cell Design

We will now design this subassembly cell using Lean principles. Figure 9 shows the Lean cell diagram. A decoupler which holds only one part is shown between each workstation.

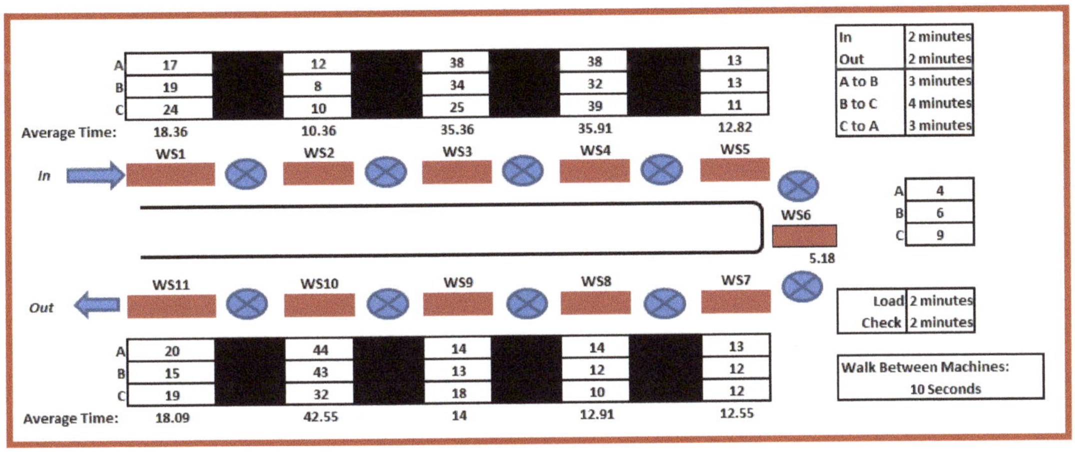

Figure 9

A Lean Mixed Model Cell

Note that all of the individual Part processing times at each workstation are less than the required Takt time of 45 seconds per part. This satisfies the first Axiom of Lean cell manufacturing given in Chapter 3.

One Worker

First, consider assigning only one worker to the cell. The cycle time per part with one worker is given by:

CT= 120 sec (walk) + 4 min (In/Out) +44 min (load/check) = 50 minutes

To this must be added the Average Changeover Time per part per cycle.

ACT = { (10*11)/ 11) min. = 10 minutes

The total average cycle time is:

CT1= 60 minutes

Since the average time to produce one part is required to be Takt = 45 minutes, this is clearly an infeasible solution. We must use two or more walking workers.

Two Workers

There are many ways that two workers could be assigned to this subassembly cell. In general, one of two strategies might be desirable: (1) Attempt to balance worker cycle times (2) Meet Takt time requirements with only one worker and assign the leftover tasks to the other worker.

Strategy 1

The cell production time per part with only one worker is 60 minutes. If Takt time of 45 minutes is to be met, then it follows that at least 15 minutes need to be eliminated per part. It might be a good strategy to assign both Station 1 and Station 2 to one worker. This strategy puts the responsibility for loading and unloading parts to the cell under the responsibility of one worker. Let us examine this strategy.

W1: Assign Workstations 1, 2 and 11 to worker 1

The average loop time per part for this assignment per part is:

W1= 50 sec (walk) + 4 min (In/out) + 12 min (load/check)

+ {(10/11)*3} setup

W1 = 19.56 minutes

W2: Assign Workstations 3-10 to worker 2.

The average loop time for this assignment per part is:

W2 = 90 seconds (walk) + 32 minutes (load/check)

+ {(0.909*8} Setup

W2= 40.67 Minutes

Note that Workstation 10 is now a bottleneck workstation for worker 2. The actual cycle time is 42.55 minutes with a bottleneck delay of 1.88 minutes per cycle.

The average time it takes to produce one subassembly is:

Max {19.56, 42.55} = ***42.55 minutes***

Strategy 2

W1: Assign Workstations 1-4 and 11 to worker 1

The average loop time per part for this assignment per part is:

W1= 100 sec (walk) + 6 min (In/out) + 20 min (load/check)

+ {0 .909*5} setup

W1 = 26.078 minutes

There is a bottleneck delay for worker 1 at workstation 4, so the actual worker 1 cycle time is:

W1= 35.91 minutes

W2: Assign Workstations 5-10 to worker 2

The average loop time per part for this assignment per part is:

W2= 100 sec (walk) +24 min (load/check)

+ {0 .909*6} setup =31.12 minutes

There is a bottleneck delay at Workstation 10, so the actual cycle time is:

W2= 42.55 minutes

The average time it takes to produce one subassembly is:

Max {35.91, 42.55} = 42.55 minutes

Assessment of the Two Strategies

It is natural to assume that Strategy 2 would be better than Strategy 1, but this may not be true. In Strategy 1, Worker 2 is busy most all of the time, and worker 1 is only busy about 50% of the time. This may not be so bad. Worker 1 is available for reassignment to material handling duties or even to another cell. In Strategy 2, both workers are busy about the same amount of time, but both finish well ahead of Takt time requirements.

We will now address some options which might be available if we choose to automate only portions of the cell. The assignments previously considered assumed that each workstation is a single cycle, walk-away-switch machine with automatic discharge to the decouplers. We have called this a *machining cell*. In reality, the cell could contain some machines which are manually operated. If this is true, the cell would be called an *assembly cell.*

Forming a Hybrid Assembly Cell

There are some possible economic advantages in not having every workstation a machining station. First, there is some additional capital investments required to automate a machine. Second, automation may involve more R&D time than management wants to invest. Third, if a worker is being paid $30 an hour, he/she is making $30 an hour over an 8 hour workday no matter how idle he/she might be. Fourth, automation tends to dehumanize and minimize the role that a worker might play at each automated workstation. In any case, let us quickly explore the impact of having some operations done manually. Notice that if any one operation or task is done manually, it simply adds that operations task time to the cycle times previously calculated: Nothing else changes. Using the two worker assignment strategy just analyzed, consider the following modifications.

Suppose we let Workstation 1 and Workstation 8 be manual operations and apply this modification to Strategy 2. This will result in the following cycle time for workers 1 and 2.

CT_1 = (26.078 + 18.36) = 44.44 minutes
CT_2 = (31.120 + 12.91) = 44.030 minutes

The resulting average time per part is: CT = max {44.24, 44.03} = 44.44 minutes with almost perfect worker balance. The subassembly cell is also well synchronized with final assembly.

Comparing Traditional Assembly Line Balancing Production with Lean Production

The traditional assembly line balancing solution required 7 Workstations. Each workstation required at least one worker. The solution obtained did not produce parts one at a time, but produced a set of 11 parts every 475 minutes. This batch production does result in improved Leveling, but frequent set=ups and changeovers can induce variance into the system. Perhaps more important, the production system for subassemblies is difficult to modify if demand patterns change.

The Lean cell overcomes most if not all of these problems. The Lean cell was designed to operate with one or more workers. Each worker clearly understands their role in producing each type of subassembly. The Lean cell operates on a unit move principle, and produces parts according to Takt time one-at-a-time. This creates a smooth, uniform flow of subassemblies to point-of-usage in final assembly. As discussed in Chapters 2-5, the Lean manufacturing strategy is flexible and versatile. It can be easily modified to add or subtract products, and responds easily to changes in demand by adding or subtracting cell workers. The Lean cell effectively uses a worker to execute both material handling and Quality control duties. By almost any metric or yardstick, the *Lean Cell Design* is superior in almost every way.

Executing Set-up and Change-over Tasks

Consider a single operator assigned to run the lean cell shown in Figure 10.

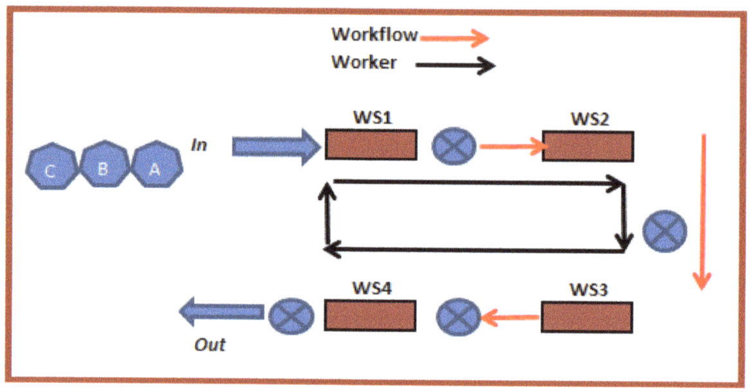

Figure 10
A Single Worker Lean cell

The cell is designed to produce three types of parts (A, B and C) one at a time. The worker will first set up machine WS1 to run part A, and then load a part A from the input station. Using a walk-away switch, the worker moves to the decoupler following WS1, which if sequences are already running, will hold a part C. Part C is then inspected. WS2 has just completed a Part B, so it is set-up for Part C; then loaded; then activated by a walk-away-switch. This procedure is repeated at WS 3 and WS 4. Upon discharging a part from WS4 (in this case part B), This could launch a Kanban signal to send another Part B to the WIP area at the Input station. In this scenario, unit flow is assumed between the feeder cell and this cell. Product flow is smooth and tightly controlled. The total WIP in this cell can never exceed 7 units. Now consider the cell shown in Figure 11, also operated by one worker.

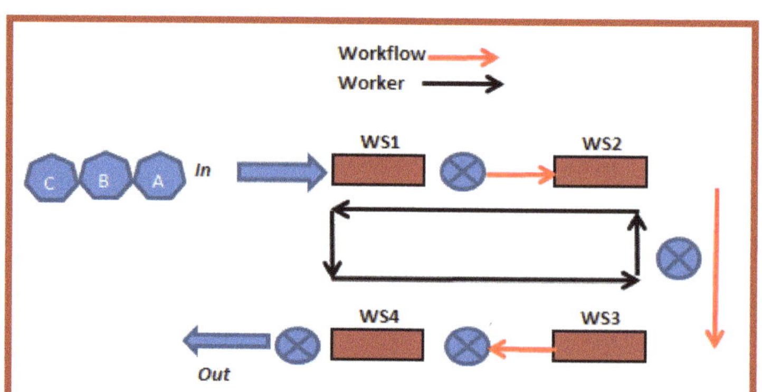

Figure 11
Another One-Worker Cell

Note that the worker order of work is now reversed from Figure 10. In Figure 11, the worker starts at WS4 and ends at WS1. The pattern of work is as follows. Starting at WS4, the worker will inspect a finished part (assume Part A) and then move it to the outbound, finished part WIP area. The worker then performs a set-up and change-over operation at WS4 for the next part in sequence (Part B). Part B is setting in the decoupler at WS4, and is inspected. It is now ready to be loaded to WS4; the machining cycle activated by a walk-away switch; and the worker proceeds to WS3. Part A is waiting in the decoupler preceding WS3, and it is inspected; loaded; and WS3 is activated by a walk-away switch. The procedure is now repeated indefinitely. If each machine does not automatically unload a part to its outbound decoupler, this additional duty is assigned to the worker and added to his/her "loop time". As usual, all machining times ad worker loop time must be less than the required cell Takt time.

The choice of whether or not to use the worker pattern shown in Figure 10 or that shown in Figure 11 is a matter of choice and preference. In reality, the choice should be based upon a prototype study by the Lean engineer. The well trained Lean engineer will design the cell for multiple products; assign one or more workers to meet Takt time requirement; and then *simulate* cell behavior to determine which work pattern will take the least time. Ideally, there should be little difference in cell behavior.

Synchronizing to Production Requirements

Regardless of which Lean cell operating strategy might be adopted, the goal is to keep the Mixed Model final assembly line or cell supplied with subassemblies for final assembly. There will be many strategies which might be adopted to achieve that goal. We have suggested that a Kanban square could be used to store subassemblies, and that an *automatic Kanban pull system* is inherently a part of this configuration. In any case, the subassembly lines should be placed very close to the Mixed Model final assembly line to minimize material handling and facilitate *just-in-time* delivery. Recall that Subassemblies are required for all parts produced. One suggestion is to place a Kanban square a next to Workstations 1 and 4 in final assembly. If the subassembly lines are close to the final assembly workstations, it might be easy to use simple power, roller conveyors to deliver subassemblies to the Kanban squares.

Managing Parts and Assembly Resources

The only thing that remains to be designed is a support system to manage and supply small parts such as nuts, bolts, washers etc at each manufacturing workstation. For illustrative purposes, consider the Mixed Model final assembly line shown in Figure 1. Note that there are four storage areas called *Supermarkets*. These are storage areas for selected raw materials. A preferred choice would be to pre-assemble kits which contain the nuts, bolts, wire harnesses, etc. The term Supermarket is being used for the storage of raw materials, but the configuration of a Supermarket is almost identical to a Kanban square. The difference is how reordering policies are executed. Raw materials often come from external suppliers, and as such a Kanban system might be difficult to use. In a supermarket, a simple reordering policy is typically used for each type of raw material being stored. There are many inventory control models which might be used to manage inventories in a Supermarket. We will demonstrate a continuous review inventory model used when demand is relatively constant, and the inventory replenishment time is also relatively constant. Define *Lead time* as the time required to fill an order from the supplier, and *Inventory On Hand* (IOH) as the current stock level.

The strategy is as follows: When the Inventory-On-Hand for any one part reaches a predetermined level (R), place an order for Q items of inventory. Figure 12 shows this reordering system graphically.

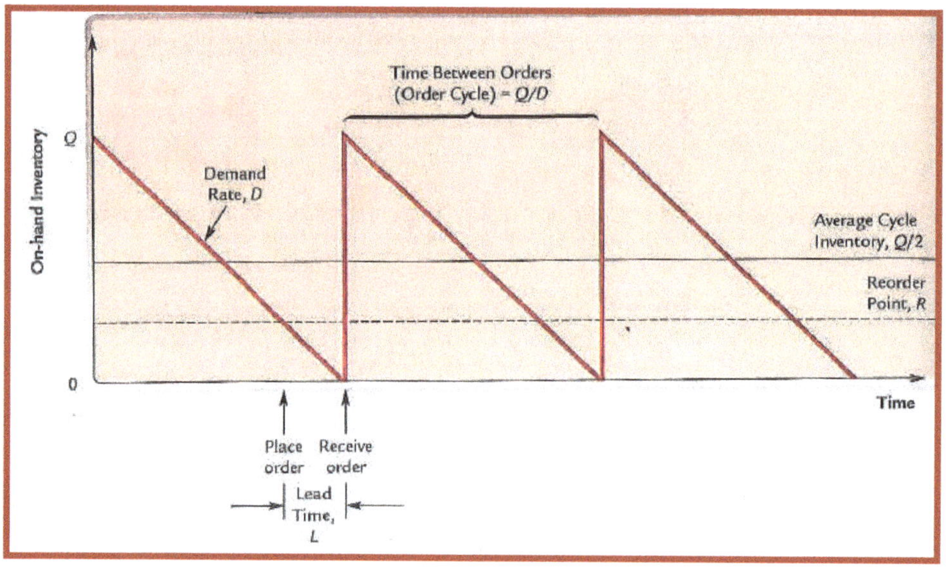

Figure 12
A Fixed Point reordering Inventory Model

Note that this type of inventory model will not allow any stock- outs of inventory. An order is placed for Q items at inventory position R. The time to place and receive an order is called *Lead time* and is designated as L. Ideally, when an order is placed, inventory-on-hand should be zero or near zero and immediately rise to a level Q. Notice that if demand is relatively constant, R should be the amount of inventory consumed during the order Lead Time. This is a simple but effective policy for Supermarkets. For illustrative purposes, suppose that it takes 1 day to place an order and be processed by the supplier. The supplier can ship and deliver an order in two days. Since a part in final assembly is produced every 41 minutes when the line is running, using a 7 hour work day (420/41)*3 = 65.33 units will be produced during the supplier's lead time. Hence, an order should be placed when the Inventory Level reaches 65 units. Although it is counter to Lean Thinking, it is generally a good idea to carry a small amount of *Safety Stock* to guard against uncertainty. In this case, ½ day inventory or (210/45) = [4.67] = 5 units shoud be sufficient. So, when inventory drops to 60 units order 65. Of course, the ordering level and ordering quantity will change if multiple stock units are used on each task; such as 10 bolts per task. It should be noted that the use of Safety Stock is to avoid raw material stock-outs and should be regarded as only a temporary fix. The Lean engineer should recognize that our old enemy variance is at work; locate the cause and eliminate if possible. This is true Lean Thinking.

The Fixed-Point Re-Order System as a Pull Model

The fixed point reordering policy just discussed is clearly a simple derivative of the classical Wilson Economic Order Lot Size model, which is based on inventory ordering and carrying costs. The model just presented does not use these costs... which in general may be difficult to accurately estimate... but instead is based upon Just-in-Time principles. In this case, the strategy is to determine a *Reorder Point* which will be determined such that from when the order is placed, the order will arrive Just-in-Time to replace inventory and resources needed as they near zero. The key is simply to estimate the *Lead time* accurately, which is usually easy to determine. The basic formula is given as:

Q = Demand (D)*Lead Time (LT) + Safety Stock (SS)

Ideally, Q is exactly equal to D*LT, but some variability in delivery time and usage rate should be buffered. This buffer is "Safety Stock"; which in a variance-free world would be zero. The demand in this case is the rate of usage at the stock point, and lead time is the amount of time required to send and receive the order. For example: if D=20 units per hour and LT =12 hours;

Q = 240 units + {SS=4} = 244 units.

The policy is a simple Kanban strategy: when the raw material inventory level reaches 240 units, reorder 244 units. Of course, this will irritate the lean engineer who does not want to carry a safety stock of 4 units. In reality, the amount of SS is dependent upon the cost of 4 units in buffer stock, the reliability and consistency of the supplier and how much variance must be buffered. Since the reorder point is 240 units, there will be no WIP buildup over the long run. Note that the reorder point cannot be set at 244 units as indicated in the basic formula.

Finally, notice that if safety stock is set at zero; Q=D*LT is simply a form of Little's Law.

$$L_{Sys} = \lambda * W_{Sys}$$

Summary and Conclusions

This case study has attempted to demonstrate how Lean, U-shaped cells can be used in Mixed model final assembly and in subassembly feeder cells. Various ways to design this system were discussed, but there are many options which might be explored.

In any case, it should be obvious that a Lean cell will in all cases dominate a traditional, batch Assembly Line balanced cell. The concepts of *Balancing, Leveling* and *Synchronization* were discussed and demonstrated.

Appendix B

Case Study II

Case Study

Improving Automobile Production

A Midwest automobile manufacturing plant is currently scheduled to produce an average of 400 cars per month or 4,800 cars per year. During the last 12 months of production, only 3,730 cars were produced. A Lean team was given the assignment to discover why planned production could not be achieved; identify improvement opportunities and increase production.

After an initial investigation, it was discovered that over the last 12 month period of time, 187 unscheduled machine downtimes occurred totaling 1897 hours. Unplanned machine downtimes cause the following things to happen.

- Machine Capacity is lost
- Production schedules and due dates are severely affected
- Workers experience unplanned idle time
- Production TAKT time cannot be met
- Workers become frustrated and blame Maintenance for the problems
- Delivery dates are compromised and late delivery occurs
- Company reputation and customer satisfaction is seriously affected

Further, it was estimated that if these unplanned delays could be 80% eliminated, a production capacity of 5,000 cars per year could be expected. The Lean team was authorized to proceed with a *Kaizen Blitz* to immediately increase production.

Pareto Analysis of Machine Downtime Events

The 187 downtime events were each plotted on a Bar Chart to determine the frequency of occurrence. From the Pareto Chart, it was found that 8 downtime events accounted for 87% of all downtimes. This is certainly an indication of which downtime events need to be eliminated. However, one must exercise caution in choosing which events should be the subject of Kaizen improvement studies. Frequency of events fail to capture three important modes of impact: (1) Cost of downtime (2) Duration of downtime and (3) Risk of the event occurring. The latter indicator should be explained. Over the last 12 months, an event which may have occurred only once or twice might be the most disruptive in terms of either lost revenue or time. Although every downtime event needs to be eliminated, further analysis is needed to determine the order in which events need to be examined.

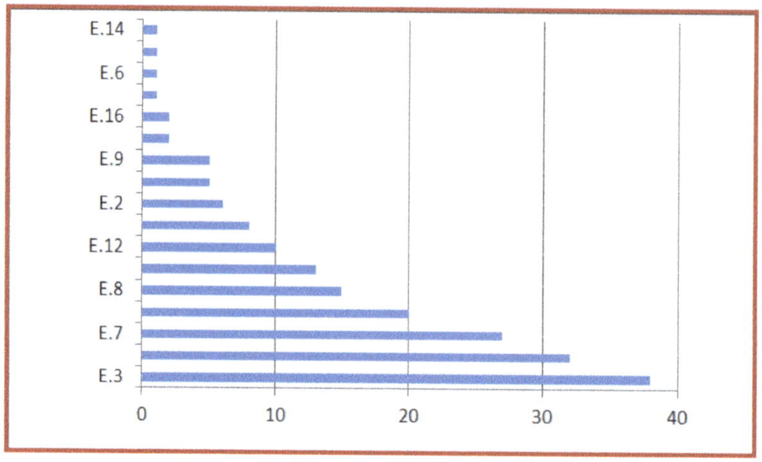

Figure 2
Pareto Analysis of Disruptive Events

The cost of a downtime event can be very difficult to determine. It is obvious that any disruptive, downtime event might involve the following costs.

- Cost of labor
- Cost of any raw materials ruined or scrapped
- Possible rework
- Cost of replacement tools or parts
- Cost of lost production
- Increased inventory costs
    - Etc.

Some of these costs can be readily estimated, but some might be impossible to accurately determine. On the other hand, the duration of downtime and the frequency of downtime is usually known or can be determined (Figure 2). It should be conceded that *any* disruptive event will cost the company money, and that all need to be eliminated. The question is: *Which are most important to eliminate?* We will use a *risk analysis* approach to rank-ordering the priority of Kaizen events.

A Risk Analysis Approach

Three importance indicators can be developed for each disruptive event, (1) The *frequency* of occurrence , (2) The *duration* of total downtime and (3) The *criticality* of failure. The frequency of failure can be summarized as follows.

Frequency Index	Yearly Disruptions
5	>= 20
4	>= 15 <20
3	>= 10 <15
2	>= 5 < 10
1	> 0 < 5

The duration of total downtimes can be captured in the following Table.

Duration Index	Duration
1	< 1 Hour
2	>= 1 Hour < 3 hours
3	>= 3 hours< 1 Day
4	>= 1 Day < 2 days
5	>= 3 days

The criticality of system downtime can be characterized in the following Table.

Criticality Index	Criticality	Impact
1	Minor	Disruption, but system recovers soon
2	Moderate	Disruption, but system slowly recovers
3	Disruptive	Disruption: Output is affected
4	Critical	Serious disruption: System never recovers
5	Very Critical	Very Serious: Line stops for a long time

A *Measure of Impact* is: **Risk Index = (Frequency Index)*(Duration Index)*(Criticality Index).** Note that the Risk Index (**RI**) can range between a minimum of 1 to a maximum of 125. A *Frequency-Duration Index Matrix* can now be constructed as follows.

	Duration				
Frequency	**1**	**2**	**3**	**4**	**5**
1	1	2	3	4	5
2	2	4	6	8	10
3	3	6	9	12	15
4	4	8	12	16	20
5	5	10	15	20	25

The ***Frequency/Duration*** cell totals were then crossed to the ***Criticality Matrix*** to determine the events with the highest impact or Criticality index.

Frequency/Duration	Criticality 1	2	3	4	5
1	1	2	3	4	*5*
2	2	4	6	8	*10*
3	3	6	9	12	*15*
4	4	8	12	16	*20*
5	5	10	15	20	*25*
6	6	12	18	24	*30*
8	8	16	24	32	*40*
10	10	20	30	40	*50*
12	12	24	36	48	*60*
15	15	30	45	60	*75*
16	16	32	48	64	*80*
20	20	40	60	80	*100*
25	25	50	75	100	*125*

Risk Analysis Criticality Indexes

Red	**High Priority Investigation**
Yellow	**Medium Priority Investigation**
Blue	**Low Priority Investigation**

A *Risk Analysis Profile Matrix* was then used to prioritize the order of Kaizen Investigations. The *Red Area* represents those disruptive events which should be immediately investigated and eliminated. The *Yellow Area* represents those events which need to be investigated and eliminated as soon as possible. The *Blue Area* represents those events which are lowest on the priority scale. Some examples of assessment and prioritization are given below.

Event Code	Event	Frequency Index	Duration Index	Criticality Index	Risk Index	Priority
E-27-4	Tool Failure	3	2	2	12	4
E-3-5	Material Handling Equipment Failure	2	4	2	16	3
Event Code	Event	Frequency Index	Duration Index	Criticality Index	Risk Index	Priority
E-127-9	Wrong Bar Stock	1	4	5	20	2

Event Code	Event	Frequency Index	Duration Index	Criticality Index	Risk Index	Priority
E-27-4	Tool Failure	3	2	2	12	4
E-3-5	Material Handling Equipment Failure	2	4	2	16	3
Event Code	**Event**	**Frequency Index**	**Duration Index**	**Criticality Index**	**Risk Index**	**Priority**
E-127-9	Wrong Bar Stock Delivered	1	4	5	20	2
E-112-5	Machine Needed Adjusting	5	1	2	10	5
E-96-4	QC Check Failure	4	2	2	16	3
E-10-4	Main Bearing Failure	2	5	5	50	1

In the actual assessment exercise, out of the 187 downtime events 12 were found to be high priority; 81 medium priority; and 94 low priority. Based upon the criticality index, the most disruptive failure in this small example was E-10-4, which was Main Bearing failures in a turning operation.

It was decided to execute a *Kaizen Blitz* on **all** of the top 12 priority events with a team of five investigators (Lean Engineer, Design Engineer, Area Foremen and two operators). The team was charged with systematically examining the remaining medium and low priority events with a Lean Engineer, an Area Foreman and one operator. During the next 6 months all 12 top priority Kaizen Blitzes were executed, and 15 medium priority events. Twenty of the low priority events were examined at a Kaizen meeting , and discussed by a team of 3 process operators. These latter Kaizen meetings proved to be very popular; area workers felt like they were being recognized and enthusiastically came up with multiple solutions.

The Kaizen Event: Investigative Procedure

Having calculated the *Risk Index* for each disruptive event and selected Bearing Failure for a Kaizen Event, the problem was then investigated. Two Magnificent 7 Lean tools were used in the investigative process: (1) A ***Fishbone analysis*** and a (2) ***Root cause analysis*** using a ***why-***

why-why procedure. The first task was to conduct a Fishbone Analysis to determine what could cause machine downtime. The following Fishbone diagram was constructed.

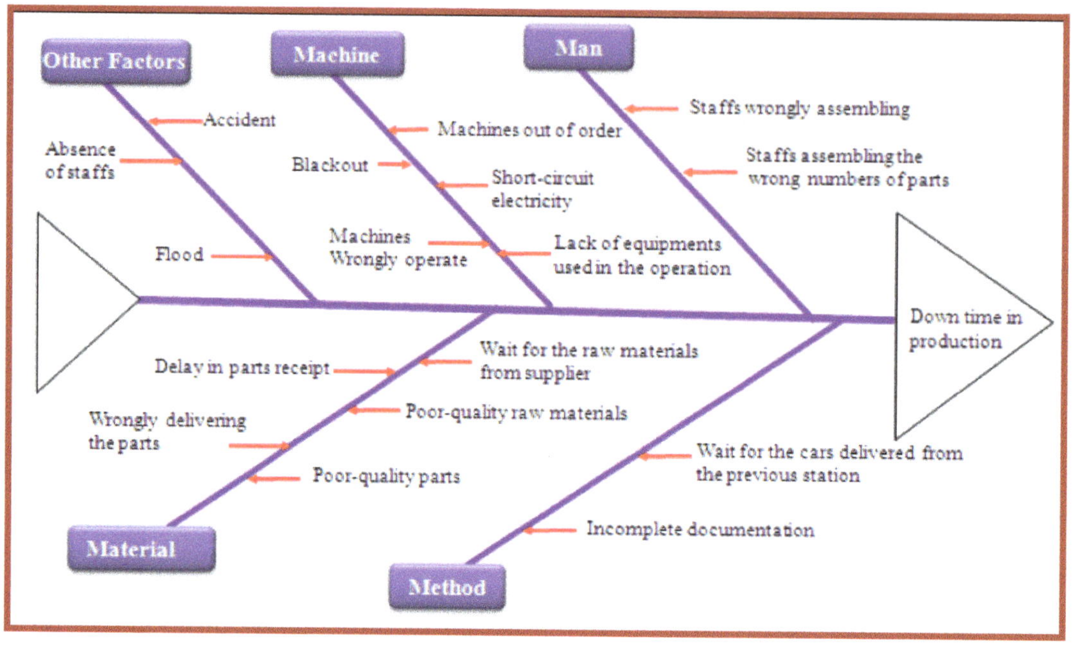

Figure 1
A Fishbone Diagram for Causes of Downtime

A typical Fishbone diagram is shown in Figure 1. The Fishbone diagram used the typical major categories that might affect machine downtime: (1) Man (2) Machine (3) Materials (4) Methods and (5) Other causes. For each of the causes identified. A Root Cause Analysis was conducted. For example, one downtime cause was that the operators sometimes followed an incorrect procedure. This caused a delay while the part was disassembled to remove improperly installed components and repeat the correct assembly.

The procedure used in Root Cause Analysis was the simple but effective *Why-Why-Why* methodology, augmented when needed by *Who, What When and Where*. For example, (a) Why was machining time being lost? .... by incorrect assembly procedures … Why? (b) The assembly instructions were sometimes incomplete…Why? .... (c) Because they had not been updated in the past 4 years. (d) Why? The assembly instructions were supposed to be updated every 6 months by the IE division, but were not …. Why? (e) In a recent cost saving move, the three Industrial Engineers who updated instructions were moved to another division to replace retired personnel. (f) Why did this prevent the update of operating procedures ? .... The IE division had to prioritize work, and it was decided that operators knew how to properly operate all machines. Operating instructions would be updated only on a yearly basis. (g) Why did this result in machine downtime? ... 9 months ago, a new Carbon Steel was chosen for many machining operations, calling for a reduction in speeds and feeds. (h) Why? .... This was a design decision. However, without reducing speeds and feeds, the old speeds and feeds with harder carbon steel caused tool chatter which damaged the surface finish and sometimes tool failure which ruined the component being installed. We could go on, but it is clear that the ***Why-Why*** procedure is to keep asking *Why* (Augmented by How, When, Where and Who) until the *root cause* is identified.

The success of this exercise was as follows:

(1) The number of cars produced rose from 3730 to 4750 during the next 12 month period.
(2) On-time delivery rose from 92% to 98% over the same time period
(3) Rework and defectives dropped from 4% to 2%
(4) Corporate profits rose 2.3%
(5) The number of disruptive events decreased from 187 to 63. Most of the new disruptive events were in the minor category.

www.ingramcontent.com/pod-product-compliance
Lightning Source LLC
LaVergne TN
LVHW071345210926
848486LV00003B/233